BATTLEGROUND
SCIENCE AND TECHNOLOGY

BATTLEGROUND
SCIENCE AND TECHNOLOGY

VOLUME 1 (A–M)

Edited by Sal Restivo and Peter H. Denton

GREENWOOD PRESS
Westport, Connecticut • London

Library of Congress Cataloging-in-Publication Data

Battleground science and technology / edited by Sal Restivo and Peter H. Denton.
  p. cm.
  Includes bibliographical references and index.
  ISBN 978-0-313-34164-9 (set: alk. paper)
  ISBN 978-0-313-34165-6 (v. 1: alk. paper)
  ISBN 978-0-313-34166-3 (v. 2: alk. paper)
  1. Science—Social aspects—North America. 2. Science—Technological innovations—Environmental aspects—North America. 3. Science—North America.
  I. Restivo, Sal P. II. Denton, Peter H., 1959–
  Q175.52.N7B38 2008
  303.48′3—dc22   2008026714

British Library Cataloguing in Publication Data is available.

Copyright © 2008 by Greenwood Publishing Group, Inc.

All rights reserved. No portion of this book may be reproduced, by any process or technique, without the express written consent of the publisher.

Library of Congress Catalog Card Number: 2008026714
ISBN: 978-0-313-34164-9 (set)
      978-0-313-34165-6 (vol. 1)
      978-0-313-34166-3 (vol. 2)

First published in 2008

Greenwood Press, 88 Post Road West, Westport, CT 06881
An imprint of Greenwood Publishing Group, Inc.
www.greenwood.com

Printed in the United States of America

The paper used in this book complies with the Permanent Paper Standard issued by the National Information Standards Organization (Z39.48–1984).

10 9 8 7 6 5 4 3 2 1

For Mr. Sanders, James Quinn, Bernard Rosenberg, Aaron Noland, Burt and Ethel Aginsky, Jay Artis, John and Ruth Hill Useem, David Bohm, and Joseph Needham, who set encyclopedic goals for me and guided me toward realizing the unrealistic; for all the generations who had the privilege of studying at Brooklyn Technical High School and the City College of New York; and in memory of my dear friends John Schumacher and Tracy Padget.

For Evelyn Nellie Powell Denton and for her great-grandchildren, Ruth and Daniel; may they live their lives with as much determination, humor, thoughtfulness, and care for other people as she has demonstrated now for 100 years.

CONTENTS

Guide to Related Topics	*xi*
Series Foreword	*xv*
Acknowledgments	*xvii*
Introduction	*xix*
Entries	
Agriculture	1
Alien Abductions	10
Art and Science	12
Artificial Intelligence	14
Asymmetric Warfare	22
Autism	31
Biodiesel	37
Biotechnology	40
Brain Sciences	45
Cancer	51
Censorship	55
Chaos Theory	59

Chemical and Biological Warfare	62
Cloning	66
Coal	69
Cold Fusion	73
Computers	76
Creationism and Evolutionism	85
Culture and Science	90
Death and Dying	101
Drug Testing	107
Drugs	109
Drugs and Direct-to-Consumer Advertising	119
Ecology	123
Education and Science	132
Epidemics and Pandemics	136
Ethics of Clinical Trials	140
Eugenics	144
Fats	155
Fossil Fuels	157
Gaia Hypothesis	163
Gene Patenting	165
Genetic Engineering	173
Genetically Modified Organisms	182
Geothermal Energy	195
Global Warming	198
Globalization	201
Green Building Design	208
Healing Touch	213
Health and Medicine	216
Health Care	224
HIV/AIDS	228
Human Genome Project	237
Immunology	241

Indigenous Knowledge	245
Influenza	249
Information Technology	253
Intellectual Property	257
Internet	260
Mad Cow Disease	271
Math Wars	273
Mathematics and Science	278
Medical Ethics	287
Medical Marijuana	291
Memory	294
Mind	298
Missile Defense	303
Nanotechnology	307
Nature versus Nurture	310
Nuclear Energy	313
Nuclear Warfare	321
Obesity	331
Objectivity	333
Off-Label Drug Use	335
Organic Food	336
Parapsychology	341
Pesticides	343
Pluto	347
Precautionary Principle	349
Privacy	354
Prostheses and Implants	357
Psychiatry	359
Quarks	369
Religion and Science	373
Reproductive Technology	382
Research Ethics	386

Robots	389
Science Wars	395
Scientific Method	398
Search Engines	400
Search for Extraterrestrial Intelligence (SETI)	402
Sex and Gender	404
Sexuality	412
Social Robotics	414
Social Sciences	419
Software	430
Space	435
Space Tourism	437
Space Travel	439
Stem Cell Research	443
Sustainability	446
Technology	453
Technology and Progress	462
Tobacco	464
UFOs	467
Unified Field Theory	469
Urban Warfare	473
Vaccines	479
Video Games	485
Virtual Reality	487
Warfare	491
Waste Management	500
Water	504
Wind Energy	512
Yeti	517
Bibliography	*521*
About the Editors and Contributors	*541*
Index	*547*

GUIDE TO RELATED TOPICS

BIOLOGY AND THE ENVIRONMENT
Agriculture
Ecology
Gaia Hypothesis
Global Warming
Green Building Design
Nature versus Nurture
Organic Food
Pesticides
Precautionary Principle
Sustainability
Waste Management
Water

DRUGS AND SOCIETY
Drugs
Drugs and Direct-to-Consumer Advertising
Drug Testing
Off-Label Drug Use
Medical Marijuana
Tobacco

ENERGY AND THE WORLD ORDER
Biodiesel
Coal

Fossil Fuels
Geothermal Energy
Global Warming
Nuclear Energy
Wind Energy

GENETICS
Cloning
Eugenics
Gene Patenting
Genetically Modified Organisms
Genetic Engineering
Human Genome Project
Stem Cell Research

MATHEMATICS AND PHYSICS
Chaos Theory
Pluto
Quarks
Space
Space Tourism
Space Travel
Unified Field Theory

MEDICINE AND HEALTH
Cancer
Death and Dying
Epidemics and Pandemics
Ethics and Clinical Trials
Fats
Healing Touch
Health and Medicine
Health Care
HIV-AIDS
Immunology
Influenza
Mad Cow Disease
Medical Ethics
Obesity
Prostheses and Implants
Reproductive Technology
Sex and Gender
Sexuality
Vaccines

MIND AND BRAIN
Autism
Brain Sciences
Memory
Mind

POSTMODERN BATTLEGROUND
Creationism and Evolution
Globalization
Intellectual Property
Math Wars
Religion and Science
Science Wars

SCIENCE
Art and Science
Culture and Science
Education and Science
Indigenous Knowledge
Mathematics and Science
Objectivity
Parapsychology
Psychiatry
Research Ethics
Scientific Method
Social Sciences

SCIENCE OUT OF BOUNDS
Alien Abductions
Search for Extraterrestrial Intelligence (SETI)
UFOs
Yeti

TECHNOLOGY IN THE GLOBAL VILLAGE
Artificial Intelligence
Biotechnology
Censorship
Cold Fusion
Computers
Information Technology
Internet
Nanotechnology
Privacy
Robots

Search Engines
Social Robotics
Technology
Technology and Progress
Video Games
Virtual Reality

WAR IN THE TWENTY-FIRST CENTURY
Asymmetric Warfare
Chemical and Biological Warfare
Missile Defense
Nuclear Warfare
Urban Warfare
Warfare

SERIES FOREWORD

Students, teachers, and librarians frequently need resources for researching the hot-button issues of contemporary society. Whether for term papers, debates, current-events classes, or to just keep informed, library users need balanced, in-depth tools to serve as a launching pad for obtaining a thorough understanding of all sides of those debates that continue to provoke, anger, challenge, and divide us all.

The sets in Greenwood's *Battleground* series are just such a resource. Each *Battleground* set focuses on one broad area of culture in which the debates and conflicts continue to be fast and furious—for example, religion, sports, popular culture, sexuality and gender, science and technology. Each volume comprises dozens of entries on the most timely and far-reaching controversial topics, such as abortion, capital punishment, drugs, ecology, the economy, immigration, and politics. The entries—all written by scholars with a deep understanding of the issues—provide readers with a non-biased assessment of these topics. What are the main points of contention? Who holds each position? What are the underlying, unspoken concerns of each side of the debate? What might the future hold? The result is a balanced, thoughtful reference resource that will not only provide students with a solid foundation for understanding the issues, but will challenge them to think more deeply about their own beliefs.

In addition to an in-depth analysis of these issues, sets include sidebars on important events or people that help enliven the discussion, and each entry includes a list of "Further Reading" that help readers find the next step in their research. At the end of volume 2, the readers will find a comprehensive Bibliography and Index.

ACKNOWLEDGMENTS

When we set out to solicit entries for these volumes, we decided to cast a wide net for participants from both sides of the Canadian–U.S. border (and beyond). Although this has meant some additional headaches (often related to e-mail servers and crashing computers), we want to thank our many authors for contributing their thoughts and expertise. We have learned a great deal from them and from each other about a wide variety of topics in science and technology and about electronically mediated collaboration.

For the origins of these volumes, we have to thank Marcel LaFollette, who introduced Sal to Kevin Downing from Greenwood Press; we appreciate Kevin's patience and professionalism in wrestling with us through the details of the big project that grew larger and took longer than any of us planned. For her organizational and editorial expertise, we owe a debt of gratitude to our Greenwood editor, Lindsay Claire, who has kept everything on track despite the multifocal nature of what we decided should be done. For their thoughtful attention to detail, we very much appreciate the work of Michael O'Connor at Greenwood and the people at Apex CoVantage. For their forbearance and support in the swirl of the past two years, we also thank Mona, Ruth, and Daniel Denton.

In the beginning, we shared editorial duties with Elizabeth Shea, a professor of rhetoric with a background in engineering and science and technology studies, who unfortunately had to drop out of the project after several months during a transition from the academy to the corporate world. We thank Elizabeth for her contributions during those hectic early stages of identifying the first set of contributors.

Finally, we wish to acknowledge the international partnership these volumes reflect, between Sal Restivo at Rensselaer Polytechnic Institute in Troy, New York, and Peter Denton, at Red River College of Applied Arts, Science and Technology in Winnipeg, Manitoba, and the support for relevant education both institutions have demonstrated for a long time. The issues we include are not parochial ones, but ones whose implications affect every member of our global society. We have done our best to make all the entries reflect a North American perspective (even if there are specific examples drawn from both sides of the border), to ensure students in Canada and the United States will benefit from them equally. We have done this without sacrificing the global reach of these volumes as we all experience the growing pains and the perils of an emerging world community.

On a personal note, our shared authorship of the Religion and Science entry is fitting; we first met in Toronto nine years ago, after we both received a 1999 International Templeton Science and Religion Course Prize, and have collaborated on various smaller projects since then. Although we bring very different skills, interests, and experiences to the table, all of the editorial decisions and commentary reflected here are mutual. We hope you enjoy and appreciate the results.

Sal Restivo and Peter H. Denton

INTRODUCTION

If you have ever walked on the site of some past battle, some historic battleground, you might have experienced an uneasy feeling. Depending on how long ago the battle was fought, there will be less and less evidence of what happened as Nature returns the site to what it was before the fighting took place. Yet there are ghosts, the realization that in this place people fought and died, and it would not be unusual to wonder what they were like, why they were here, and what was so important that it led to a battle on this spot. Widening the circle of questions, you wonder about the larger reasons there was a war (for there is rarely just one battle), the motivations of the groups who sent people to this place to fight, who won and who lost, and whether in the end the war was worth what it cost.

It is in this way that we would like you to read our volume in the Battleground series. For all of the topics selected here, there are battlegrounds. Some are historic, some describe conflicts taking place at the moment, and others sketch places where battles have yet to be fought. Yet as we selected and edited the entries for these volumes, the relationships between these battles began to emerge, giving us a sense that there were larger reasons for the fighting and of why particular conflicts are taking place. In the end, to understand any particular battle, one needs to understand the war, who was fighting and for what reason, and—if there ever was a winner—what was won or lost.

Although conflicting—and opposing—perspectives are found on the various topics presented, we have tried to maintain the larger, global perspective urgently required for finding solutions to the problems confronting our generation. At this juncture in human history, other ways need to be found to represent differences, choices and debates than models that entail not only fighting, but also winners and losers. There are no winners if problems such as nuclear war, environmental degradation, or food and water shortages overtake us all. When a child dies in a refugee camp in Darfur, no one gains; in the words of the seventeenth-century English poet, John Donne, we are all diminished when the bell tolls because something of our humanity also dies. While there may be some self-satisfaction in breathing the last breath of clean air or chopping down the last tree or shooting the last lion or earning the last dollar, from any perspective that values life itself, human or otherwise, this sense of accomplishment is utterly perverse.

Before we continue, we would like to situate this project in the world of encyclopedias. It is impossible to contemplate and carry out the preparation of an encyclopedia without considering the experiences of one's predecessors. Consider Pliny the Elder (Gaius Plinius Secundus, 23–79 C.E.), who was among the individuals credited with writing the first encyclopedia. Pliny wrote at least 75 books and almost 200 unpublished notebooks, but the only one of these works for which he is remembered is his 37-volume *Natural History* in which he set out to present in detail the contents of the whole world. Pliny was so inclusive that his encyclopedia included references to dog-headed people and headless people with eyes in their shoulders. Even someone historians describe as "humble," Isidore of Seville (c. 560 to 636 C.E.), also credited with writing the first encyclopedia, set out to include everything known in his time. We have tried to be both more modest than Isidore and certainly more realistic than Pliny. Somewhat immodestly, but without any pretensions to what Denis Diderot and Jean le Rond D'Alembert achieved in their great encyclopedia project of the eighteenth-century French Enlightenment (published between 1751 and 1777 in 32 volumes), we might align ourselves with the words that Diderot used to introduce the project. "This is a work," he wrote, "that cannot be completed except by a society of men of letters and skilled workmen, each working separately on his own part, but all bound together safely by their zeal for the best interests of the human race and a feeling of mutual good will."

The authors we have invited to contribute to this project include distinguished scholars, colleagues, highly talented graduate students, writers, and thinkers. They have all worked hard to represent their topics in ways that students and the general reader can understand. At the same time, they (and we) have not shied away from introducing some challenging ideas, technicalities, and terminologies. We hope each entry will provoke further inquiry, whether it is to learn more about a topic or simply to look up a word you have encountered for the first time. All of the entries are starting points for your own research and thinking, not the final word.

We need to begin with some thoughts on the subject matter of these volumes, to sketch something of the larger perspective that guided our choices for what to include. What is the battleground in science and technology that leads to the title of this encyclopedia?

Technology certainly might be considered to be unfolding on a battleground of unintended consequences. The U.S. Congress once had an Office of Technology Assessment that was supposed to monitor possible environmental and broader social and cultural impacts of new technologies. It is easy to see that new technologies affect different sectors of the population (for example, rich and poor, educated and uneducated, men and women, old and young) differently, leading to differing opinions about whether the new technologies—or the changes they bring about—are good or bad. The building of a nuclear power plant nearby may be welcomed by people in some communities and strenuously resisted by people in other communities, and for very different reasons; the waste products of nuclear energy, in particular plutonium, are extremely dangerous, are difficult to dispose of with current technologies, and can be targeted by a nation's enemies even if the chances of an accident can be eliminated. Life-saving medical technologies may be more readily available to the wealthier members of our society and out of the reach of those who have less money, raising questions about whether to spend public funds on treatments that do not benefit everyone. Large-scale technologies may pose dangers to our environment or be associated with long-term risks that may be weighed differently in different communities (and in different countries) when who benefits from these technologies is taken into account.

There are many more examples of how different people weigh the costs and benefits of a technology differently, leading to conflicts about whether the technology should be developed or used. Science, however, may not initially appear to be something that creates such conflicts. Science does not "move forward" in the wake of belching smoke stacks, polluted lakes, and damaged ecosystems. Nonetheless, battlegrounds in science have indeed emerged in two ways. First, the links that historians and social scientists have uncovered over the last 50 years or so between science and technology have blurred the once transparent distinction between science and technology; this is reflected, for example, in the introduction of the term *technoscience*. Second, identifying science as a social institution, not just a method of problem solving, has given us a clearer notion of the extent to which it is intricately intertwined with the norms, values, and beliefs of the ruling classes, particularly in the Western culture to which it owes its origin. Thus, conflicts in technology in some sense are also conflicts in science.

These two volumes are designed to give you a better appreciation of the complexities involved in what might be called "the science-technology-society nexus" (ST&S nexus). Science and technology—separately or together—are human activities, taking place within the boundaries of human society and culture. Whether they are used, for example, as tools to solve problems or are the source of problems themselves, they figure prominently in the social, cultural, and environmental choices we make as individuals and as members of a global society.

Any approach to problem solving in human settings must start, however, with some image of what it means to be human. Critiques of capitalism, communism, or any political economy as destructive and alienating or as beneficial to all humanity make no sense unless they are based on a consensus of what it means to be human. If we accept that it is acceptable to use child labor to mine coal, that it is reasonable to minimize housing and food resources for such children in the interest of saving money, if we believe that profits are more important than people and the environment itself, any critique of capitalism collapses.

If we want to defend communism or socialism, we have to point to realistic examples; contrary to popular opinion, the closest examples to which we could point are the ancient human settlements that appear to have been based on egalitarian systems in which power was equitably distributed across sex and gender. Primitive communism is a realistic historical condition; advanced communism has effectively not been tried (the Soviet Union, China, and Cuba notwithstanding). The point here is that the very idea of a "battleground" suggests value judgments, ethical stands, things over which some struggle is required. The struggle in our time is to strike a sensible balance between belief and faith and hope on the one hand and grounded reason or science on the other. Some form of pervasive realism—material and ethical—seems now to be a prerequisite not merely for the survival of local, regional, or even subcultural populations but of the planet and the human species itself. The entries in this encyclopedia seek to enroll you in this struggle and to ground belief, faith, and hope in such realism; we neither shirk from a realistic depiction of the problems nor encourage despair as a viable response.

Because our present circumstance is at the end of an historical trajectory, some of the points along that trajectory need to be identified. After all the ups and downs of wars, technological achievements, and cultures rising and falling throughout history, the Newtonian revolution in science eventually produced an exaggerated confidence in the scientific worldview. This confidence led eighteenth- and nineteenth-century thinkers to adopt Newtonian mechanics as the model for everything from jurisprudence to social science. Toward the end of the nineteenth century, some philosophers and scientists were convinced that science had solved all the big mysteries of the natural world, this on the eve of the revolutions wrought by non-Euclidean geometries, relativity theory, and quantum mechanics. These revolutions did not so much shake confidence in the ultimate power of science as much as they complicated our understanding of the nature of the foundations of science—perception, observation, experimental method, and logic itself were all transformed in the years leading up to the Great War of 1914–18 (World War I).

The Great War itself shook confidence in many different areas unrelated to the devastating effects of warfare on an industrial world. The generations lost in the trenches of the Western Front were accompanied by a loss of moral direction; whether one agreed with the direction in which the moral compass of Western culture was pointing or not, at least previous to the Great War, there had been some direction. Afterward, the structures of meaning—like the political and economic structures of the pre-War period—were in ruins. The fear was frequently expressed that the science and technology developed and used in the new civilization had outpaced the moral capacity to use it wisely; new and more deadly weapons had been handed to the same old savage, who was now able to wreak havoc on a scale hitherto unimaginable. Although technology had always been thought to have a dark side, this was now accompanied by the idea that the collapse of classical materialism and the rise of relativity theory in physics were mirrored in the collapse of moral structures and the rise of moral relativism.

A different sort of change was in store for science and technology in the wake of World War II. Hiroshima and Nagasaki demonstrated that science (represented in the image of the impish Albert Einstein) and technology (represented in the image of the mushroom cloud) could combine to destroy worlds as well as to transform bread mold into the lifesaving medical breakthrough of penicillin. The physicist J. Robert Oppenheimer (1904–67), known as "the father of the atomic bomb," reflecting on the atomic destruction of the two Japanese cities, famously said (quoting from the ancient Hindu text the *Bhagavad Gita*), "If the radiance of a thousand suns were to burst at once into the sky, that would be like the splendor of the mighty one. Now I am become Death, the destroyer of worlds." The physicists, he went on to say, "have known sin."

If the atomic bomb can be considered the first sin of the scientists and technologists, the Vietnam War (1959–75) might be considered their second sin. In a crude way, and in the spirit of Oppenheimer, we might say that science—represented in the famous equation $E = mc^2$—was at the root of the first sin, and technology—in the form of sophisticated high-tech weaponry and the dioxin-based chemistry behind Agent Orange—was at the root of the second sin. Frederick Su's novel *An American Sin* gives us an Oppenheimer-like perspective on the Vietnam War. There was something novel about the way scientists, as both critics of and participants in the war machine, reacted to science and technology in that war and the impact this had on the public understanding of and attitudes toward science and technology. Scientists became unusually reflective and critical about their own sciences, and these attitudes and understandings were magnified among the intellectual and activist critics of the war. With communication and information technologies new to war and society, the media brought the American and other nations' publics closer to the everyday realities of warfare than ever before. This war was one of the most fundamental provocations for the development of the radical science movement and thus has a place in the history of the emergence of interdisciplinary science and society programs at universities all over the world during the late 1960s and throughout the next two decades. If it was the physicists who knew sin in the flash and noise of the atomic bomb and the future it foretold, it was biologists and chemists who became the villains—who knew sin—as we learned more and more throughout the middle part of the twentieth century about the unintended consequences of their discoveries and inventions, from radiation sickness to cancer and environmental damage to the loss of ecological diversity and the extinction of species.

Yet it was the image of The Bomb, in its different guises, that dominated the horizon. In the period following the end of World War II, many of the physicists associated with the Manhattan Project, the setting for building the first nuclear weapons, turned to political activities and new scientific pursuits in the life sciences. Manhattan Project scientists established the Federation of American Scientists after the war and began publishing the *Bulletin of the Atomic Scientists* in 1945. The magazine is still in circulation and continues to inform the general public and policy makers alike of the nature and dangers of nuclear weapons and nuclear war, the political economy of missile defense systems, arms control policies, and related issues.

On July 5, 1955, Bertrand Russell and Albert Einstein released the Russell-Einstein Manifesto calling for scientists to address the issue of nuclear weapons (and, at least by implication, other potential weapons of mass destruction). The Canadian industrialist Cyrus Eaton, who had known Russell since before the outbreak of the war, offered to finance a conference on the Manifesto in his hometown, Pugwash, Nova Scotia. After other offers (such as one from Prime Minister Jawaharlal Nehru to host it in New Delhi) were discussed, Eaton's offer won out. The first Pugwash conference, with the Russell-Einstein Manifesto as its charter, was held in July 1957.

The United States was by now in the midst of the Cold War, and World War II and the Cold War saw significant developments in science and technology and in science and technology policy. Prior to 1940, the U.S. government's involvement in science and technology was relatively minor in contrast to what it would be following the war. The period was marked by the development of the Land Grant Colleges, the work of the Coast and Geodetic Survey, and weapons development in the Departments of War and Navy. During World War II, scientists and engineers were mobilized to support the national defense. This initiated defense research and development as a centerpiece of American policy for science and technology. In the years immediately following the end of the war, the United States established the National Science Foundation and founded a number of agencies devoted to military research and development. In the 1950s, the major developments involving the government-science-technology nexus included civilian programs in atomic energy (including civil defense efforts) and aerospace technology as well as many other programs. It was during this period that the White House Science Office was created.

The upshot of all these developments was that government interests were now shaping science and technology in more visible and demonstrably significant ways than in the past. Government influence was not new, but the sheer degree of that influence has led to one of the important battlegrounds in science and technology; shaping science and technology to national interests may contradict the interests of those scientists and engineers whose results are not consistent with national security or other political agendas of the government of the day. The findings of various presidential commissions, based on good scientific methods, thus have often been ignored. During the George W. Bush administration, the contradictions between good science and the science promoted from within the White House have led to what has been labeled "junk science." Cartoonist Gary Trudeau recently introduced his fans to Dr. Nathan Null, the president's situational science advisor. Trudeau has his character define situational science as a matter of respecting both sides of a scientific controversy, "not just the one supported by facts." This is an allusion to the rhetoric of junk science versus sound science that has become common in contemporary political and legal disputes. Many of those disputes help define and shape the battlegrounds discussed in our two volumes.

The 1960s were a significant watershed for our understanding of the ST&S nexus in two respects. First, the social and cultural revolutions of that period enveloped science and technology, fostering concerns about ethics, values, and social responsibility in science and technology. Many scientists and engineers, some of international repute, put aside all or some of their science and technology to pursue political agendas. This led to the emergence of the Science for the People movement in the United States and the Radical Science Movement in England, both of which eventually expanded into other countries or influenced comparable movements elsewhere. In the late 1960s and early 1970s, a new academic discipline influenced in part by the radical scientists and engineers began to emerge. Science policy programs at universities such as Sussex in England and science and society programs such as the Science Studies Unit at the University of Edinburgh in Scotland were soon followed by the development of science and society, technology and society, and science policy programs at universities in the United States. In the early 1970s, the Massachusetts Institute of Technology instituted a Technology Studies faculty seminar, and Rensselaer Polytechnic Institute found external funding from such organizations as the Lilly Foundation, the National Science Foundation, and other private and public foundations to support a Center for the Study of the Human Dimensions of Science and Technology. A science and technology battleground was beginning to come into focus as scientists and engineers battled the demons of the atomic bomb, cancer, and environmental degradation and the calls for bringing ethical and value concerns into the center of an arena traditionally cloaked in disinterestedness and detached objectivity. Science and society was taking on the look of science versus society.

In the midst of this, an embryonic science and technology studies field of teaching and research visible at Sussex, Edinburgh, MIT, Rensselaer, Cornell, and elsewhere crystallized in the founding of the Society for Social Studies of Science at the 1975 meeting of the American Sociological Association. A year later, the Society's first meeting was held at Cornell University, already a leading center for the sociology of science. A few sociologists and anthropologists were beginning to study scientific laboratories using the classical methodologies of the ethnographer, methodologies traditionally applied in the study of so-called primitive societies, traditional and usually non-Western societies.

One of our students read a section on postmodernism in a textbook and asked the following question: "Why would it ever occur to anyone to criticize science?" He was surprised to learn that celebrated thinkers have found reason to criticize science: Jonathan Swift, Rousseau, Goethe, Nietzsche, William Blake, and more recently, scholars including Theodore Roszak and Paul Feyerabend.

Feyerabend, an influential twentieth-century philosopher of science, wrote that two central questions have to be addressed in any discussion of science. One is "What is science?" The second is "What's so great about science?" For most of the last three hundred years, the question "What is science?" has been answered by scientists themselves, philosophers and historians, and biographers and journalists. Their answers have tended to be heroic and hagiographical, and—in the case of scientists reflecting on their past achievements—subject to problems of memory loss, memory distortion, and self-aggrandizing goals. What happened when sociologists of science started to consider this question led to a revolution in our understanding of science as a discourse and a practice.

The second question, like our student's question, can be taken seriously only if we separate science as a term we use to refer to the basic forms human reason takes across societies from modern science as a social institution. Once sociologists began to study science as a social institution, it was only a matter of time before science would be analyzed for its contributions to improving our individual and collective lives, but also for the ways in which it contributes to alienation, environmental degradation, and the deterioration of ecological niches on local, regional, and global scales. Let's begin by looking at what science is. There is already a mistake in the way we ask the question "What is science?"

There is a tendency in and out of science to think and talk about science in the grammar of the ever-present tense—science *is*, or science per se. This reflects the idea that science emerged at some point—in the midst of the so-called Greek miracle of the ancient world or during the so-called European scientific revolution beginning in the seventeenth century—and then held its shape without change up to the present. Traditional students of science thus gave us universal, ubiquitous, ever-present features in their definitions of science. Science is a method; science is mathematized hypotheses, theories, and laws. The definition of science may vary from one philosophical school to another, but the answer is always given in the grammar of the ever-present tense. Classically, science is defined in terms of "the scientific method." This method, which one can still find outlined in school science textbooks at all levels, involves observation; hypothesis formation; experimental design and testing; and interpretation of results. At a more sophisticated level, a group of philosophers collectively known as the Vienna Circle initiated a program between the two world wars to establish the Unity of Science. Like all efforts that have been undertaken by philosophers to identify one set of criteria for "science," the Unity of Science movement failed. These efforts are themselves problems in the sociology of knowledge. The question they raise is this: why are people so interested in questions of unity and universalism? The sociological reasons for this take us into the realm of ideology, ritual, and solidarity.

What happened when sociologists joined philosophers and historians as analysts of science? In the beginning, starting in the 1930s, they followed the lead of the philosophers and historians by analyzing the institutionalization of science in seventeenth-century western Europe. By the time sociologists entered the picture, it was generally believed that once it had become institutionalized, science was on its own. It became an autonomous social system, independent of historical, cultural, and social influences. The focus of interest in the fledgling sociology of science quickly turned to the study of science as a social institution, or more abstractly as a social system. At the same time, sociologists celebrated science along with their colleagues in philosophy and history as a flower of democracy. It is easy to understand this celebratory orientation once you realize that the sociology of knowledge and science emerged at the same time as fascism, with all that fascism entailed. Given that ideologically charged context, it is easier to see why science and democracy have become interwoven into the very fabric of our society and why issues in science can very easily become issues in how the universe is understood or what it means.

Science is a compulsory subject in our schools. Parents can decide whether to have their children instructed in this or that religious tradition; they can decide against any religious instruction whatsoever. Their children must, however, learn something about the sciences. Moreover, scientific subjects are not subject to substitutions; you cannot opt out of science in favor of magic, astrology, or legends as alternatives. Science has become as much a part of the web of social life as the Church once was. As a nation, the United States is committed to the separation of state and church, but state and science are intricately intertwined. Feyerabend claimed that we are Copernicans, Galileans, and Newtonians because we accept the cosmology of the scientists the same way we once accepted the cosmology of the priests.

Perhaps Feyerabend is right in some way but does he go too far in his rhetorical battle against the priests of science? Certainly, Feyerabend loved and respected science as a mode of inquiry and as a form of life. The fact is that we can sustain our view of science as our preferred way of inquiry and at the same time recognize that as a modern social institution it is implicated in the social problems of our modern political economy. We can then adopt a social problems approach to science.

This is not a new idea. It occurred to the sociologist C. Wright Mills, who drew our attention to the "Science Machine." Mills helped us to get behind the curtains of ideologies and icons to the cultural roots and social functions of modern science. He distinguished between science as an ethos and an orientation on the one hand and a set of "Science Machines" controlled by corporate and military interests and run by technicians on the other. He echoed Marx's and Nietzsche's conceptions of modern (bourgeois) science as alienated and alienating and Veblen's critique of modern science as a machine-like product of our matter of fact techno-industrial society.

These are not simply the conclusions drawn by radical social critics and theorists. On January 17, 1961, conservative Republican, former five-star General of the Army, and 34th president of the United States Dwight D. Eisenhower delivered his farewell address from the Oval Office in the White House. He famously introduced into the American vocabulary the phrase "military-industrial complex," cautioning that "in the councils of government, we must guard against the acquisition of unwarranted influence, whether sought or unsought, by the military-industrial complex. The potential for the disastrous rise of misplaced power exists and will persist." He also warned that we must be alert to the potential of federal employment, project funding, and money in general to dominate the agenda of the nation's scholars; at the same time, he warned, we must avoid the "equal and opposite danger that public policy could itself become the captive of a scientific-technological elite." (Remember, however, that even though Eisenhower warned about the dangers of the military-industrial complex, he believed it was a necessary component of the American agenda.)

From the vantage point of Mill's "Science Machine" idea, modern science could be described as an "instrument of terror," a social activity driven by an assault on the natural world, profit motives, and the pursuit of war and violence. Let's remind ourselves again that critics of modern science as a social institution are generally, like Mills and Feyerabend, advocates of science as an ethos, an orientation, a mode of inquiry, a form of life. The sociology of science has helped us to understand why it has occurred to so many well-known and respected scholars to criticize science. It has pressed the seemingly trivial idea that scientists are human and that science is social toward generating a new understanding of the very nature of science.

This same line of reasoning works for technology too, insofar as technology can be separated analytically or otherwise from science. As we have already suggested, it is a little easier to make the case for why we should worry about whether we can assume that all technologies are automatically beneficial and progressive for all humanity. These two volumes show why we need to be critically vigilant about everything from drugs and vaccines to computers, information technologies, cold fusion, and nanotechnology. Again, we want to caution against any tendencies to adopt anti-science and anti-technology strategies based on the historical vicissitudes of the technosciences. Rather, exploring the battlegrounds of science and technology should help you assess, evaluate, criticize, perhaps apply the precautionary principle, and not assume that everything technoscientific is automatically for the best.

Globalization, multiculturalism, and postmodernism have been important features of the world's intellectual, cultural, and political landscape for many decades. These are complicated terms, but we can suggest their essential significance briefly. Globalization is the process of linking more and more of the world's peoples as transportation, communication, and cultural exchange networks expand across the planet. This process is fueled cooperatively through activities such as tourism and trade and conflictually through wars and takes place within the truly global context of sharing the same planet. Multiculturalism is the awareness of and support for the variety of ways of life within and across cultures. Postmodernism is a set of intellectual agendas all of which have been fueled by a century of growing awareness that our teaching and learning methodologies have limits and are flawed. This has led some intellectuals to adopt an extreme relativism and to give up on the search for objective truths. Others have responded to the loss of certainty by recognizing the complexity of truth and truth-seeking methodologies. In this context, we have come to recognize that what Americans and Europeans view as universal science and technological progress may be functions of a narrow cultural vision.

Multicultural theorists introduced the idea of ethnoscience to focus attention on the scientific practices of non-Western cultures. For example, the local knowledge of a culture about its plants and herbs was referred to as ethnobotany. It was just a matter of time before the idea of ethnoscience was applied to Western science itself, transforming it from universal science to just another ethnoscience. We have already alluded to some of the reasons this might make sense, especially the idea of science as a social institution. "Universal science" and "technological progress" were seen as products of Western imperialism and colonialism. The old notion that the cross follows the sword could be expanded as follows: the laboratory follows the cross follows the sword. First, we conquer militarily; then we impose a moral order through the efforts of missionaries or other moral agents; and then we impose a culture of laboratories, lab coats, and "the" scientific method. This has been, to put it in schematic form, the way the West has spread its ethnoideologies around the world.

We do not want to imply that Western science as an ethnoscience is on a par with all other ethnosciences, only that comparisons and judgments need to be made about the nature of its claims and conclusions and not simply assumed. It is also important to keep in mind that the very idea of "Western science" is corrupted given that as the West emerged as the dominant political economy on the world scene, starting as early as the thirteenth century in the Italian city states, it became heir to the scientific contributions of more and more of the world's cultures, as well as their interpreter.

Any project, even an encyclopedic one, has its limitations. These limitations include the amount of space (e.g., word counts) we are allowed and the amount of time we have to put the volumes together, dominated by deadlines. This means that we cannot hope to cover every possible topic or even every possible significant topic. Our choices reflect topics, issues, and problems (battlegrounds) that are on the front pages of today's newspapers and lead stories on the evening news. They necessarily also reflect the perspectives of the editors and of the publisher. Given these inevitable constraints and biases, and the arbitrary decisions that affect the final outcome, we believe we have identified key battlegrounds, key ideas, concepts, issues, troubles and problems at the nexus of contemporary science, technology, and society. In particular, we believe our choices reflect the battlegrounds on which the future of our species, our planet, and our global culture is being played out.

Interestingly, despite the range of topics and the disparate authors, and despite the tendency of encyclopedias to have the structure of a list, there is a narrative thread in these two volumes. That thread tells the story of the conflict between science and religion as it plays itself out on the largest stage in history, Earth itself, which has become a battleground in its own right. The game is afoot, as Sherlock Holmes might say, and we would like to think that these two volumes will not only serve as an initial guide to some of the crucial social and cultural battlegrounds in science and technology, but also show how they are related to the human quest for meaning in life, the universe, and everything. We have tried to offer a realistic picture of the battlegrounds and their potentials for glory or disaster, along with some ideas as to how the disasters might be avoided in order to ensure the future of life, human and otherwise, on the only planet we know that supports it.

HOW TO READ AND USE THE BOOK

The "further reading" sections at the end of each entry include any of the sources on which the author has relied as well as additional materials we think are useful; any numbers or statistics are taken from one of the sources listed at the end of the entry. We want you to use these volumes as resources for thinking as opposed to shortcuts for research.

We have added editorial sidebars to some of the entries, either to supplement the ideas or information presented by the author or to indicate other ways in which the topic might have been approached. Our sidebars are thus not intended to be corrections, but elaborations; the presence or absence of a sidebar indicates only the alignment of the planets, not the author's alignment with what the editors really wanted to see in a particular entry.

Peter H. Denton and Sal Restivo

A

AGRICULTURE

Agriculture has been at the center of human society since the birth of civilization and at the center of disputes and debates ever since. Most economists, historians, and cultural anthropologists would argue that permanent communities became possible only once people discovered means of producing regular amounts of food in one location and then were able to produce enough surplus food to allow some of the population to undertake nonagricultural work.

Current debates center on the means of agricultural production; the social and cultural contexts of agriculture; specialized farming; the spread of disease; and the implications and consequences of the introduction of novel crops and organisms (particularly through biotechnology).

The ancient civilizations of the Middle East and Mesoamerica arose after humans learned to farm sustainably, and permanent towns, villages, and cities were established. Both the Egyptian and Mesopotamian civilizations arose in fertile river valleys that contained excellent farmland; for many centuries, civilizations remained almost exclusively in temperate areas, which allowed for productive farming with primitive methods and the few crop species available, such as in India, China, the Mediterranean, and Central America. Agricultural issues, images, and ideas were so central to the lives of ancient civilizations that they dominated the language of the religions that arose in the ancient world. The Judeo-Christian book of Genesis, for example, begins with stories of humans being given control of the plants and animals of the world to use for their sustenance and ends with stories of Joseph, who reached fame and acclaim in Egypt by creating a grain storage system during years of plenty that minimized the effects of a devastating drought that followed.

Even though farming and food production have been carried out for thousands of years, these are areas of intensive scientific development and often-furious debate, with new issues arising regularly as scientific and technological innovation creates new situations. Agriculture may have given birth to civilization, but it is a parent that is developing with the speed of an adolescent, producing new questions, issues, and challenges all the time.

Farms and farm machinery do not just get bigger over time. They also become more specialized as agricultural technology becomes more advanced, leading to important agronomic and social consequences for farmers and for their nations that can be seen in the farmland that spreads across the interior of the United States.

Typical of the changes in farming methods and machinery that have occurred in recent decades is the situation on the Great Plains, which spread from the Midwest north to the Canadian border and west to the Rocky Mountains.

Through to the 1980s, the most common form of farming was for farmers to till (to turn over and churn up with machinery) the soil repeatedly during the year, in order to kill weeds before and after the crop was grown, relying on chemical pesticides to control weeds growing inside the crop. In the drier parts of the plains, farmers would often leave fields bare of any crop for a full summer so that all the rain that fell that year could soak in and be available for the next year's crop. During such "fallow" years, farmers would till the cropless soil so that weeds could not develop and drain the moisture. A negative consequence of this approach of tilling and fallow was that the churned-up soils were vulnerable to the harsh prairie winds. Much of the topsoil—the nutritionally rich first few inches—would blow away. In a drought situation the problem could become chronic and extreme, which is why the years of the 1930s became known as the "dirty thirties," and the 1980s brought back bad memories of the 1930s for thousands of farm families.

By the 1980s, a new approach to farming (known as "minimum till" or "zero till" farming) was embraced by thousands of farmers who wanted to avoid losing their topsoil and did not want to leave fields unproductive during fallow years. This approach relied on new methods, tools, and machinery. Farmers killed the weeds on their fields before seeding by spraying chemical pesticides and then used seeding equipment that would insert seed and fertilizer into the soil without tilling it, like using a needle rather than a rake. When the crop was harvested, the stalks were left in the soil, providing a root base that not only kept the soil from becoming loose and blowing but that also allowed the soil, because it contained plant matter, to become much softer and less likely to compact. Because there was less tillage, the soil did not dry out nearly as much, so most fields that had been occasionally fallowed could be put into yearly production because of the better moisture conservation. This method relies on pesticides because the age-old farmer's method of tilling to kill weeds was forsaken. Proponents say minimum-till farming is better for the environment because the soil is preserved, and soil moisture is used most efficiently.

At the same time, organic farming began to develop an alternate approach. Rather than embracing new machinery, tools, and pesticides, organic farming attempts to produce crops in the same conditions but with no synthetic pesticides or fertilizers. Instead of using radical new farming methods, it tries to rediscover and reincorporate the farming methods of the days before chemical fertilizers and pesticides were available. This is done not only in order to produce food for which some consumers are willing to pay a premium price but also because some farmers believe it is better for both farmers and the environment to rely less on expensive new technological solutions such as pesticides and specialized equipment. They believe that the chemical-dependent form of agriculture is just as bad for the soil as tillage agriculture because pesticides often kill most of the naturally occurring systems that make both the soil healthy and crops productive. Organic farmers focus on embracing the natural systems that exist in the soils and among crops in order to farm without needing to rely on recent technological advances, which they see as being as much of a curse as a blessing.

Although these two farming methods may seem to be poles apart because of their radically different methods and perspectives, they are united by the desire of the farmers who practice them to farm in a less environmentally destructive manner and have control over the health of their crops. While minimum-till farmers believe they are saving the soil by leaving it mostly undisturbed, organic farmers think they are saving the soil by not molesting its natural processes with chemicals. In recent years these two poles have been moving toward each other. Much organic research has focused on finding ways to ensure that organic fields do not leave tilled soil vulnerable to wind erosion, and many advancements in the use of cover crops have been made, often involving organic-specific equipment. Some minimum-till research has focused on developing "minimum input" agriculture, in which farmers can embrace the key concepts of minimum tillage while reducing their reliance on pesticides and other inputs.

In the end, much of the difference in approach between these methods (and between the farmers who embrace them) can be summed up as common difference between people who prefer high-tech or low-tech methods. Minimum-till farmers tend to like high-tech methods that allow them new possibilities for dealing with age-old problems. Organic farmers tend to prefer low-tech methods, in which tried and true methods and a holistic approach are appreciated.

Most people in the United States have an idealized picture of the American family farm. They envision a few fields of crops, some beef cattle, a pen full of pigs, and a milk cow called Daisy, along with a multigenerational family that lives in the same house. This image is common in movies, books, and TV commercials. Perhaps the most famous example of a fictional family farm is Dorothy's in *The Wizard of Oz*. Although there is hardship and struggle on the farm, and the land of Oz is wondrous, exciting, and magical, Dorothy longs to return to the simple life and good people in Kansas. After all, "there's no place like home." The family farm is the mythical home that many Americans feel is an important element of their nation's identity. In countries such as France, the psychological importance of the family farm is arguably greater still.

Perhaps because of that, making fundamental changes to the nature of the farm in America can become heated and intense. As most of those changes occur because of technological and scientific innovations, the social and political debate often becomes one of debating various production practices and farming approaches.

This is true of the general move away from small mixed farms to large specialized farms. Although it is often couched in terms of a debate of "family farms versus factory farms," or "family farm versus corporate farm," the reality is far more complex and challenging to summarize. Is a small farm that has some livestock and a few fields really a family farm if all it supports full-time is one farmer, with the other spouse working off the farm in town and the children moved away to college or jobs in the town or city because there is not enough work on the farm? Is a large, multi-barn hog operation just a factory and not a family farm anymore because of its huge scale and specialized nature, even if it provides enough work to employ a number of members of one family and allows the family to live in the countryside? Is it a farm at all if an agricultural operation is owned by a corporation that produces farm products merely as a business venture focused on making a profit? There are no easy answers to these questions. In the case of Dorothy's farm, the farm appears to be owned by her family, so in that sense it appears to be a family farm. Much of the labor, however, appears to be supplied by the three hired hands—surprisingly similar to the scarecrow, the cowardly lion, and the tin woodsman of Oz—so by some definitions it might not be a family farm!

Regardless of how one defines "family farm," "corporate farm," or "factory farm," there is no question that most farms in America have become far more specialized in the past century. Very few farmers now, even if they raise both livestock and crops, would consider attempting to supply all their family's food needs by having a few each of a dozen types of animals and a large vegetable garden, which was common a century ago. This may be partly due to farmers understandably no longer being willing to work from sunrise to sunset in order to simply support themselves. Many farmers still do work from sunrise to sunset—or beyond—but they generally restrict themselves to doing the kind of work that brings them the best results. For many crop farmers in the past century, that has meant spending more and more time on simply growing crops and not spending much time on other less productive areas of agriculture. The same applies to livestock production: an expert hog raiser will not necessarily be an excellent cotton grower. As with most jobs in society, after the initial wave of immigration and settlement, individuals have increasingly focused on doing what they do best. Even among farmers who still have mixed farms, such as the corn-soybean-pig farmers of the Midwest, production is focused on a semi-closed circle: the corn and the soybean meal are often fed to the pigs.

This specialization has been aided by the blossoming of new technologies that allow farmers to be far more productive and efficient than in the early years of the twentieth century. This comes with a cost, however. New technologies tend to be expensive and demand far more skill of the farmer than operating a horse-drawn plough or relying on an outdoor hog pen. This has encouraged farmers to concentrate ever more closely on the specific area of production in which they feel most comfortable. Over the decades, this has produced a U.S. farming industry that is highly productive and efficient, producing huge amounts of relatively cheap food for a burgeoning U.S. population.

Although this may seem a source of joy to U.S. society, with average families now able to earn enough to pay their yearly food bill in less than six weeks, it often leaves farmers feeling that they are trapped on a treadmill, climbing as fast as they can to simply stay in the same place. If they stop moving upward, they will go down and be crushed, they fear. While food is cheaper for consumers, and farmers are producing far more on every acre or with every pig or cow than they did in previous decades, they feel no more secure. Farm population numbers have been falling at a fairly steady rate for decades, and even though most farms are much larger now than they were in the past, many are still only marginally financially viable.

This is mainly the result of steady technological advances. Bigger and more efficient tractors and combines allow crop farmers to produce larger crops more cheaply. The farmers who have this technology can afford to sell each bushel for a little less than those relying on smaller equipment, putting pressure on the farmers with the older equipment. A farmer who raises 50,000 hogs probably does not need much more labor than a farmer who raises 10,000 hogs, but the farmer with 10,000 hogs would have to make five times as much profit per pig to enjoy the same income as the bigger farmer. It is now possible to raise far larger numbers of livestock per worker because modern production methods are much more efficient, but the less efficient farmer is in danger of being driven out of business. If he expands his operation, he may survive longer but never prosper. If he expands by borrowing a lot of money but then has poor production results, he may very rapidly be driven out of farming. Technological advances may help those who first embrace them and use them well, but for most farmers they are simply necessary in order to survive.

Average farm sizes have vastly expanded, but many farmers do not believe their profits have kept pace. While their profits may seem stagnant, their exposure to risk has greatly increased: most crop farmers owe hundreds of thousands of dollars for farm equipment they have bought and fertilizers and pesticides they have used; owners of large hog barns or cattle feedlots often owe hundreds of thousands or millions of dollars because of construction costs.

Farmers have responded to this cost-price-exposure squeeze in a number of ways. Some minimize their debt and try to get by with what they can produce with a modest land or facility base and spread out their risks by growing a wide number of crops or raising a number of species of livestock. Others specialize in producing only one product and take on whatever debt is required to give them the most efficient production system possible. Others try to find small-scale, low-debt ways of producing something unusual—a "niche market" product such as pasture-raised chickens or wild boars—that will provide their farm with enough income to be viable, but not so much debt that a bad year will bankrupt them.

Regardless of the approach taken, one factor seems common to most farms all the way back to Dorothy's fictional farm in Kansas: farming is a challenging business that constantly confronts farming families with difficult situations created by the inexorable flow of technological progress. The public may have a romanticized view of farming, but farming families know there is often little romance in the struggle to survive, whatever route they choose.

Food has always been vulnerable to contamination by pests and diseases. For thousands of years, people around the world have been cautious about pork because if the animal is raised in infected premises or fed contaminated feed, its meat could harbor a dangerous microorganism called trichna. (This disease has now been almost eradicated in the United States, Canada, and Europe.) Many people's grandmothers and aunts know of the dangers of leaving egg salad and mayonnaise-based foods out in the heat during long wedding celebrations. Hamburger fans know that if they do not cook the patties thoroughly, they can be struck by "hamburger disease," which is caused by the *E. coli* bacterium.

In recent years, public concern has become much greater as a result of the outbreaks of new diseases such as bovine spongiform encephalopathy ("mad cow disease"), foot-and-mouth disease (FMD), and avian influenza ("bird flu"). Millions of people have become terrified of these diseases, and governments have introduced regulations and formed crisis plans to deal with possible future outbreaks. Although disease panics are not new, incredibly fast changes in agricultural production methods have caused many to challenge whether farmers, food processors, or governments might be responsible for making the dangers worse than they need to be. Debate has erupted over whether new—or old—farm practices are part of the cause of the recent outbreaks. As with many topics in the overall debate about agricultural innovation, technology innovation proponents generally believe new developments will help solve the problems and eliminate the causes, whereas skeptics fear that technological changes both may have caused the problems and may make their solutions much more difficult.

Mad cow disease became a major worldwide panic after hundreds of thousands of cattle in the United Kingdom became sick with the disease in the late 1980s and early 1990s. Early in the epidemic, government scientists and agriculture department officials assured the public that the United Kingdom's beef supply was safe for humans to consume. Within a few years, it became clear that mad cow disease could jump the species barrier and infect humans with a form of the disease. Although it is unknown how many humans will eventually die from the human form of mad cow disease (it may reach only a few hundred rather than the millions once feared), the British and European public's confidence in government scientists, food regulators, and the food production system was badly damaged, causing a lasting state of skepticism among millions of citizens.

The spread of the disease from a few infected animals to hundreds of thousands of cattle and more than 100 people has generally been accepted as being the result of material—especially brain and spinal material—from infected animals being mixed into animal feed and fed to healthy animals. This produced protein products that were spread out to hundreds or thousands of other cattle in a circular process that caused the number of diseased cattle to spiral, leading many people to distrust the industrialized food production and processing system in general.

The rapid spread of foot-and-mouth disease (FMD) across the United Kingdom in 2001 also shocked the public in that nation and in the European Union (EU). Animals from a few infected flocks in the north of England spread the disease across the country because of the livestock production and marketing system (common to most industrialized countries) that caused millions of animals to be transported long distances between farms and feeding facilities and slaughter plants each year. Critics of industrialized agriculture argued FMD revealed how vulnerable nations become when they allow agriculture to become a nationwide and international business, in which one or a few infected animals can wreak havoc on millions of other animals and cause damages costing millions and billions of dollars.

The worldwide outbreaks of avian flu have also provided ammunition to the proponents of industrialized livestock production, who have argued that small, unconfined flocks of chicken and other poultry on small farms create reservoirs of vulnerable birds that could permanently harbor that disease or many others. A large confinement barn, in which all the birds are kept inside and not allowed to mingle with the outside world, can make for easy elimination of an outbreak within the flock through extermination of all the birds. The facility can then be sanitized with little chance that the disease will be reintroduced, proponents of confinement agriculture argue. The situation is the opposite for unconfined flocks that are allowed to be outside: the domestic fowl can come into contact with wild birds that have diseases such as avian flu, and even if they are all killed to control the infection, replacement birds will always be vulnerable to contact with infected wild birds.

So, as with many agricultural debates, the argument over whether new technologies and industrialized production methods have made consumers, farmers, agricultural products, and the environment safer or more threatened is divisive and unlikely to be easily resolved.

Perhaps no area of current agriculture is as rife with debate as biotechnology. The word alone—*biotechnology*—reveals the nature of this divisiveness. Until recent history, technology was seen as mainly something to do with tools and machines or methods employing specialized tools and machines. Plants and animals were generally seen as being part of the natural or biological world, even if they had been bred and developed to suit human needs. Biology and technology seemed to exist in different realms. In the past few decades, however, the science of plant breeding has evolved beyond the relatively simplistic methods employed since the dawn of civilization to employ the most cutting-edge laboratory technology possible. This includes the splicing of a gene or genes from one species into the genetic code of another species. Similar approaches are being developed in animal breeding and production. Just as *biotechnology* is a word combining two spheres not usually thought to be compatible, so too do biotechnological innovations bring into sometimes-jarring combination the natural and the scientific.

For some people, biotechnology has been a wonderful revolution, producing much more food for consumers at a lower cost, making farming simpler for farmers, and offering the promise of future developments that will provide animal and plant-based products that can cure health problems and provide an abundance of food for a rapidly expanding world population. For others it has been a frightening birth of an industrial-technological food system that has undermined a wholesome and natural food production system on which human society is based. Proponents say it produces better and cheaper food for consumers, better results for farmers, and better hope for the future of the planet, which has a limited amount of agricultural land. Critics say it threatens the health of consumers with radical new products, robs farmers of their independence, pollutes the environment with unnatural genetic combinations, and endangers the natural equilibrium that exists in farmers' fields.

The debate over glyphosate-resistant crops is typical of the overall debate. In the 1990s, Monsanto, a chemical company, helped develop a way of splicing a soil bacteria–based gene into crop species that allowed those crops to be sprayed with glyphosate and survive. Glyphosate, known most commonly by the trade name Roundup, kills virtually any living plant to which it is applied. When the naturally occurring gene is spliced into plants, however, creating a genetically modified organism (GMO), those plants become immune to the pesticide, allowing farmers to spray their crops with glyphosate and kill almost all the weeds, but leave the crop growing. Fans of the technology say it produces more crops per acre at lower cost than either conventional pesticide approaches or organic methods. Some farmers like it because it makes growing a crop much easier, with fewer or less complicated pesticide sprayings required. Some grain companies and food processors like it because it tends to provide harvested crops that have lower amounts of weed seeds and can produce grains that are more uniform. Critics and opponents, however, believe that the radical technique of genetic modification is so new that long-term tests on the safety of consuming the GMO crops are needed. Proponents of GMO crops argue that because the end product—the vegetable oil, the flour, or any other substance—is virtually identical to that made from non-GMO varieties of the crop, it really does not matter what specific method of plant breeding was used to develop the crop that produced the product. Critics counter that because the technology is so new, unexpected dangers may lurk within it somewhere, and it should not be allowed to create food products until more is known.

Similar arguments occur over the issue of the environmental impact of glyphosate-resistant and other GMO crops. Critics say there is a danger that the overuse of these crops will eventually, through natural selection in the field, produce weeds that are also immune to glyphosate or that are able to spread into and supplant natural plants. These "superweeds" will then pose a grave danger to farmers and the environment because they will be difficult to eradicate. Proponents of the technology say there are many other pesticides available to kill plants and that weeds develop resistance to any overused pesticide, regardless of how the crop is developed at the breeding stage. Not using glyphosate-resistant crops would force most farmers to return to using more pesticides, which would not be better for the environment.

The issue of farmer control is also one of keen debate with regard to GMOs. Because they work so well, farmers overwhelmingly embrace these crop varieties, giving them much of the acreage in the United States for soybeans and corn.

In order to buy the seeds for the GMO crop, however, farmers sometimes have been required to sign a license with companies that commits them to selling all of their seed for processing at the end of the growing season, preventing them from saving some of the seed to use for planting in future years. Some farmers say this robs them of a crucial right of farmers to save seeds that they have produced on their own fields, a practice that farmers have maintained for thousands of years. Companies have argued that farmers do not have to grow the GMO crops in the first place—many non-GMO varieties still exist—so if they want the ability to grow a crop that the inventors developed at great expense, they will have to agree to certain conditions, such as allowing the company to control the use and production of the seed stocks and to make money from selling the seed. Critics also say the extra crop produced by the GMO technology depresses world prices, yet farmers are forced to pay fees to grow it, damaging the farmer's income from both the revenue side and the expense side. Defenders say patent protection and other forms of intellectual property protection last only a few years, and eventually these new innovations will be free for anyone to exploit. For example, glyphosate lost its protection in the 1990s and is now widely available from a number of manufacturers, generally at a much lower price than during its years of one-company control.

The biotechnology debate has gripped the world since the mid-1990s and has caused disputes between the United States and European Union. The United States quickly embraced GMO technology and saw much of its farm production switch to biotechnology-boosted crop and animal products. The European Union was far more cautious, placing bans on the production, sale, and import of many GMO crops and animal products and only very slowly lifting restrictions. This difference in approach has caused repeated trade battles between the EU and the United States and has caused trade tensions between countries around the world. The United States has tended to see the EU restrictions as simply trade barriers in disguise; the EU has tended to see the American push for all countries to embrace biotechnology as an attempt to gain an advantage for its companies, which have created many of the biotechnological innovations.

At its core, the passion of the debate over biotechnology appears to reveal both the excitement and the uneasiness that accompanies any major technological advance. Proponents and visionaries see golden opportunities developing. Opponents and critics worry about unforeseen dangers and damaging consequences. Proponents seem to see the opportunities offered by the new technology as its main element and the harmful consequences as secondary and manageable. Therefore, to them, the technology should be embraced. Critics tend to see the dangers as being potentially greater than or equal to the benefits and think caution is wiser than excitement.

One thing is certain: the biotechnological rabbit is out of the hat and running, so this debate is unlikely to disappear any time soon.

See also Biotechnology; Genetically Modified Organisms; Mad Cow Disease; Organic Food; Pesticides.

Further Reading: Brouwer, Floor. *Sustaining Agriculture and the Rural Environment: Governance, Policy and Multifunctionality.* Northhampton, MA: Edward Elgar Publishing,

2004; Cochrane, Willard W. *The Development of American Agriculture: A Historical Analysis.* Minneapolis: University of Minnesota Press, 1993; Duram, Leslie A. *Good Growing: Why Organic Farming Works.* Lincoln: University of Nebraska Press, 2005; Hillel, Daniel. *Out of the Earth: Civilization and the Life of the Soil.* New York: The Free Press, 1991; Hurt, R. Douglas. *American Agriculture: A Brief History.* West Lafayette, IN: Purdue University Press, 2002; Kimbrell, Andrew, ed. *The Fatal Harvest Reader: The Tragedy of Industrial Agriculture.* Washington, DC: Island Press, 2002; McHughen, Alan. *Pandora's Picnic Basket: the Potential and Hazards of Genetically Modified Foods.* New York: Oxford University Press, 2000.

Edward White

ALIEN ABDUCTIONS

The heart of the controversy surrounding alien abductions tends to be whether or not they actually occur. Proponents of abduction, either abductees themselves or those who interview them, give a strong impression of the abductees having been through some sort of ordeal. Self-identified abductees often point to the uncertainty of scientific work as the largest piece of evidence in their favor, though paradoxically it is the lack of evidence to which they point. Skeptics, both self-identified and labeled as such by "UFO" communities, often point to the monumental barriers to space travel itself, much less the enormity of the conditions that would make extraterrestrial life possible. Both the vast distance entailed in space travel and the unique biochemistry of life are reasons to doubt the presence of alien contact with Earth.

As with many debates of this nature, there is a role here for social scientists too often left on the sidelines. The role of the social scientist is not to come down on one side or the other, but to further understand the social origins of the debate itself. Where do abduction narratives come from? Why do they follow such a common format, one that has become almost universal in its structure and formula?

If we examine the common abduction narrative, typically, the abductee is home alone or with only one other family member. The abduction usually takes place at night or in a secluded, remote spot such as a field. Abductees get a "feeling" that they are not alone or are being monitored in some way; many abductees use language that suggests their thoughts are being read or invaded. It is usually at this point that aliens manifest in some way, and almost universally they are small in stature, with large heads and eyes. Their skin tone is usually described as gray, green, or dull white. When the aliens arrive, abductees are usually paralyzed or "frozen." They are then experimented on, either at the abduction site or sometimes in an alien locale, such as a spacecraft. This experimentation usually involves the removal of bodily fluids, the installation of implants (often for the purpose of mind control or monitoring), or both.

What is worth noting about this narrative are all the elements it has in common with the mythos of vampires, to the point of being almost identical: a creature appears late at night in the window of a solitary victim, who is transfixed and frozen as the creature steals precious bodily fluids. In both narratives the victim often has no memory of the event and recollects it only under the influence of hypnosis or other "psychological" therapy.

The psychoanalysis component that both narratives share is a major key to their underpinnings. Both the modern vampire narrative and the alien abduction narrative have emerged since Freud's time and have an implicit, though folk, understanding of contemporary psychological theory. Popular conceptions of the subconscious mind and dream imagery are required to fully appreciate the narratives, and as such they are both dramas of the modern psyche. Narratives regarding monsters have existed in virtually every known human culture, and they almost always revolve around setting social boundaries: boundaries between adults and children, women and men, group and non-group members, and the living and the dead. All of these elements have social taboos in common, and monsters represent either the taboo itself or the consequences of its violation. Alien abduction, following this logic, is the late twentieth-century incarnation of the monster/taboo tale.

If we examine the idea that aliens are the quintessential modern monsters, what does this narrative express? For one, the space age brought a remarkable new amount of information that needs to be incorporated into our view of the human universe. There is a lot of new knowledge about our solar system, the Milky Way galaxy, and the newfound enormity of the rest of the universe. Astronomy with telescopes has been practiced for only about 400 years, and although that may seem like a long time, in the grand scheme of things, human culture has a lot of catching up to do. Consider as well that a manned mission to the moon took place only 40 years ago and that the modern alien abduction narratives have also existed for about that time period. The "space race" was set against a backdrop of Cold War tensions, clandestine government programs, the civil rights conflict, and a collective national identity that regarded technological progress as a presumed way of avoiding a worldwide "disaster." It is not surprising that monster stories, already incorporating social taboos and anxiety as their fuel, would emerge around a modern, space-age narrative that incorporates secret government programs, unidentified flying objects, alien technologies, and modern methods of psychic suppression. The common depictions of aliens represent everything Western culture regards as advanced: large, evolved brains (often allowing for "psi" powers); advanced technology; and a secret agenda that is not beholden to any one government.

Another common denominator of modern life in the twentieth and twenty-first centuries is wireless communication technology, specifically radio and television waves. An often overlooked but major component of space age technology is the satellite, allowing for instant global communication via relayed signal transmissions. There is a cultural association between the power of such signal transmissions (especially wireless Internet, which is a new form of radio transmission) and socio-technical progress. It is thus not surprising that official pursuits of extraterrestrial life, such as the search for extraterrestrial intelligence (SETI) program, involve the search for radio waves of external origin. SETI scientists are keen to criticize self-identified abductees and their researchers as participating in "junk science," sometimes offering psychological explanations of trauma or mental illness as their actual malady. Amateur radio astronomers who are certain that extraterrestrial life exists are also met with criticism from SETI. Conversely, abductees and amateur researchers criticize SETI for being exclusive and narrow in their scope.

See also Search for Extraterrestrial Intelligence (SETI); UFOs.

Further Reading: Freud, Sigmund. *Totem and Taboo.* Mineola, NY: Courier Dover, 1998; Fricke, Arther. "SETI Science: Managing Alien Narratives." PhD diss., Rensselaer Polytechnic Institute, 2004; UMI no. 3140946.

Colin Beech

Alien Abductions: Editors' Comments

Monster narratives are one of the resources humans have to deal with trauma. It is likely that alien abductions are in fact a victim's way of dealing with and grounding traumatic sexual or other assault experiences. It is also interesting to note how the character of aliens changes in line with broad social and cultural changes. In the 1950s, aliens often visited Earth to help us manage our out-of-control nuclear weapons and other Cold War escalations or as representations of the nuclear threat that hung over all our heads. In the film *The Day Earth Stood Still*, Klaatu, an alien in human form, and Gort, a robot, warned earthlings to get their act together, or a peaceful confederation of planets would destroy Earth as a potential threat. By the 1980s, the alien narrative had transformed to fit with a cultural period in which childhood trauma narratives were getting more attention, and the dangers around us were perceived to be more personal and less global. Depression is now closer to us than destruction by nuclear war. A combination of the popularity of conspiracy theories, especially concerning UFOs, and a very poor folk and professional understanding of psychosocial processes fueled the alien abduction narratives.

ART AND SCIENCE

The distinction between the sciences and the arts has not always existed, yet as professions have become more specialized and less generalized in the centuries since the Renaissance, the worlds of artists and scientists have drifted apart and now occupy much different spheres that do not seem to have anything in common with each other.

Leonardo da Vinci, one of the most famous painters of all time, would have been described as a "natural philosopher." For da Vinci, who was well known for his skills as a painter, sculptor, architect, musician, and writer, as well as an engineer, anatomist, inventor, and mathematician, there was no distinction between his roles as scientist and artist. In fact, the word *scientist* did not even exist until the 1800s, more than 300 years after da Vinci lived.

Historians have noted, however, that since the Renaissance, the aims of the arts and sciences have long been unknowingly intertwined. Both artists and scientists highly value creativity, change, and innovation. Both use abstract models to try to understand the world and seek to create works that have universal relevance. Furthermore, artists and scientists seem to borrow from each other on a regular basis: scientists often talk of beauty and elegance in reference to equations and theories; artists who draw the human figure must acquire an intricate knowledge of the human body. Abstract ideas and aesthetic considerations are fundamental to both groups' higher goals.

Given such similarities, it is no wonder artists and scientists find common ground through technology. Even though technology is most often associated with items that have a utility purpose (such as a hammer) or high-tech devices (such as high-definition televisions), technology, in the core sense of the word, is central to both artist and scientist alike. It is difficult to imagine a painter without a brush or musician without an instrument, just as it is difficult to imagine an astronomer without a deep-space telescope or a biologist without a microscope. Emerging technologies, which are technologies that are on the verge of changing existing standards, have come to be used by both artists and scientists and have brought the two groups closer together.

Twenty-first-century art and science will involve technology more than ever, and it seems that artists and scientists are both learning from each other. Artists have used modern technology to push past the boundaries of technological utility, and scientific breakthroughs have cast new light on old issues in the art world.

One new technology, cave automatic virtual environments (often referred to as CAVEs), has proved a fertile developing ground for scientists and artists alike. Originally developed for research in fields such as geology, engineering, and astronomy, CAVEs are 10-foot cubicles in which high-resolution stereo graphics are projected onto three walls and the floor to create an immersive virtual reality experience. High-end workstations generate three-dimensional virtual worlds and create the sounds of the environment. Special hardware and software track the positions and movements of a person entering that virtual environment, changing the images in the cave in a way that allows the visitor to feel immersed in the virtual space.

CAVEs allow for cutting-edge research to be conducted in the sciences by allowing scientists to test prototypes without physical parts, but they also provide a new type of canvas for artists, who can create interactive artwork like never before and allow their audience to actively engage art. CAVEs serve as a striking example of how technology can feed new developments simultaneously in both the arts and the sciences, as well as an example of how artists have taken the original purpose of a technology to new and unexpected levels.

Just as technology creates new fields in which artists and scientists can collaborate, so too have rapid developments in natural sciences created new questions for artists to tackle. Biological and medical researchers continue to discover more facts about the human body and enable radical new possibilities; artists will play a vital role in helping society come to grips with these novel discoveries.

It is not just scientists who need artists to help interpret new technological developments. The art world has taken hold of many scientific techniques for historical and conservation purposes. As digital art has proliferated, new fields, such as information arts and image science, have emerged.

With art sales having turned into a multimillion-dollar enterprise, art historians now use scientific analyses to date and authenticate pieces of art. One recent case involving art dating, however, highlighted the wide gap that still exists between the philosophies of the arts and sciences. A number of paintings, believed to be by the artist Jackson Pollock, were recently discovered in a storage locker, wrapped in brown paper. An art historian, who was an expert on Pollock, deemed the works to be authentic based on the artist's distinct drip and splatter style and the artist's relationship with the man who had kept the paintings. Yet a chemical analysis of the paint the artist used revealed that the paint was neither patented nor commercially available until after Pollock's death in 1956. Neither side of the argument has fully been able to prove its claim, but both stand by their trusted techniques.

The Pollock case may never be resolved, but a new generation of virtual splatter-painters has emerged that may erode such differences. Collaboration, not separation, may ultimately prove to be the greatest advantage for both science and art.

See also Virtual Reality; UFOs.

Further Reading: Ede, Siân. *Art and Science.* London and New York: I. B. Tauris, 2005; Jones, Stephen, ed. *Encyclopedia of New Media: An Essential Reference to Communication and Technology.* Thousand Oaks, CA: Sage, 2003; Wilson, Stephen. *Information Arts: Intersections of Art, Science, and Technology.* Cambridge, MA: MIT Press, 2002.

Michael Prentice

ARTIFICIAL INTELLIGENCE

The term *artificial intelligence,* or AI for short, refers to a broad area of applied scientific research dealing generally with the problem of synthesizing intelligent behavior. Usually this means building or describing machines that can perform humanlike tasks. Sometimes the work is highly theoretical or philosophical in nature; often computer programming is involved. In any case, a theory or example of artificial intelligence must necessarily be built upon some idea of what we mean by human "intelligence." Defining this concept is no simple matter because it touches on contentious philosophical subjects such as the idea of the consciousness and the spirit or soul; intelligence is in some sense what defines a human as opposed to a nonhuman animal, a volitional process as opposed to a process of nature.

The classic image of an artificial intelligence can be seen over and over again in popular and science fiction; the HAL-9000 computer from *2001: A Space Odyssey* is the quintessential example. This is a machine that can interact with people on their own terms; it is capable of speaking and of understanding speech; it appears to express emotions and opinions and can even demonstrate an appreciation of art and music. Of course, the computer also has superhuman capabilities in solving logical problems in chess, arithmetic, and electromechanical diagnostics. This fictional character is worth mentioning because it so clearly represents a prototypical object of AI research: a machine that is capable of humanlike behavior, at least in those capacities attributed to the mind.

Humans and their minds are incredibly complex creations, among the most complex in the known universe, capable of a very wide range of behaviors indeed. The actual means by which the mind (supposing it is an entity at all) performs its amazing feats of reasoning, recognition, and so on remain largely mysterious. Research in AI is usually focused on emulating or synthesizing one or another mental capability, and opinions differ widely on which capabilities are most exemplary of the state we call "consciousness" or "intelligence." One agreed-upon idea is that intelligence is made up of the ability to solve various problems; so the study of intelligence usually equates to the study of problem solving, and there are as many approaches to AI as there are types of problems to be solved.

One of the basic controversies in AI concerns how to decide what an AI is and how you know when you have one and, if AI is a matter of degrees, to what degree an instance of AI conforms to some ideal notion or definition of AI. The possibility of human technology attaining some definition of volition or consciousness is vehemently defended as a real one by prominent scientists such as Ray Kurzweil, whose book *The Singularity Is Near* includes a chapter titled "Responses to Criticism." There is strong popular and expert resistance to the idea that machines might be made to "do anything people can do," and these include some criticisms that are based in religious or other strongly held beliefs. Such philosophical and metaphysical debates are fascinating in their own right, but most practical AI research avoids them (for now) by considering only limited categories of problem-solving programs.

Within the actual practice of AI research is another, more technical arena for debate. There are essentially two approaches to programming intelligence, which derive their philosophical motivation from the computationalist and connectionist models of the mind in cognitive psychology. Connectionism, which studies intelligent behavior as an emergent property of many simple autonomous components, is the newer trend and is related to the use of neural networks and agent-based models in AI research. Computationalism envisions the brain as an abstract information processor and inspired the earlier, logic- and rule-based AI models for abstract reasoning.

The study of problem solving by machines has a longer history than one might expect. The mechanization of arithmetic processes was first accomplished by the ancient Chinese invention of the abacus. Europeans in the seventeenth century built clockwork machines that could perform addition and subtraction; notable examples were produced by Wilhelm Schickard in 1623 and Blaise Pascal in 1642. More sophisticated engines capable of performing inductive reasoning and algebraic analysis were theorized but not successfully constructed by Gottfried Wilhelm von Leibniz (1646–1716) and Charles Babbage (1791–1871).

General-purpose electronic computers are clearly the result of a long progression of technological innovations. Equally important, however, are the philosophical and theoretical inventions that laid the groundwork for modern theories of computation. In the seventeenth century, several complementary lines of thought were formulated and developed. One was the empiricist or rationalist doctrine in philosophy, exemplified by the writings of René Descartes and Thomas Hobbes, which essentially regarded the mind as an introspective computational force that could be considered distinct from the body.

Parallel to the development of the philosophical groundwork for AI research was the growth of a mathematical foundation for the science of computation. Descartes and others created analytic geometry, which correlated the study of tangible geometric entities with that of abstract algebraic ones.

The logician George Boole, in the mid-1800s, made a fundamental contribution to future sciences with the formulation of his system for logical arithmetic. This system essentially consists of the values 1 and 0 (or "true" and "false," sometimes notated as T and $\bot$) and the operators AND (* or $\wedge$), OR (+ or $\vee$), and NOT ($\neg$). This last operation inverts the value of a variable, so $\neg 1 = 0$ and $\neg 0 = 1$. The other two operators behave the same as in regular arithmetic. They preserve the arithmetic properties of commutativity ($x + y = y + x$), associativity ($x + (y + z) = (x + y) + z$), and distributivity ($x * (y + z) = (x * y) + (x * z)$).

More complex logical operators can be defined in terms of these fundamentals. For example, *implication* ($x \rightarrow y$) is equivalent to $\neg x \vee y$ and can be expressed semantically by the sentence "if x then y," the variables x and y referring to statements that can be evaluated as true or false. Boole demonstrated that such constructions, using his three simple operators, could be used to express and evaluate *any* logical proposition, no matter how complex, provided only true/false statements are concerned in its premises and conclusions.

In 1937 an engineer named Claude Shannon demonstrated in his master's thesis that electromechanical relays could represent the states of Boolean variables and that, therefore, circuits composed of relays could be used to solve Boolean logic problems by analogy. This insight formed the basis for the systematic development of digital circuit design and modern digital computing. Furthermore, a theoretical basis had been established for electronic circuits that were capable of addressing logical problems of arbitrary complexity, or in a very specific sense, machines that think.

Arguably the most important figure in the history of AI was the British mathematician Alan Turing. Turing was responsible for the development of several early computer systems as part of his work as a code-breaker for Allied intelligence during World War II. Even before that, however, he revolutionized the theory of computing by his description in 1937 of what is now known as a Turing machine. This "machine" is an abstraction, which essentially consists of a long (potentially infinite) sequential memory of symbols and a mechanism to read, write, and move forward or backward within the memory. (Technically, a Turing machine is a type of mathematical construct known as a finite-state automaton.)

Turing's breakthrough was to demonstrate that such a simple device could, given enough memory and the correct set of initial symbols (or program), compute any describable, computable function that maps one string to another. Because this description can be considered to include written language and mathematical thought, Turing thereby demonstrated, well ahead of his time, the universality of the digital computer as a problem-solving device.

Turing's formal mathematical work on computability was fundamental because it suggested that logical statements—that is, propositions from the world of ideas—could be expressed in discrete symbolic terms. Furthermore, the reducibility of the symbols themselves was shown to be extreme; any discrete sequence of logical actions could be expressed in terms of a very few basic actions, just as George Boole and his successors had shown that all mathematical knowledge could be expressed (albeit in a clumsy and convoluted fashion) as the outcome of his fundamental logical operators.

As crucial as Turing's theoretical work proved to be, his name is perhaps most widely recognized today in conjunction with a relatively speculative paper concerning the notion of intelligent machinery (Turing 1950). In it, he makes a historic attempt at a practical, empirical definition of intelligence in the form of a thought experiment he called the "imitation game," which is now popularly known as the Turing test.

The basis of the test is very simple: a human and a computer each must answer questions put to them remotely by a human judge. It is assumed that the questions and answers are in the form of text. The judge's task is to identify the computer by this means alone. If the computer can successfully pass itself off as human, it can be fairly said to have demonstrated intelligent behavior.

Turing's hypothetical test informally set the tone for a generation of AI researchers. He neatly sidesteps the philosophical problems inherent in the very proposition of a synthesized intelligence, casually deploying the term *human computer* as a reference point for judging the performance of his hypothetical intelligent machines. This implies a belief in the fundamentally discrete and quantifiable nature of human thought processes, and the imitation test itself lays out a working definition of "intelligence" or "consciousness" based on the ability to manipulate an entirely abstracted set of symbols. As a general description of intelligence, this has been widely challenged, particularly by the science of embedded AI, which considers physical survival as a primary goal of all autonomous entities.

The old school of AI research describes logic-based or rule-based conceptions of the mind, based on computationalist theories from cognitive psychology. The computationalist theory describes the human brain as a function that transforms a set of information.

In this approach to AI, intelligence is conceived of as an algorithm by which information (about the state of the world or the premises of a problem) is coded into symbols, manipulated according to some set of formal rules, and output as actions or conclusions. As Herbert A. Simon and Alan Newell write, "a physical symbol system is a necessary and sufficient condition for general intelligent action" (Newell and Simon 1976). The methods used to arrive at this kind of model of AI are derived from the framework of formal logic built up by George Boole, Bertrand Russell, Alfred Tarski, and Alan Turing (among many others).

A "traditional" problem-solving program proceeds to calculate a solution by applying rules to a set of data representing the program's knowledge of the world, usually at a high level of abstractions.

This kind of intelligence essentially equates to an aptitude for *search* over large and complex spaces. Its effectiveness is perhaps best demonstrated in the application to specialized and formal problems, and the playing of games such as chess is a notable example. Much work has been done on the development of game-playing programs, and in 1997 a chess-playing computer famously defeated reigning world champion Gary Kasparov.

A procedure for playing a winning game of chess can be abstractly represented as a search for an optimal (winning) state among a great number of possible outcomes derived from the current state of the board. A computer is uniquely capable of iterating searches over huge sets of data, such as the expanded pattern of possible positions of a chess game many, many moves later. It should be pointed out, though, that the "brute force" computational approach to winning at chess used by a computer program does not claim to emulate a human being's thought process. In any case, Deep Blue (successor to Deep Thought), currently the world's best chess-playing computer, does not rely on brute force but on the technique of "selective extensions." Computer chess programs search more or less deeply. Deepness measures the number of moves ahead that the computer searches. Six moves is the minimum depth, 8 or so moves is considered average depth, and maximum depth varies but is typically 10–20 moves. Selective extension adds a critical dimension to the computer's search algorithm.

It is no coincidence that the study of AI originally developed in conjunction with the formulation of game theory in the 1950s. This movement in mathematics studies the formal and probabilistic process of playing games of chance and decision making. It has been applied to models of human activity in economic, military, and ecological contexts, among others.

Another notable application of rule-based AI is in the study and emulation of natural language. Text "bots" provide convincing artificial interactions based on applying transformative rules to input text. These language-producing programs' obvious limitation is their lack of flexibility. All language bots thus far produced are strictly limited in the scope of their conversation.

The program SHRDLU, written in 1971, clearly and easily discusses, in English, an extremely simple world consisting of several blocks of various shapes and sizes, a table, and a room. SHRDLU also has a virtual "hand" with which it can manipulate the objects in its world.

Research into language processing by computers has since grown to encompass an entirely new field, that of computational linguistics. Chat bots rely on specific models of the representation of human knowledge and the working of language, and computational linguists use simulations to explore and evaluate such models. Machine language processing also has its practical applications in such tasks as content search and automatic translation.

The more recent chat bot ALICE has a somewhat extended range of conversation and is further extensible by the use of a specialized markup language that defines the program's set of recognizable patterns and responses. However, the program still relies on a top-down, pre-specified set of rules for its behavior and does not include the crucial facility for training and self-development that would allow its behavior to become truly intelligent. It is thought that more fundamental models of language and grammar development might be brought to bear on the problem.

Rule-based AI has had another strong application in the creation of computer expert systems. These are programs in which the knowledge of a human specialist (in, say, tax law, medicine, or popular music) is encoded in a database and used in a process of inference to guess the best solution to a particular question. Expert systems have been used with considerable success to aid in such tasks as diagnosing medical conditions and predicting a person's taste in music or film.

In recent years, the design of expert systems has been strengthened by probabilistic techniques such as fuzzy logic and Bayesian networks. These techniques encode degrees of uncertainty into the database used for inference.

Fuzzy logic adds continuous variables to the toolset of classical logic. Rather than declaring that a statement is or is not true, fuzzy logic allows for some in-between degree of truthfulness. Conclusions are reached by comparing input data to various ranges and thresholds. Fuzzy logic can be performed by neural networks (which are discussed later).

Bayesian networks, named after the Reverend Thomas Bayes, an eighteenth-century mathematician, use conditional probability to create models of the world. Bayes's innovation was to describe a formal representation of the real-world process of inference. In his description, actions are associated with effects with some degree of strength. Once such associations are defined, it is possible to estimate a probable sequence of events based on the resultant effects. Bayesian networks automate this procedure and provide a strong and efficient facility for revising conclusions based on updates in real-world data. They are used in risk analysis and a variety of diagnostic tasks.

The newer school of AI research creates agent-based intelligent behavior models. These are concerned less with an entity's ability to solve formal problems in an abstract problem space and more with the idea of survival in a physical or virtual environment. This general movement in AI owes its existence to twentieth-century revolutions in the sciences of psychology and neuroscience. The goal of these researchers is to investigate emergent behavior, in which the system is not told what behavior is expected of it (in the form of a set of rules and scripted responses) but rather produces interesting behavior as a result of the particular structure of interconnections among its parts.

The information-processing systems of agent-based AIs are often patterned on connectionist models from cognitive psychology, and the most typical of these models is the neural network. A neural net is built up of a large number of simple processing units, coupled to input variables (or sensors) and output variables (or actuators). These units are neuronal cells in a biological neural network and similarly behaved programming constructions in artificial neural nets. Each essentially records its level of activation and influences the level of activation of units to which it is connected.

In addition, a neural network as a whole is equipped with some facility for development and learning. Development may be as simple as randomly mutating the strengths of connections between neurons. Learning is implemented as some kind of reinforcement of successful behaviors. Neural networks excel at pattern recognition tasks, in which a given set of input is tested against a target state, or set of states, to which it has been sensitized. This facility has made the implementation of neural networks an important element in such tasks as e-mail spam filtering and the digital recognition of faces and handwriting.

An important feature of many kinds of artificial intelligences is feedback. Input data is processed and fed back to the output part of the agent, the means by which it influences the world: for example, the movement capabilities of a robot or the physical expression of a genome. The output then affects the agent's next perceptions of the world, and the program alters its input/output mapping, tuning itself into an optimal state.

Not all agent-based AIs use neural networks or directly concern themselves with data manipulation. Approaches to synthesizing intelligent behavior vary widely in their degree of abstraction. One particularly interesting example is that of Valentino Braitenberg, a neuroscientist whose 1984 book *Experiments in Synthetic Psychology* describes a series of very simple but increasingly complex robots that consist entirely of direct couplings of sensors and motors. For example, the initial, "type 1" vehicle consists of a single sensor whose output affects the speed of a single wheel, positively or negatively. Type 2 vehicles have two sensors and two motors, which allows for several more variations in the possible arrangements of linkages, and so on.

What is interesting is that such a simple design paradigm can give rise to a wide variety of behaviors based on the specification of the sensor–motor linkages. A type 2 vehicle, equipped with light sensors such that brightness in the left "eye" inhibits the speed of the left wheel, and likewise for the right, will tend to orient itself toward light sources. If the coupling is reversed, with the left eye inhibiting the right wheel and vice versa, the robot will avoid light sources. Much more complex behaviors can be achieved by introducing more sensors, various types of sensors, and multiple layers of linkages; for example, the right eye may inhibit the right wheel but speed up the left wheel, to whatever degree. Some of the more complex vehicles exhibit markedly lifelike behavior, especially with multiple vehicles interacting, and have been described by observers as behaving "timidly" or "altruistically."

The Braitenberg vehicles demonstrate some important points raised by embedded intelligence research. The neural substrate of the vehicle is no more important than the types of sensors it possesses and their relative positioning. There is no reasoning facility as such; that is, there are no internal decisions being made that correspond to human-specified categories of behavior. Describing the mechanisms of the robot deterministically is a simple task, but predicting the robot's behavior can be virtually impossible. These robots demonstrate, in a moderately abstract way, how progressively complex brains might have evolved.

The prototypical example of emergent intelligence to be found in nature is that of selective evolution in organisms. One cornerstone of the theory of natural selection is that traits and behaviors are not abstractly predetermined but arise from the varied genetic characteristics of individuals. The other component of the theory is the potentially unintuitive fact that the survival value of a trait or behavior exerts selective pressure on the genes that give rise to it in an individual's descendants. The process of selection produces organisms without predetermined design specifications beyond the goal of survival.

An important branch of AI and robotics research attempts to harness the somewhat unpredictable power of evolution by simulating the process in a virtual environment, with the survival goals carefully specified. The typical approach is to begin with a trivial program and produce a generation of variations according to some randomizing process and some set of restraints. Each individual program in the new generation is then evaluated for its fitness at performing some task under consideration. Those that perform poorly are discarded, and those that perform well are subjected to additional mutations, and so on. The process is repeated for many generations.

The most critical element in this kind of simulation is the selection of a function for evaluating fitness. In the real world, the only criterion for selection is survival, which is determined by the effects of a hugely complex system of interactions between the individual organism and its entire environment. In artificial evolution, the experimenter must take great care when designing the fitness function for the particular task at hand. Considerable success has been achieved in evolving solutions for many practical problems, ranging from mechanical design to new algorithms for data compression. Ultimately, evolutionary approaches to creating AI suffer from limitations of scope similar to those that haunt other approaches. The problem is the inherent impossibility of designing a fitness function that matches the complexity of real-world survival demands.

A particularly interesting example of artificial evolution can be seen in the virtual creatures created by Karl Sims (1994). These are structures of blocks whose size, shape, and arrangement are dictated by growth patterns encoded in genotypes. The resultant morphologies then compete at various tasks such as swimming, walking, and jumping, in a virtual world that includes a detailed physics model. Some of the resulting creatures resemble biological shapes and perform somewhat familiar gestures. Others are quite novel, such as creatures that move by rolling themselves over, or are shaped asymmetrically. These experiments thereby illustrate the phenomenon of emergent, unforeseen behaviors, which is considered a key feature of artificial intelligence.

Within the field of AI research, as in many branches of science, there is considerable variation and debate over what levels of abstraction to use in tackling a problem. At the highest level, computers can be useful aids in abstract decision making when provided with both extensive and applicable sets of rules. At the lowest level, specifications might be created for growing artificial brains from the simplest components. At either extreme, there are serious limitations; rule-based AI can encompass only limited domains of expertise, and agent-based and evolutionary approaches typically lack the power to solve complex problems.

Some integration of the various historical techniques is underway and will likely be pursued more fully in the future. Hierarchical architectures such as the subsumption architecture incorporate layers of symbolic logical reasoning and sub-symbolic reactive processing. Future evolutionary approaches may take cues and starting points from existing biological structures such as the brain (which, recent research suggests, actually undergoes selective "rewiring" in individuals during development).

Significant successes have already been achieved in using machines to solve problems that were recently considered only manageable by humans. The cutting-edge research projects of 10 or 20 years ago can now be found diagnosing heart conditions, controlling car engines, and playing videogames. In some cases AI techniques are now indispensable, such as in communications network routing, Internet searches, and air traffic control. Many of the techniques previously categorized as AI are now considered just "technology," as the bar for machine intelligence continues to be raised.

See also Brain Sciences; Memory; Mind; Robots; Social Robotics.

Further Reading: Brooks, Rodney. "Elephants Don't Play Chess." *Robotics and Autonomous Systems* 6 (1990): 3–15; Jackson, Philip C. *Introduction to Artificial Intelligence.* 2nd ed. New York: Dover, 1985; Kurzweil, Ray. *The Singularity Is Near: When Humans Transcend Biology.* New York: Penguin, 2005; Luger, G. F. *Artificial Intelligence: Structures and Strategies for Complex Problem Solving.* 5th ed. London: Addison-Wesley, 2005; Newell, Alan, and Herbert Simon. "Computer Science As Empirical Enquiry." *Communications of the ACM* 19 (1976): 113–26; Pfeiffer, Rold, and Christian Scheir. *Understanding Intelligence.* Cambridge, MA: MIT Press, 2001; Simon, Herbert. *The Sciences of the Artificial.* 3rd ed. Cambridge, MA: MIT Press, 1996; Sims, Karl. "Evolving Virtual Creatures." *Computer Graphics (Siggraph '94 Proceedings)* (1994): 15–22; Turing, Alan. "Computing Machinery and Intelligence." *Mind* 59, no. 236 (1950): 433–60.

Ezra Buchla

Artificial Intelligence: Editors' Comments

The relation of artificial intelligence research to current models of brain function is not coincidental. Viewing the brain as an organic computational device and depicting human experience as the rendering into mind of the computational implications of electrochemical sensory stimulation provides some powerful and persuasive tools for understanding what, to this point, has always been a mysterious process. That religious experience does not involve contact with another level of reality, spiritual dimension, or divine entity is a conclusion inherent in such an approach; thus, while the application of what are essentially mechanistic metaphors to brain function may create powerful analytical tools, what these tools tell us about the brain—or about religious experience—remains a highly disputed area, laden with the presuppositions of both those who are looking for God and those who regard God as a sociocultural construction.

ASYMMETRIC WARFARE

Asymmetric warfare is a term whose definition evokes as much conflict as its examples represent. Although not a new idea—often the point is made that it goes back to at least the biblical account of David and Goliath when a small boy using a sling and stone defeated a giant armor-clad warrior—it has come to characterize various conflicts in the twentieth and twenty-first centuries.

Definitions of asymmetric warfare have ranged from simple differences between two opposing forces (thus a lack of symmetry) to one side "not playing fair." These and other definitions add little to the overall understanding of the term because they are not easily tied to specific examples.

Perhaps the best definition that illustrates the complexity of the concept was offered by Steven Metz:

> In military affairs and national security, asymmetry is acting, organizing and thinking differently from opponents to maximize relative strengths, exploit opponents' weaknesses or gain greater freedom of action. It can be political-strategic, military-strategic, operational or a combination, and entail different methods, technologies, values, organizations or time perspectives. It can be short-term, long-term, deliberate or a default strategy. It also can be discrete or pursued in conjunction with symmetric approaches and have both psychological and physical dimensions.

To frame his definition, Metz includes unconventional or guerrilla operations (operational asymmetry), Mao Tse-tung's "People's War" (military-strategic asymmetry), and the North Vietnamese actions during the Vietnam War (political-strategic asymmetry). Additional terms such as *unconventional, guerrilla,* and *irregular* to describe warfare are effectively corollaries to the term and yield parallel examples. Asymmetric warfare is not bound by a specific time frame, nor does the employment of asymmetric tactics preclude a combatant from also using symmetric tactics as well, perhaps simultaneously.

An early example of asymmetric warfare is found in a conflict that occurred between two empires in the third century B.C.E. during the Second Punic War. Charged with defending Rome against Hannibal and his Carthaginian forces, the Roman General Fabius Maximus soon realized that he could not afford to confront this great army in a direct battle. To counter Hannibal's advantage, Fabius adopted a protracted approach in his engagements with the Carthaginians. Fabius determined when and where his inferior forces would conduct small-scale raids against his opponent; he avoided Hannibal's proficient cavalry and always favored withdrawal over a decisive encounter. Through the use of such tactics, Fabius not only wanted to inflict casualties upon the Carthaginian force but also hoped to affect their morale. Fabius enjoyed some initial success, but this strategy required time to be effective, time that the Roman Senate did not believe they had. They removed Fabius, and his replacement chose to engage Hannibal in a decisive battle that proved disastrous to the Roman forces. Fabius's impact is still visible today in what is known as a "Fabian strategy"—that is, when an inferior force uses a similar approach.

History is replete with other examples of a weaker force adopting Fabian tactics. Generals George Washington and Nathanael Greene relied on a similar strategy during the American Revolutionary War. Realizing that their Continental Army lacked the capability and experience of the British Army, they avoided large-scale direct clashes. Both Washington and Greene attempted to wear down the British through a series of small-scale clashes against rear and lightly defended areas, combining guerrilla-type attacks with limited conventional force battles. Washington understood victory would not come on the battlefield alone. Maintaining pressure on the British Army and continuing to inflict mounting casualties, Washington hoped that Britain would not want to continue to pay such a high cost for achieving its goals in the colonies. The asymmetric tactics employed by Washington and Greene successfully brought about American independence and the withdrawal of all British forces.

Although these examples of asymmetric conflicts fit Metz's definition, they are not likely parallels to asymmetric warfare in the twenty-first century. Recent wars in Iraq and Afghanistan would, no doubt, dominate any discussion of the topic. Well before the continuation of the war in Iraq, however, or before that fateful late summer day in September 2001 when terrorist attacks on the World Trade Center and Pentagon precipitated the war in Afghanistan, a blueprint for asymmetric war was born in China during the first half of the twentieth century.

China was a nation in a state of flux. Shortly after the abdication of its last emperor in 1912, China entered into a long period of internal struggle that was exacerbated by the Japanese invasion of Manchuria in 1931. In 1925 Chiang Kai-shek became the head of the dominant Chinese political party (the Kuomintang), which at the time was receiving assistance from Russian Communist advisors. In 1927 Chiang split with the Communists and attempted to consolidate China under a Nationalist government, thereby starting the long civil war between the Nationalists and the Communists. In the ensuing civil war that lasted throughout the Japanese invasion of Manchuria and World War II, the leadership of the Communist Party would eventually be assumed by Mao Tse-tung. One of

FOURTH-GENERATION WARFARE

Another term that has been associated with asymmetric warfare is *fourth-generation warfare*. This type of warfare is characterized as an evolved type of insurgency that uses whatever tactics are necessary to convince the enemy that their goals are either unobtainable or too costly. It prefers to target an opponent's political will over its military strength. According to Thomas X. Hammes, first-generation warfare evolved from medieval times and was symbolized by the type of warfare fought by nation states with large armies as witnessed in the Napoleonic era. The amount of resources needed to train and equip such a large army could only be gathered at this time by a nation-state. By contrast, second-generation warfare centered on fire and movement, with emphasis on indirect fire by artillery (like what was experienced on the Western Front at the beginning of the Great War of 1914–18). Third-generation warfare was epitomized by the German Blitzkrieg (usually translated as "Lightning War") during World War II, with an emphasis on combined arms operation, the use of the radio to direct forces in the field, and an emphasis on maneuver rather than static warfare. Hammes observes that changes in the type of warfare in each case were not just a result of new technology but were also brought about by changes in political, economical, and social structures.

the main themes of Mao's approach during the civil war was to make it a People's War based on the millions of peasants that he would organize into a military and political movement, a strategy that eventually led the Communists to victory over the Nationalists.

In 1937 Mao wrote his famous work *On Guerrilla Warfare.* In this book, Mao outlines his philosophy on how guerrilla warfare should be conducted. A key passage states that if guerrillas engage a stronger enemy, they should withdraw when he advances, harass him when he stops, strike him when he is weary, and pursue him when he withdraws. Mao identified three phases that along with the preceding reference became the basis for what can be seen as his approach to asymmetric warfare. In the first phase, insurgents must build political support through limited military action designed to garner support of the population. Actions such as politically motivated assassinations would be appropriate during this phase. Phase 2 sees the insurgent/guerrilla expand his operations with the goal of wearing down his opponent, which is usually government forces or an occupying power. In this phase, insurgents increase their political footprint and attempt to solidify control over areas in which they have made gains. In the final phase, asymmetry meets symmetry as insurgent activity is combined with more conventional action against the opponent. This phase should not occur prior to the insurgent movement achieving a favorable balance of forces.

In his philosophy, Mao did not set out a timeline for the shift from phase to phase or for fulfilling the overall strategy. In addition, he not only believed that an insurgency might shift from phase 3 back to a phase 2 or 1 depending on the situation, but also observed that the insurgency did not have to be in the same phase throughout the country at the same time. Mao's approach to his insurgent movement reflects many elements of the definition of asymmetric warfare later developed by Metz.

Unlike Mao, who for the most part employed his asymmetric strategy in a civil war against other Chinese forces, Ho Chi Minh had to contend with not one but two foreign powers during his struggle to unite Vietnam. At the end of World War II, France attempted to reestablish its colonies in Southeast Asia but was opposed in Vietnam by Ho Chi Minh and General Vo Nguyen Giap (Ho's military commander). Both Ho and Giap were Communists who had spent time in China during World War II and were influenced by Mao's theories on guerrilla warfare. Ho drew on Mao's three-phase approach to guerrilla warfare against the French and then against the Americans. Throughout the conflict with the French, Giap did not attempt to engage them in a decisive battle for he knew that if he massed his forces against an enemy with superior firepower, he would likely be defeated. Instead, Giap conducted small-scale raids on French forces that spread throughout the country. When the French attempted to mass and conduct an offensive operation, Ho's forces, known as Viet Minh, would disperse and avoid contact.

As the Viet Minh were engaging French forces, Ho Chi Minh consolidated his support amongst the Vietnamese population and, in accordance with phase 2 of Mao's doctrine, attempted to establish control over various parts of the country. In an effort to draw the Viet Minh into a major battle, the French penetrated deep into Viet Minh territory and tried to seize control of Dien Bien Phu. The

Viet Minh refused to become engaged and instead continued to strike at French forces in other areas of the country. To reinforce these areas, the French drew from troops committed to Dien Bien Phu. Eventually, the balance of forces at Dien Bien Phu swung in favor of the Viet Minh, and Giap, in line with Mao's phase 3, attacked the depleted French forces. The French were routed and began their withdrawal from Vietnam in 1954.

Vietnam at this time was divided along the 17th parallel (with Ho's forces in the north) and became a key region for Cold War politics, with Ho supported by China (united since 1949 under Mao) and the Soviet Union and with South Vietnam supported by the United States. Ho's struggle to unite Vietnam would eventually lead him into conflict with the United States. Until the mid 1960s, aid from the United States consisted mainly of economic and military hardware as well as some military advisors. At the beginning of 1965, the U.S. military became directly involved in the conflict, with the initial deployment of air and marine forces. Throughout the year, the United States' commitment drastically increased, and they became involved in offensive operations against the Viet Minh forces in the north and Viet Cong in the south. Although the Viet Cong was an insurgent group fighting against the U.S.-backed South Vietnamese government, arms and key leadership were supplied by North Vietnam. Just as they had done against the French, Ho and Giap avoided large-scale clashes with the United States, preferring to hit them where they were weaker and then fade away into the countryside. Ho realized that just as time had been on his side against the French, it would once again prove an ally against the Americans. This was a characteristically asymmetric insight.

Despite the increasing casualty figures, the message to the American people from the political and military leadership was that the Americans were winning the conflict in Southeast Asia. One of the biggest turning points in the Vietnam War occurred when Ho and Giap decided to switch to a more symmetric strategy and launched the Tet Offensive in January 1968. Although this decision proved fatal for the Viet Minh and Viet Cong as U.S. and South Vietnamese troops crushed the offensive, the fact that they were able to mount such a large-scale attack came as a shock to most Americans.

It was at this point that many in the United States started to believe that the Vietnam conflict was not going to end quickly. Protests against the U.S. involvement in the war grew, and Ho took advantage of this opportunity. He attempted to show, through the media, how his forces were struggling against a corrupt government in the south and more importantly how that corrupt government was willing to allow thousands of U.S. soldiers to die so that it could survive.

Ho was able to use Mao's "three phase" approach to help him eventually defeat the U.S. and South Vietnamese forces. Ho shifted between the various phases depending on the status of his insurgency. He utilized both asymmetric (hit-and-run) and symmetric (Tet Offensive) strategies and was able to exercise political control over contested territories. In addition, Ho was able to introduce a new facet into the asymmetric conflict; using mass media, he targeted the morale of the U.S. population. He capitalized on the fact that time was on the side of the insurgent and that asymmetric conflicts were no longer just fought with guns and bullets.

While Ho was winning the public relations war abroad, at home, the character of the conflict increased his support among the population. There was no possibility of Viet Minh and Viet Cong forces striking out against Americans in their homeland. The Vietnamese were fighting the Americans (as they had fought the French) in a struggle for survival.

Other recent examples of asymmetric warfare are not hard to find. Throughout the 1990s and into this century, asymmetric conflicts dominated the Middle East as the Palestinians struggled against the Israelis for a homeland. Unlike the first *Intifada* (Arabic for "uprising") from 1987 to 1993, the second *Intifada* that commenced in 2000 saw the Palestinians adopt the use of suicide bombers. Suicide bombers were not a new tactic, but they were used with increasingly lethal effects against the civilian population. In response, Israel launched conventional force actions that also resulted in the loss of civilian life. In addition, globally, the impact of terrorism appeared to be on the rise. International and domestic terrorism used asymmetric approaches in attempts to both draw attention to and advance their causes. All these examples combined, however, did not do as much to bring the concept of asymmetric warfare into the headlines as did the events of September 11, 2001 (9/11).

Although there was much more planning and preparation involved for the attacks on 9/11, it was represented by the hundreds of analysts who appeared on numerous news programs as an asymmetric attack of untold proportions—19 terrorists armed with box cutters killed thousands of innocent civilians. The asymmetry in this kind of attack was more far reaching than just the events of that day. The attacks shut down the busiest airspace in the world; its impact on financial markets was enormous; and trillions of dollars have been spent since September 2001 on prosecuting what became known as the Global War on Terror.

Following the attacks on the World Trade Center and Pentagon, responsibility was laid at the feet of Osama bin Laden and his al Qaeda terrorist network. It was the second time that al Qaeda had attacked the World Trade Center; the first occurred in 1993 when a truck laden with explosives blew up in the underground parking area. At the time of the attacks on the United States, bin Laden was living in Afghanistan where al Qaeda had, with the support of the Taliban-led government, established a number of terrorist training camps.

In a speech before Congress on September 20, 2001, President George W. Bush demanded that the Taliban hand over bin Laden and close all the terrorist training camps. The Taliban rejected the demands and on October 7, 2001, less than one month after the attacks, the United States and Britain launched air strikes against Afghanistan. By December 2001, with the fall of Kandahar in the south, the majority of Taliban and al Qaeda forces, reportedly including bin Laden, had fled to the mountainous regions of Afghanistan or into the border area of Pakistan. Before leaving Afghanistan in early December, the leader of the Taliban, Mullah Omar, stated that the Taliban would regroup and conduct a guerrilla campaign against the U.S.-led forces.

The capability of the Taliban and al Qaeda to carry out the type of asymmetric attacks that Mullah Omar spoke of had been highlighted even before September 2001. On October 12, 2000, in what was suspected to be an al Qaeda–supported attack, two men in an explosive-laden small boat approached the USS *Cole* while it was in the port of Aden, Yemen, and detonated their craft. It was an incredible feat considering that the *Cole* was one of the most sophisticated warships afloat, one designed to protect against air, surface, and subsurface attack and built at a cost of about $800 million. Although the ship was saved, the U.S. Navy lost 17 sailors, and another 39 were injured. On the ground, asymmetric attacks combined with hit-and-run tactics would later form the basis of the Taliban's reaction to coalition forces in Afghanistan.

The insurgency that formed after the defeat of Mullah Omar's government was more than just remnants of al Qaeda and the Taliban. Historically, a foreign force on Afghan soil has been a rallying point for numerous indigenous groups, some of which were loosely affiliated with the Taliban, while others may have even fought the Taliban when they ruled Afghanistan. Regardless of their affiliations prior to the arrival of the U.S.-led coalition, these groups joined in the resistance. From 2002 until 2006, the insurgents conducted numerous small-scale attacks on coalition forces through the use of rocket attacks, improvised explosive devices, and suicide bombers. Despite the coalition's greater mobility and advanced technology, insurgents were able to inflict mounting casualties during this time period. Just like Mao and Ho, the Taliban do not have a set time frame for their struggle—it has been said with regard to the methodical approach of the Afghan insurgents that NATO has all the watches, but the Taliban have all the time.

Since the beginning of 2006, there has been a spike in insurgent attacks, especially in southern Afghanistan. Although still trying to avoid large-scale engagements with coalition forces, insurgents are becoming bolder. They are still using improvised explosive devices to inflict the vast majority of casualties, but in a sign of their growing strength and resolve, they have been engaging larger numbers of coalition forces. The use of suicide bombers has dramatically increased since the beginning of 2006, and of particular concern to the coalition forces and international community, some of these attacks are occurring in areas of the country such as the west and north that have been relatively free of insurgent activity. The suicide bomber has always been viewed as one of the ultimate asymmetric weapons not only because he or she is usually accurate but also because the unpredictable nature of such an attack also instills fear in the population. Coalition forces are now more attentive to anyone who approaches their convoys and have fired on innocent civilians as they employ more aggressive force-protection measures. In such an environment, it is hard to win hearts and minds.

The Taliban, just like Ho Chi Minh in Vietnam, and other insurgent groups are attempting to convince certain nations that the goals they might want to achieve in Afghanistan are not worth the price their citizens will have to pay, and they are doing this through asymmetric means. A surge in kidnappings is just one of the methods employed by the insurgents, as evidenced by the July 2007 kidnapping of 23 South Koreans traveling from Kabul to Kandahar. During their six weeks in captivity, two hostages were freed and two were executed. The remaining 19 were released when the South Korean government agreed to withdraw their approximately 200 troops by the end of 2007. Notwithstanding that the mandate for the South Korean troops was due to end in 2007, this incident highlights insurgent attempts to affect the morale of troop contributing nations. When the hostages were released, a Taliban spokesman claimed that kidnapping was a useful and cost-effective strategy, one that would demonstrate to coalition partners that the United States was not able to guarantee security within Afghanistan. Tactically speaking, the loss of 200 troops was not a significant issue for the international coalition; however, the psychological and visual impact of having your citizens kidnapped and executed may make governments think twice before committing troops in the future, to say nothing as to how events such as this affect the thousands of aid workers attempting to bring some sort of normalcy to the everyday lives of average Afghans.

In early 2003 the world was told the United States and the coalition of the willing would engage in a campaign of "shock and awe" in an attempt to liberate Iraq from Saddam Hussein's rule. Through the use of air strikes and ground forces, the United States wanted to secure a quick, high-tech conventional victory—and they did, as Iraqi armed forces were quickly routed within one month. In a very high-profile event on May 1, 2003, President Bush announced on board USS *Abraham Lincoln* that major combat operations in Iraq were over. In one sense they were, but what followed and continues at the time of this writing is a violent asymmetric insurgent campaign.

Since 2003, coalition casualties have increased under a relentless attack from numerous insurgent elements. The types of attacks ongoing in Iraq are similar to those being conducted in Afghanistan. Ambushes, improvised explosive devices, and suicide attacks continue to impede the return to a relatively normal life in Iraq. Insurgents are not engaging coalition forces in large-scale battles, preferring instead to inflict as many casualties as possible before fading away, just like Ho Chi Minh's forces did in Vietnam. Slowly and steadily, the number of coalition dead continues to rise. In addition to the attacks on coalition forces, the insurgents are also targeting elements of the newly formed Iraqi security forces. As in Afghanistan, there have been numerous kidnappings followed by gruesome executions as the insurgents call for the withdrawal of all coalition forces. These kidnappings and executions have been highly publicized for the consumption of audiences in troop-contributing nations.

The asymmetric attacks used by the insurgents are designed not only to weaken the morale of the soldiers but also to target public opinion back in the United States, Britain, and other nations that have troops serving in Iraq. Al Qaeda is also sponsoring and inspiring asymmetric attacks directly against the nations with troops in Iraq. On March 11, 2004, in Madrid, Spain, a series of explosions ripped through several trains, killing 191 and injuring more than 1,800, just three days before Spain's national election. In an unexpected outcome, Spain elected a Socialist prime minister who had said during the election campaign that he would remove Spanish troops unless the United Nations took charge of the Iraqi mission. As promised, the newly elected Spanish prime minister pulled out Spain's combat troops by the end of April 2004 and the last of the support troops left one month later. The cause and effect of the bombings and the election results can be a source of debate, but the fact remains that the Socialist Party was not expected to win the elections prior to the bombings.

Another tactic being used very effectively in Iraq is political assassination. As described earlier, political assassination is mentioned in phase 1 of Mao's phases to guerilla war as a means for an insurgency to build political support. In September 2007, just 10 days after meeting with President Bush, a key Iraqi ally, Abdul-Sattar Abu Risha, was assassinated. Abu Risha was seen as an important figure in the Iraqi province of Anbar, where during the first part of 2007 a growing opposition to al Qaeda in Iraq had developed. Local residents had been opposed to al Qaeda's attempt to coerce citizens of Anbar to join their movement. Senior members of President Bush's administration cited on numerous occasions the developments in Anbar and Abu Risha's leadership as examples of how the conflict in Iraq was turning around. Only four days prior to this assassination, the top U.S. soldier in Iraq, General Petraeus, had singled out the province of Anbar for its successful campaign against insurgents. This type of asymmetric attack affects events not only in Iraq but also back in the United States; combined with the numbers of coalition soldiers dying as the war in Iraq approaches its fifth year, the message from the insurgents is they are there for the duration.

After the preceding lengthy discussion of examples of asymmetric warfare in the twenty-first century, it might seem strange now to ask if labeling warfare as asymmetric is appropriate. *Asymmetric warfare* is a phrase that is used to describe a multitude of conflicts both past and present; to some it is a term that is often thrown around without due consideration of just what is meant when a conflict is labeled asymmetric. In fact, one might say that *all* wars could in one way or another be classified as asymmetric.

In a recent article, Peter H. Denton states that the key issue in twenty-first-century conflicts is not symmetry versus asymmetry but rather a disparity between opposing forces. He claims that there is little value in using the term *asymmetric*; instead, what we need to look at is the difference in the kind of force each side brings to the fight and the disparate objectives of the opposing sides. Denton claims that it is unproductive to consider that an apple is asymmetric compared to a screwdriver when there is no common frame of reference; by extension, he states that the same can be said for conflict in the twenty-first century.

The overwhelming majority of the literature on current warfare is centered on U.S. conflicts, which by their vary nature are asymmetric. The United States now finds itself involved in many regional conflicts defending what are perceived to be its national interests—for example, global security or free-market access to oil in the Middle East. American dominance in manpower and technology naturally leads to conflicts having asymmetric aspects. Therefore, labeling so many conflicts as asymmetric may be pointless; unless the United States finds itself in combat against a major power such as Russia or China, all conflicts in which it participates will be asymmetric.

So is it still appropriate to use the term *asymmetric warfare* to describe conflicts in the twenty-first century? Depending on what author you pick, the answer will be different. What we do know is that if a conflict is labeled as asymmetric in the twenty-first century, the use of that term itself will be the subject of debate. Simply put, if a conflict is labeled as asymmetric, then that term needs to be defined and put in context if it is to have any value. Asymmetric warfare has been around for thousands of years, and some would say that it will dominate warfare for the foreseeable future; others say that the term is currently overused and no longer adds value to the description of conflicts. This debate is not one that will end soon.

The examples of asymmetric warfare provided in this entry are just that; they are not meant to be all-inclusive but rather serve to demonstrate some of the common characteristics of asymmetric warfare. Determining whether a conflict is an example of asymmetric warfare in the twenty-first century is a contentious subject. Whether present-day conflicts in Afghanistan and Iraq are examples of asymmetric warfare will continue to be debated for many years to come. That being said, the importance of understanding asymmetric warfare and the role it has played throughout history and will play in the future cannot be overstated.

See also Chemical and Biological Warfare; Nuclear Warfare; Urban Warfare; Warfare.

Further Reading: Cassidy, Robert M. "Why Great Powers Fight Small Wars Badly." *Military Review* (September–October 2002): 41–53; Denton, Peter H. "The End of Asymmetry: Force Disparity and the Aims of War." *Canadian Military Journal,* Summer 2006, 23–28; Hammes, Thomas X. *The Sling and the Stone.* St. Paul, MN: Zenith Press, 2006; Lambakis, Steven J. "Reconsidering Asymmetric Warfare." *Joint Force Quarterly*, no. 36 (Winter 2005): 102–8; Liddell Hart, Basil Henry, Sir. *Strategy.* 2nd rev. ed. New York: Praeger, 1972; Mack, Andrew. "Why Big Nations Lose Small Wars: The Politics of Asymmetric Conflict." *World Politics* 27, no. 2 (1975): 175–200; Metz, Steven. "Strategic Asymmetry." *Military Review,* July–August 2001, 23–31; Petraeus, David H. *Report to Congress on the Situation in Iraq.* September 10–11, 2007. http://www.foreignaffairs.house.gov/110/pet091007.pdf; Tse-tung, Mao. *On Guerrilla Warfare.* Trans. Samuel B. Griffith. New York: Praeger, 1967.

William MacLean

AUTISM

Arguments over autism tend to revolve around causes and practical treatments. There is increasing (though disputed) evidence that mercury in pollution and vaccines may cause autism. Medical doctors and researchers generally advocate treatments to make autistic individuals more socially acceptable, whereas alternative approaches tend to highlight personal comfort and environmental influences. The rate of autism among children grew from 1 in 2,500 to 1 in 166 (an increase of 15-fold) over the 1990s. Both advocates of mainstream medical approaches and advocates of complementary and alternative medicine (CAM) regard this increase as shocking and urge immediate action in response.

An individual diagnosed with autism typically exhibits abnormal social, communication and cognitive skills. No child is diagnosed with autism at birth.

On average, diagnosis occurs when the child is around 44 months old; autism is several times more common in boys than in girls. Slowness or inappropriate behavior in social interactions (unprovoked tantrums, screaming, objection to physical closeness, not making eye contact); withdrawing into isolation from others; problems with touch, taste, and smell; problems with hearing and language; and sleep problems can characterize an autistic individual. The manifestation of symptoms varies drastically among autistic people, however. Severe cases may cause an individual to exhibit self-destructive tendencies, constant bodily motion of some sort, and insensitivity to pain. In regressive autism, children begin to develop normally and then suddenly and catastrophically reverse their normal skill development. On the other end of the autistic spectrum, some individuals diagnosed with autism are indistinguishable from their peers at school and work. Autistic savants, individuals who exhibit extraordinary abilities in a certain area (music, mathematics, etc.), are relatively rare. Autistic individuals have the same life expectancy as normal individuals; families caring for autistic kin therefore face ongoing problems finding practical approaches to daily living.

Until the 1970s, because of the theories of Dr. Bruno Bettelheim and others, parents were often blamed for their child's autistic condition. This created cycles of guilt and doubt for families attempting to establish supportive and loving environments. Even after autism was accepted as a neurological (brain and nervous system) disorder in the mainstream medical community, many caregivers still suffered a stigma in both public places and doctors' offices. Although medical experts generally insist that autism is strictly a brain disorder, caregivers consistently report that children diagnosed as autistic all have physical disorders in common. These include food allergies and sensitivities, asthma, epilepsy, sleep disorders, inflammatory bowel disease and other digestive disorders, and persistent problems with both viral and yeast infections. Although it is generally agreed that more autism-trained professionals are essential and that a much better understanding of dietary and environmental triggers is needed, battles over funding priorities and immediate concerns often separate mainstream medical and alternative approaches to autism. Everyday caregivers and medical researchers often hold different research priorities that reflect their respective interests in autism. Medical doctors and researchers emphasize the importance of funding genetic research and developing new drug treatments. Parents of autistic children emphasize the need for better care options using more complete understandings of interacting physical, social and environmental factors.

Autism was an unknown condition before 1943. It was first diagnosed in 11 children born in the months following the introduction of thimerosal-containing vaccines given to babies in 1931. Thimerosal is a mercury-based preservative used to preserve vaccines between the time they are made and the time they are injected. Mercury is a known neurotoxin (brain poison), and its effects are most damaging to children's still-growing brains and nervous systems. Its effects are also more damaging to persons who, for one reason or another, are less able to flush mercury from their bodies. In 1991 the U.S. Centers for Disease Control and Prevention (CDC) and the U.S. Food and Drug Administration (FDA) began recommending that more vaccines containing thimerosal be added to routine childhood vaccinations, even within hours of birth. At that point, doctors and pharmaceutical companies had not considered that these additional injections would expose children to levels of mercury that were 187 times greater than "safe" exposure levels set by the U.S. Environmental Protection Agency (EPA). Dr. Neal Halsey, pediatrician and chairman of the American Academy of Pediatrics committee on infectious diseases from 1995 to 1999, was an outspoken and lifelong advocate of vaccination programs and accordingly of these additional vaccines. Dr. Halsey had a drastic change of opinion in 1999, when documents alerted him to thimerosal exposure levels that could account for the skyrocketing autism rate. At his urging, the American Academy of Pediatrics and the Public Health Service released a statement strongly recommending that vaccine manufacturers remove thimerosal from vaccines and that doctors postpone the thimerosal-containing hepatitis B vaccine given to newborns.

Many medical doctors and researchers, government organizations, and pharmaceutical companies do not readily accept blame for increased autism incidence through advocating thimerosal-containing vaccines, however. They generally make two arguments against this assertion. First, safe levels as established by the EPA are for methyl mercury, whereas thimerosal contains ethyl mercury, which may behave slightly differently in the body. Although studies and dangers of methyl mercury are available in published literature, there are fewer studies available on the toxic effects of ethyl mercury. Second, they assert that, although the 15-fold increase in autism incidence coincides with the years in which more thimerosal-containing vaccines were given to infants, the increase can be explained by development of more effective diagnostic methods. In other words, they assert that autism rates increased because doctors could better identify and diagnose autism.

The National Academy of Sciences' Institute of Medicine (IOM) reported in 2004 that there was no proven link between autism and vaccines containing thimerosal, frustrating researchers and caregivers who insist the link is indisputable. Members of Congress, such as Florida representative David Weldon, also a physician, criticized the IOM report for using studies that were poorly designed and flawed. Under pressure from Congress, the IOM agreed to review their first report.

Ongoing studies are establishing new links between mercury poisoning and autism and continue to evoke charged arguments. Thimerosal was being phased out, however, with vaccine stores being shipped to developing countries and used up by 2005 in the United States. Whatever the final answer, anecdotal evidence is troubling. After American manufacturers introduced thimerosal-containing vaccines in China, the autism rate leapt from virtually zero to over 1.8 million cases in seven years.

Mercury in thimerosal-containing vaccines is not the only connection between poisoning and autism. The symptoms of autism are similar to those of heavy metal poisoning. Exposure to environmental pollution, especially from electricity and plastics production, is increasingly implicated in the rise of autism rates. Dr. Claudia Miller, professor of family and community medicine at the University of Texas, published a study in the March 2005 issue of the journal *Health and Place* that linked mercury pollution from coal-fired power plants to increases in autism incidence in surrounding communities. This corroborated findings of a Florida study released in 2003 that linked mercury emissions from incinerators and power plants to toxic buildups in surrounding wildlife. Toxic mercury buildups in fish and other wildlife have been shown worldwide to cause brain degeneration and nervous system disorders in native peoples who depend on wildlife as food resources. Environmental groups continue calling for more stringent emissions regulations. They, along with parents in communities that surround known mercury and other toxic pollution sources, argue that affordable technology exists to produce electricity without such dangerous emissions. The electric power industry, through such research groups as the Edison Electric Institute and the Electric Power Research Institute, argues that wind carries mercury pollution generated in a single area to all over the world. They say this means that just because a power plant has mercury emissions does not mean that the mercury pollution from that plant will stay in the local area. Research suggests that genetic susceptibility to autism would frequently go undetected if not triggered by an environmental influence, however. Dr. Jill James published results in the April 2005 issue of *Biology* reporting a biochemical condition found in blood samples of autistic individuals. The particular condition rendered all 75 autistic individuals tested less able to effectively detoxify poisons, especially heavy metals such as mercury, in their bodies. Out of 75 individuals tested who were not autistic, none showed this biochemical trait. Genetic traits may have little effect on children until they are exposed to high levels of toxins with which their bodies are less able to cope; the resulting buildup of toxins could then result in autistic symptoms.

Although this research is promising, many parents and caregivers of autistic children emphasize that genetics alone cannot provide practical answers to treatment questions and everyday problems. They are more concerned with practical treatment issues such as dietary triggers; fungal, viral, and bacterial infections; allergies; management of the monetary costs of care; and communication. Autistic individuals tend to have digestive problems traceable to undiagnosed food allergies and yeast (fungal) proliferation. Some debate exists as to whether chronic allergies and yeast problems can be a contributing cause of autism or a result of autism. Some high-functioning autistic individuals have experienced a reversal of autistic symptoms and consequently are not recognized as autistic after changing their diets and dietary supplements to alleviate allergy and yeast problems. Others have experienced an alleviation of symptoms or lessened symptom severity. Dietary changes can include eliminating flours, sugar, and starches such as corn and potato and taking over-the-counter caprylic acid and vitamin supplements. The medical establishment tends to advocate prescription anti-yeast, antiviral, and antibiotic medications. Although many parents and caregivers have found these to be of great help, others witness too many side effects from prescription pharmaceuticals and opt for CAM strategies.

Two other reasons families and caregivers cite for trying CAM treatments include the insufficient number of medical experts trained to treat autistic symptoms and the exorbitant costs of American medical care. Because the behavioral and cognitive symptoms of autism are accompanied by chronic infections and allergies, the numbers of visits to physicians and medication prescriptions are high among people with autism—over $6,000 a year on average for one autistic individual. Behavioral and cognitive testing and training add to the regular costs of care. Most autistic people require supervision around the clock, making regular employment for the caregiver difficult. The lack of available, appropriate, and low-cost day care for autistic individuals also increases the burden of care on families.

Autism-trained professionals are essential for the care and effective treatment of autistic individuals, many of whom may learn, to different degrees, to compensate in social situations for their autism. Autistic people have difficulties maintaining conversational rhythm, in addition to difficulties with classifying what is important in different social situations. They usually look toward other rhythms, through their own bodily senses, to make their everyday world meaningful. Alexander Durig suggests understanding autistic individuals on their own terms, instead of the more common practice of trying to make autistic individuals act and appear "normal." He points out that everyone has difficulties in accomplishing certain types of tasks and difficulties fitting into certain types of social situations. Whatever the cause or the options for treatment, the increasing prevalence of autism in children is a troubling and costly statistic.

See also Brain Sciences; Immunology; Memory; Vaccines.

Further Reading: Allen, Arthur. *Vaccine.* New York: Norton, 2006; Durig, Alexander. *Autism and the Crisis of Meaning.* Albany: State University of New York Press, 1996; Grandin, Temple. *Thinking in Pictures, Expanded Edition: My Life with Autism.* New York: Vintage Press, 2006; Liptak, Gregory S., Tami Stuart, and Peggy Auinger. "Health Care Utilization and Expenditures for Children with Autism: Data from U.S. National Samples." *Journal of Autism and Developmental Disorders* 36 (2006): 871–79.

Rachel A. Dowty

BIODIESEL

Biodiesel is a diesel substitute made from biological materials that can be used directly in a diesel engine without clogging fuel injectors. It is the product of a chemical process that removes the sticky glycerines from vegetable and animal oils. Because it is made from biomass, biodiesel is considered to be a "carbon neutral" fuel. When burned, it releases the same volume of carbons into the atmosphere that were absorbed when the biomass source was growing. Controversy exists, however, in particular over the use of high-quality grains for biodiesel because land is taken out of food production and devoted to the production of fuel.

In comparison with diesel, biodiesel has reduced particulate, nitrous oxide and other emissions and emits no sulfur. Biodiesel is used as a transportation fuel substitute, either at the rate of 100 percent or in smaller percentages mixed with diesel. It mixes completely with petroleum diesel and can be stored safely for long periods of time. Biodiesel is biodegradable and does not contain residuals that are toxic to life forms. It has a higher flash point than diesel and is safer to ship and to store. Biodiesel is mixed with kerosene (#1 diesel) to heat homes in New England. It has been used as an additive in aircraft fuel, and because of its oil-dissolving properties, it is effective as a nontoxic, biodegradable solvent that can be used to clean oil spills and remove graffiti, adhesive, asphalt, and paint; as a hand cleaner; and as a substitute for many other petroleum-derived industrial solvents. Biodiesel is appropriate as a renewable alternative to petrochemical diesel because it can be produced domestically, lowers emissions, and does not cause a net gain in atmospheric carbon. The overall ecological benefits of biodiesel however, depend on what kinds of oils are used to make it.

Biodiesel is made from the oils in seeds, nuts, and grains or animal fats. Oil sources for biodiesel production are called biomass "feedstock." Agricultural crops are specifically grown to be utilized for biodiesel production. The crops vary according to region and climate; in the northern hemisphere, biodiesel is most often made from soybean, sunflower, corn, mustard, cottonseed, rapeseed (also known as canola), and occasionally, hempseed. In tropical and subtropical regions, biodiesel is made from palm and coconut oils. Experiments have been conducted in extracting oils from microorganisms such as algae to produce the fuel. These algae experiments have raised hopes of converting sunlight more directly into a renewable fuel to be used with existing diesel machinery. Biodiesel offers an efficient use for waste oils that have already been used for frying or otherwise manufacturing food for human consumption. Waste grease biodiesel is made from the oils left over in the fryer and the grease trap, as well as from the animal fats and trims left over from the butchering process.

The fuel is manufactured from both fresh and waste oils in the same way, through a chemical reaction called transesterification, which involves the breaking up, or "cracking," of triglycerides (fat/oil) with a catalytic agent (sodium methoxide) into constituent mono-alkyl esters (biodiesel) and raw glycerin. In this process, alcohol is used to react with the fatty acids to form the biodiesel. For the triglycerides to react with the alcohol, a catalyst is needed to trigger a reorganization of the chemical constituents. Most often, a strong base such as sodium hydroxide (lye, NaOH) is used as a catalyst to trigger the reaction between the triglycerides and the alcohol. Either methanol (CH_3OH, or wood alcohol, derived from wood, coal, or natural gas) or ethanol (C_2H_6O, known as grain alcohol and produced from petrochemicals or grain) is used as the alcohol reactant. The chemical name of the completed biodiesel reflects the alcohol used; methanol makes methyl esters, whereas ethanol will produce ethyl esters. Most frequently, methanol is the alcohol used.

Triglycerides are composed of a glycerine molecule with three long-chain fatty acids attached. The fatty acid chains have different characteristics according to the kind of fat used, and these indicate the acid content of the oil. The acid content must be taken into account in order to get the most complete reaction and thus the highest yield of biodiesel. To calculate the correct proportions of lye and methanol needed to transesterify a sample of oil, the acid content of the oil is measured through a chemical procedure called titration. Waste vegetable oil has higher fatty acid content and requires higher proportions of lye catalyst than fresh oil. The titration results determine the proportion of lye to combine with the methanol or ethanol to form a catalytic agent that will complete the reaction fully.

To make biodiesel, the oil is placed in a noncorrosive vessel with a heating element and a method of agitation. The mixture of lye and alcohol is measured and mixed separately. Usually the amount of methanol or other alcohol needed amounts to 20 percent of the volume of oil. The amount of lye depends on the acidity of the oil and is determined by titration and calculation. When the oil reaches 120–30 degrees Fahrenheit (48–54 degrees Celsius), the premixed catalytic solution is added. The oil is maintained at the same heat, while being gently stirred for the next 30 to 60 minutes. The stirring assists in producing a complete reaction. The mixture is then left to cool and settle; the light yellow methyl esters or ethyl esters float to the top and the viscous brown glycerin sinks to the bottom. The esters (biodiesel) are decanted and washed free of remaining soaps and acids as a final step before being used as fuel.

Although transesterification is the most widely used process for producing biodiesel, more efficient processes for the production of biodiesel are under development. The fuel is most simply made in batches, although industrial engineers have developed ways to make biodiesel with continuous processing for larger-scale operations. Biodiesel production is unique in that it is manufactured on many different scales and by different entities. It is made and marketed by large corporations that have vertically integrated supplies of feedstocks to mesh with production, much the same way petrochemical supplies are controlled and processed. There are also independent biodiesel-production facilities operating on various scales, utilizing local feedstocks, including waste oil sources, and catering to specialty markets such as marine fuels. Many independent engineers and chemists, both professional and amateur, contribute their research into small-scale biodiesel production.

Biodiesel can be made in the backyard, if proper precautions are taken. There are several patented and open source designs for biodiesel processors that can be built for little money and with recycled materials. Two popular designs include Mike Pelly's Model A Processor for batches of waste oils from 50 to 400 gallons and the Appleseed Biodiesel Processor designed by Maria "Mark" Alovert. Plans for the Appleseed processor are available on the Internet for free, and the unit itself is built from a repurposed hot water heater and valves and pumps obtainable from hardware stores. Such open source plans and instructions available online have stimulated independent community-based biodiesel research and development. In some cases, a biodiesel processor is built to serve a small group of people who decide to cooperate on production within a region, taking advantage of waste grease from local restaurants. The biodiesel fuel is made locally and consumed locally, reducing the expenses of transporting fuel.

The use of vegetable oil in diesel engines goes back to Rudolf Diesel, the engine's inventor. The diesel engine demonstrated on two different occasions in Paris during the expositions of 1900 that it could run on peanut oil. (The use of peanut oil was attributed to the French government, which sought to develop an independent agriculturally derived power source for electricity and transportation fuel in its peanut-producing African colonies.)

As transportation fuel, biodiesel can be used only in diesel engines in which a fuel and air mixture is compressed under high pressure in a firing chamber. This pressure causes the air in the chamber to superheat to a temperature that ignites the injected fuel, causing the piston to fire. Biodiesel cannot be used in gasoline-powered internal combustion engines. Because its solvent action degrades rubber, older vehicles running biodiesel might need to have some hoses replaced with those made of more resistant materials. The high lubricating capacity of biodiesel has been credited with improving engine wear when blended at a 20 percent rate with petroleum diesel.

The benefits of burning biodiesel correspond to the percentage of biodiesel included in any formulation. The overall energy gains of biodiesel are also assessed according to the gross consumption of energy required to produce the oil processed into fuel. Biodiesel processed from waste grease that has already been utilized for human food consumption has a greater overall energy efficiency and gain than biodiesel produced from oils extracted from a virgin soybean crop grown with petrochemical-based fertilizers on land previously dedicated to food production.

Biodiesel's emissions offer a vast improvement over petroleum-based diesel. Emissions of sulfur oxides and sulfates (the primary components of acid rain) are eliminated. Smog-forming precursors such as nitrogen oxide, unburned hydrocarbons, and particulate matter are mostly reduced, although nitrogen oxide reduction varies from engine to engine. The overall ozone-forming capacity of biodiesel is generally reduced by nearly 50 percent. When burned, biodiesel has a slightly sweet and pleasant smell, in contrast to the acrid black smoke of petroleum-based diesel.

Biodiesel has the additional and important advantage of carbon neutrality, in that it is produced from the energy stored in living organisms that have been harvested within 10 years of the fuel's manufacture. During their growing cycle, plants use carbon dioxide to process and store the energy of the sun in the form of carbon within their mass. When plants are converted to fuel source and burned, they can release into the atmosphere only the amount of carbon consumed and stored (through photosynthesis) during their life cycle. When petroleum fuel is burned, carbons are released into the atmosphere at a much faster rate. The atmospheric release of the fossilized carbons of petroleum fuel places an impossible burden on existing living biomass (trees and plants) to absorb the massive quantities of carbons being released.

The mass production of biodiesel from biomass feedstock grown specifically for fuel has not been proven to produce a net energy gain because of the energy inputs needed in current industrial farming methods. These include the inputs of petroleum-derived fertilizers and herbicides, fuel for farm machinery, and the energy needed to pump water and transport the fuel. Concerns have also been expressed about taking land and other agricultural resources previously devoted to food production for the production of biomass for fuels such as biodiesel.

See also Fossil Fuels; Global Warming.

Further Reading: Official Site of the National Biodiesel Board. http://www.biodiesel.org/markets/mar; Pahl, Greg. *Biodiesel: Growing a New Energy Economy.* Burlington, VT: Chelsea Green, 2004.

Sarah Lewison

BIOTECHNOLOGY

Although biotechnology can be defined broadly to include any technological application that uses a biological system or organism, the term has become synonymous with the use of modern technology to alter the genetic material of organisms. The ability to recombine DNA across species has created significant social controversy over the creation of biohazards, "terminator" genes, genetic pollution, "playing God," and the ethics of altering the lives and appearance of animals.

Biotechnology may be considered as any technological application that uses biological systems, living organisms, or their derivatives. The term *biotechnology* covers a broad range of processes and products and can be understood from at least two perspectives. From one perspective, biotechnology (a) is the process of using (bio)organisms to produce goods and services for humans. The use of yeast in the processes of fermentation that make bread and beer and the historical domestication of plants and animals are examples of this kind of biotechnology. From another perspective, biotechnology (b) is the process of using genetic technologies to alter (bio)organisms. This perspective is illustrated by the hybridization of plants, the cloning of sheep, and the creation of genetically engineered food crops. Although both perspectives are debatable endeavors, biotechnology type (b) is inherently more problematic than type (a). Most ethical, moral, and religious criticism of biotechnology focuses on type (b) biotechnology. The United Nations (UN) definition, then, focuses on the history and problems associated with type (b) biotechnology.

Biotechnology type (b) began in the late nineteenth century as the rise of the science of genetics established a basis for the systematic and conscious practice of breeding plants and animals. In 1944 Oswald Avery identified DNA as the protein of heredity. In 1953 James Watson and Francis Crick discovered the structure of DNA. Biotechnology blossomed in the late 1960s and early 1970s with the development of recombinant DNA technology and the birth of the biotechnology industry. In 1997 the human genome was mapped and sequenced. Since the 1990s, an increasing number of techniques have been developed for the biotechnological reproduction and transformation of organisms. An examination of controversies associated with biotechnology includes at least the biotechnological modification of microorganisms, of plants, and of animals.

In the early 1970s, researchers across the world began exploring recombinant DNA (rDNA) technology, or the technology of joining DNA from different species. rDNA technology is performed either by a gene-splicing process, wherein DNA from one species is joined and inserted into host cells, or by cloning, wherein genes are cloned from one species and inserted into the cells of another.

In 1972 biochemist Paul Berg designed an experiment allowing him to use rDNA technology to insert mutant genetic material from a monkey virus into a laboratory strain of the *E. coli* bacterium. Berg did not, however, complete the final step of his experiment because he and his fellow researchers feared they would create a biohazard. Because the monkey virus was a known carcinogen, and because the researchers knew that *E. coli* can inhabit the human intestinal tract, they realized their experiment might create a dangerous, cancer-inducing strain of *E. coli*.

Berg and other leading biological researchers feared that, without public debate or regulation, rDNA technology might create new kinds of plagues, alter human evolution, and irreversibly alter the environment. Berg urged other researchers to voluntarily ban the use of rDNA technologies and sent a letter to the president of the National Academy of Science (NAS). The NAS responded by establishing the first Committee on the Recombinant DNA Molecules. In 1974 that committee agreed to the temporary ban on the use of rDNA technologies and decided that the issue required the attention of an international conference. Scientists worldwide were receptive to the voluntary ban and halted their work on rDNA experiments.

In February 1975, Berg and the NAS organized the Asilomar Conference on Recombinant DNA. Lawyers, doctors, and biologists from around the world convened in Monterey, California, to discuss the biohazard and biosafety implications of rDNA technology and to create a set of regulations that would allow the technology to move forward.

This conference provided a meaningful forum for discussing both whether scientists should use rDNA technologies and how to safely contain and control rDNA experiments. The Asilomar participants were able to identify proper safety protocols and containment procedures for some of these experiments, and they also prohibited some experiments, such as Berg's experiment involving cloning of recombinant DNA from pathogenic organisms.

The Asilomar conference resulted in the first set of National Institutes of Health (NIH) guidelines for research involving recombinant DNA. These guidelines are still the primary source of regulation of recombinant DNA research and have been periodically updated by the NIH.

The Asilomar conference also stimulated further controversy involving rDNA technologies. On one side, concerned citizens and public interest groups that had not participated in the conference began to demand a voice in the regulation of recombinant DNA technologies. The city of Cambridge, Massachusetts, exerted its power to control the rDNA research conducted in its universities, creating the Cambridge Biohazards Committee to oversee DNA experiments. The environmental organization Friends of the Earth even brought a lawsuit demanding that the NIH issue an environmental impact statement on rDNA research. On the other side, biological researchers opposed the inclusion of the public in the rDNA discussion. These researchers feared that public participation in the matter might restrict and compromise the freedom of scientific research.

Humans have for centuries used selective breeding and hybridization techniques to alter food-producing plants. The introduction of recombinant DNA technologies, however, has allowed humans to genetically cross plants, animals, and microorganisms into food-producing plants. There are two basic methods for passing genetic traits into plants. First, biologists can infect a plant cell with a plasmid containing the cross-species genes. Second, biologists can shoot microscopic pellets carrying the cross-species genes directly through the cell walls of the plants. In either case, biologists are reliably able to move desirable genes from one plant or animal into a food-producing plant species.

For instance, scientists have already spliced genes from naturally occurring pesticides such as *Bacillus thuringiensis* into corn to create pest-resistant crops and have genetically altered tomatoes to ensure their freshness at supermarkets.

Genetic technologies allow an increase in properties that improve nutrition, improve the capacity to store and ship food products, and increase plants' ability to resist pests or disease.

In 1994 the Food and Drug Administration approved the first genetically modified food for sale in the United States. Now genetic modification of food supplies is pervasive, particularly in staple crops such as soy, corn, and wheat. Cross-species gene splicing has, however, created at least two significant controversies.

One controversy arose over the use of "terminator" gene technologies. When biotechnology companies began to produce foods with cross-species genes, they included terminator genes that sterilized the seeds of the plants. This terminator technology served two functions: it kept the plants from reproducing any potential harmful or aberrant effects of the genetic engineering, and it also ensured that farmers who purchased genetically modified plants would need to purchase new seeds from the biotechnology companies each year.

The use of terminator technologies caused an international social debate, especially when biotech companies introduced their genetically modified foods into developing countries. Because farmers in developing countries tend to reseed their crops from a previous year's harvest, the terminator technology created a new and unexpected yearly production expense. Civil and human rights groups urged banning the introduction of genetically modified crops in developing countries, arguing that any potential nutritional or production benefits offered by the genetically modified foods would be outweighed by the technological mandate to purchase expensive, patented seeds each year. In response to this, Monsanto (the biotechnology company that owns the rights to the terminator gene patents) pledged not to commercialize the terminator technology. Human rights groups continue to work toward implementing legal bans on the use of the technology, however.

Another controversy arose with a concern about genetic pollution. Although biologists are reliably able to splice or physically force gene sequences from one species into another, they are not always able to control the reproduction and spread of the altered plants. This has created serious debate over the introduction of genetically modified food from laboratories into natural ecosystems.

One concern is that the genetic alterations will pass from the food-producing crop to weeds that compete for nutrients and sunlight. One good example occurs with pesticide-resistant crops. Some biotechnology companies have modified crops to resist the application of certain pesticides. This allows farmers to apply pesticide to their fields while the modified crop is growing, thus reducing competition from weeds and attacks by pests. However, biologists cannot always control whether the pesticide-resistant gene will stay confined to the food-producing crop. Sometimes the pesticide-resistant gene migrates to surrounding plants, thus creating "super weeds" that are immune to the application of pesticides.

Another concern is that the genetic alterations will unintentionally pass from the modified food-producing crop into another natural strain. Here the concern is that the uncontrolled movement of cross-species genetic alterations may alter evolutionary processes and destroy biodiversity. For example, one controversy focuses on whether the introduction of genetically modified corn has led to the cross-pollination of native Mexican strains of maize. There is also a concern about introducing strains of genetically modified potatoes into areas of Peru, where subsistence farmers safeguard many native strains of potatoes.

The final and perhaps most important social concern is the safety and quality of the food produced by genetically altered plants. There has been a general inquiry into the safety of genetically modified foods. Because few tests have been conducted into the safety of these foods or the long-term effects on human health, there is a strong movement, particularly in western European countries, to ban "Frankenfoods." There has been an even stronger reaction over the labeling and separation of genetically modified foods. Moving genes from one species to another food-producing crop can raise serious allergy and safety concerns. When, for example, one company began splicing desired genes from Brazil nuts into soybeans, it became apparent that the resulting modified soya plant would induce allergic reactions in any person with a nut allergy. However, because food distribution systems, especially in industrialized countries, tend to collectively amass and distribute staple crops, if no labeling or separating requirement is installed, there is no way to tell which foods contain genetically altered plants. This raises concerns about the ability to recall products should scientists discover a problem with genetically altered foods.

As with agriculture, humans have long practiced forms of animal biotechnology by domesticating animals and through practices of animal husbandry. However, the use of rDNA technology has allowed humans to clone animals and produce transgenic animals. Scientists now genetically insert genes from cows into chickens to produce more meat per animal, genetically alter research laboratory rats to fit experiments, and genetically modify pigs to grow appropriate valves for use in human heart transplant procedures.

Although all the concerns about plant biotechnology, and particularly the concern about genetic pollution, apply to the genetic manipulation of animals, there are several controversies unique to the application of biotechnology to animals.

The first, and perhaps most fundamental, controversy over the application of biotechnology to animals is the moral reaction against "playing God" with recombinant DNA technologies. Many religious and ethics groups have chastised biologists for violating fundamental limits between species that cannot, without major evolutionary changes, otherwise breed. This has brought a serious debate over whether the biotechnological mixing of species is unnatural or whether it merely demonstrates the arbitrary segregation of our scientific categories of kingdoms and species.

Another controversy unique to applying biotechnology to animals concerns the rights and welfare of genetically modified animals. Genetic technology has, for example, allowed great advances in xenotransplantation (the use of pigs as sources of organs for ailing human beings) and in genetically altering laboratory rats. This enables scientists to "pharm" medical products and laboratory subjects from genetically altered animals. However, this ability to extract resources from animals comes into direct conflict with a growing awareness of ethical duties toward animals and of animal rights. Although few critiques suggest that these ethical duties require us to abandon the practice of applying biotechnology to animals, they have raised serious questions about how genetic modifications alter the lives of animals and what sorts of safeguards or standards should be employed in animal biotechnology.

See also Gene Patenting; Genetic Engineering; Genetically Modified Organisms; Nanotechnology; Precautionary Principle.

Further Reading: Metz, Matthew. "Criticism Preserves the Vitality of Science." *Nature Biotechnology* 20 (2002): 867; Patterson, D. J., et al. "Application of Reproductive Biotechnology in Animals: Implications and Potentials." *Animal Reproductive Science* 79 (2003): 137–43; Quist, David, and Ignacio Chapela. "Transgenic DNA Introgressed into Traditional Maize Landraces in Oaxaca, Mexico." *Nature* 414 (2001): 541–43; Rodgers, Kay. *Recombinant DNA Controversy.* Washington, DC: Library of Congress, Science and Technology Division, Reference Section.

Celene Sheppard

BRAIN SCIENCES

Neurological studies generally focus on answering questions about how the brain governs relationships connecting brain cells, body, and environment. Social sciences, such as psychology, anthropology, and sociology, generally focus on answering questions about how social interactions influence the brain and behavior. When social scientists and brain scientists (neuroscientists) work together, their respective methods and approaches require them to use cross-disciplinary categories for understanding both the brain and cognition.

One common strategy neuroscientists and cognitive scientists (such as psychologists) use involves combining brain imaging technologies together with standardized educational and psychological tests. In linking standardized testing and brain imaging technologies, such as functional magnetic resonance imaging (fMRI), many disciplines must come together and, in the process, adjust disciplinary categories and definitions to accommodate diverse areas of expertise. In fact, special training programs in fMRI have increasingly been established in an effort to bridge the many disciplines involved. This lack of standards and common categories led to the publication of a number of studies that were later retracted because of methodological flaws. Yet frequently by the time conclusions were retracted, other researchers were already building on the flawed methods, and the erroneous information already had been disseminated to the general public through publications, media, and advertising.

Use of a combination of standardized testing and fMRI characterizes many brain studies that aim to answer questions about brain and cognition. In such projects, standardized testing categories are used alongside categories for brain function in localized brain regions. For example, reading comprehension is both a standardized testing category and a category of brain function localized in the brain region called Wernicke's Area. So when researchers scan the brain of a research subject while asking him or her questions from standardized reading comprehension tests, they can connect and compare the scores from the test with the level of brain activation shown occurring in Wernicke's Area while the research subject is asked and answers the question. Their analyses can then include comparisons with any number of fMRI or standardized test results from other studies in various disciplines, helping to bridge the boundaries between studies of cognition, mind, brain, and learning.

In deciphering the meaning of brain scans, health tends to be considered in terms of the absence of disease and less frequently as disease being the absence of health. One reason for this is the lack of consensus about which criteria to use and how to use them to identify a normal, healthy brain. Many problems establishing what constitutes a healthy brain have to do with individual variation (intersubject variability). A person may be considered healthy and normal but have a drastically different brain structure and different patterns of brain activation from another person considered just as healthy and normal. Still, descriptions of a "representative subject" are widespread throughout the neuroscience literature.

To distinguish between healthy and diseased brain scans, neuroscientists established measurement coordinates that form the criteria for what are commonly referred to as "standardized brains." One of the most common coordinate systems used for finding sites of brain activation and correlating those activations with brain functions is the Talairach coordinate system. This system uses the locations identified in a particular elderly woman's brain as a standard for localizing structures and functions in fMRI studies. Such coordinates are the most useful labels for averaging fMRI data between subjects (comparing neuroimaging results). But before brains can be compared using Talairach coordinates, the brain images must literally be deformed ("normalized") such that structures align properly with the standard brain against which it is compared.

Once the brain is deformed properly and aligned with the standard brain, it is divided up into separate areas, including around 70 areas called Brodmann's areas. Brodmann, a German neurosurgeon, published his categories for brain areas between 1905 and 1909. Before Brodmann, there was no standardized, or even widely agreed upon, structure of the cerebral cortex. Neuroscientists have added categories to further subdivide Brodmann's areas into many smaller regions. Our understanding of brain structure and function is based on these smaller subdivisions of Brodmann's areas in (a deformed image of) an elderly alcoholic woman's brain at autopsy. The old systems of classification and measurement are still at work when cognitive neuroscientists use the Talairach coordinate system with cutting-edge fMRI brain scans. Fewer social and cultural investigations elaborate on the invisible work required for the success of fMRI, presumably at least in part because the technology has not been around as long as standardized testing tools.

In contrast to difficulties in classifying what constitutes a normal brain scan, scoring criteria on standardized educational and cognitive tests have a rich history in debates about distinguishing scores that are normal from those that are not. For this reason, many studies using fMRI depend on standardized tests to link cognition to blood and oxygen flows in the brain. That is, because standardized cognitive and educational testing has a more established record of use than fMRI, combining the scientific authority of the tests with the technological authority of fMRI gives the brain sciences greater power in their pursuit of funding and helps to validate their research achievements.

Generally, standardized tests and fMRI are combined according to the following basic protocol: A subject lies still in the fMRI scanner, while holding some sort of device that allows him or her to press buttons (usually described as a "button box"). Each button corresponds to a multiple-choice answer to a test question. The subject sees the test question and possible answers projected into the scanner, usually via a projection screen and mirror(s), and presses a button that corresponds to whichever answer—(a), (b), (c), and so on—he or she chooses. Meanwhile, technicians (who can be from any one of many disciplines such as physicists, computer programmers, engineers, radiologists, psychologists, neurologists, and graduate students) control the scanner that magnetically alters the very spin of hydrogen molecules within cells such that the computer can register these changes as changes in blood flow and blood oxygen levels in the brain. Computerized images of the subject's brain thinking the answer to a standardized test question are generated through a complicated, many-peopled process that is contingent on the timing of measurements taken: how much time, after the question is displayed for the subject, it takes for brain blood flow to be considered as reacting to the test question; the amount of time the subject spends thinking about the answer; the time until the subject presses a button to answer; and the time before the next question is displayed to the subject, and so on. The detected brain's blood and oxygen flows are then correlated to brain maps (such as Talairach coordinate system maps), which tell the scientist what the parts of the brain receiving blood and oxygen do in terms of cognition. Along with the test scores (number correctly answered), all this information is used to draw conclusions about the scientific hypotheses being put forth by neuroscientists. Of course, this is an extremely simplistic overview of the process. Disciplines, technological proficiency, statistical classification, definitions of normalcy, and theoretical applications are diverse; the interactions among them are complex.

Other brain scientists prefer methods that involve more interpersonal interaction to study the human brain and cognition. Dr. Oliver Sacks is known for his ability to see things other brain doctors miss by emphasizing the importance of the patients' perspective. He studies brain damage in people by observing their posture, gaze, and gestures around people and in environments with which they are familiar to help him try to see the world from their point of view. If they are capable of speech, he listens carefully to their descriptions of their problems and experiences.

Dr. V. S. Ramachandran also tends to first interact with subjects and develop low-tech solutions that help people use their brains to overcome cognitive and physical disability. Many of his studies have been of people with "phantom limbs," that is, people who have lost one or more limbs and who still have vivid sensations and sometimes excruciating pain in the lost ("phantom") limb(s). Sensations in limbs that are no longer attached to a person's body are common because of very real changes that take place in the brain as well as the body when a limb is missing. He explains that the areas of the brain that control sensations in a missing limb are still there, but relatively drastic changes in what brain regions correlate with what specific function can take place in brain areas surrounding the damaged area. For example, perhaps by stroking the jaw or foot, one can elicit a response in what was formerly the missing limb's brain area that tells the person she or he feels specific sensations in some part of the missing limb. Ramachandran studied people who have pain in a phantom arm or otherwise feel it in an uncomfortable position. He and his colleagues found that the pain and discomfort can frequently be relieved when the person is shown a mirror reflection that is designed to superimpose the reflection of the person's existing arm such that the person is able to see an arm where there is no arm. This enables the person's visual cortex region in the brain to "fill in" an arm, at which point he or she can then exercise some control over sometimes uncontrollable actions of phantom limbs.

Science and technology studies scholar Susan Leigh Star highlights reasons to not take for granted how neuroscientists correlate a location of the brain with a specific bodily or cognitive function. She reviews many social, historical, and political pressures that brain scientists in the late nineteenth and early twentieth centuries had to navigate in their work. With respect to the cases of Talairach coordinates and interdisciplinary classification problems, she identifies ways that the work of these early brain scientists echoes approaches of the brain sciences today. Many neuroscientists are still looking for how specific brain regions control specific functions, although mounting evidence shows that these localizations of brain functions can be drastically changed at any age (such as with phantom limbs). Research into brain function and its relation to cognition and to "mind" remains popular today, and just as it did a century ago, it uses anatomical and physiological techniques, now with an emphasis on brain imaging studies, to explore its subject.

See also Autism; Memory; Mind; Nature versus Nurture.

Further Reading: Ramachandran, V. S., and Sandra Blakeslee. *Phantoms in the Brain: Probing the Mysteries of the Human Mind.* New York: Harper Collins, 1999; Rose, Steven, and Dai Rees, eds. *The New Brain Sciences: Perils and Prospects.* Cambridge: Cambridge University Press, 2004; Sacks, Oliver. *The Man Who Mistook His Wife for a Hat.* London: Picador, 1986; Star, Susan Leigh. *Regions of the Mind: Brain Research and the Quest for Scientific Certainty.* Stanford: Stanford University Press, 1989.

Rachel A. Dowty

Brain Sciences: Editors' Comments

Looking to the brain for explanations of human behavior is more of a cultural prejudice than a scientifically grounded approach. What do you see when you look at brain scans taken while a person is playing chess? Is the brain playing chess? Is the person playing chess? It becomes immediately evident that there is something odd about separating brains from persons. Indeed, some of the more recent studies of the brain are beginning to offer evidence for the problematic nature of long-taken-for-granted dichotomies such as mind–brain, mind–body, brain–body, and brain–mind. Increasingly, it appears that in order to deal with some or all of the anomalies emerging in the brain sciences, we may need to reconceptualize the entire brain–body system. The mind may be more of an artifact than a "real" entity, a natural kind. It may indeed be nothing more than a secular version of the soul, an entity posited in order to identify something unique about humans relative to other animals.

The traditional students of mind in philosophy and psychology have been hampered in their work by an individualist cognitive approach to the person. Brains and persons tend to be treated as freestanding entities in these fields. Sociologists of mind have of course paid more attention to social processes, social institutions, and social constructions. The most interesting development in this arena is that social scientists are now taking the brain as an appropriate if non-obvious social object. As early as 1973, the anthropologist Clifford Geertz proposed that the brain was a social thing. Since then there has been mounting evidence that both connectivities in the brain and the very structure of the brain are influenced by social practices. Thinking and consciousness are increasingly being viewed as things that socially constructed bodies do in social contexts, not things that individuals or individual brains do. Neuroscientists such as Steven Rose as well as sociologists of mind and brain such as Sal Restivo are persuaded that we need to abandon our conventional classification and categories regarding brain, mind, self, and body. We may be moving in the direction of a model that eliminates mind as an entity and brings the brain, central nervous system, and body into a single informational structure.

CANCER

Knowledge and understanding of cancer, the leading cause of death in the United States and worldwide, has grown exponentially in the last 20 years. Investment in research and technology has greatly reduced the effects of cancer through advances in prevention, detection, and treatment. Survival rates have never been greater; in 2003 the rate of cancer deaths dropped in the United States for the first time since 1930. Radically different approaches to prevention and treatment, despite their successes, however, continue to divide the medical and scientific communities.

Developments in cancer research stretch across the medical spectrum. From identifying new drugs to developing new screening tests and implementing more effective therapies, breakthroughs occur every day. Each of the 100 different types of cancers affects the body in unique ways and requires specific prevention, detection, and therapy plans. Understanding the complexities of the disease that afflicts over half of all men and a third of all women in the United States is vital to the medical health of the nation.

The causes of cancer are becoming better understood. Genetics and lifestyle both can contribute to a person's susceptibility to cancer. For example, diet can greatly affect a person's chances of getting cancer. Certain lifestyle choices, such as having excess body fat, eating red meat, not engaging in physical exercise, or consuming alcohol, all increase the likelihood of developing cancer. Many cancers tend to be caused by long-term exposure to cancer-causing agents, such as environmental toxins, rather than by a single incident. Environmental factors and lifestyle choices, however, do not always predict the appearance of cancer; instead, they should be taken as indicators of a higher risk. Understanding how these things interact with genetic factors over the course of a person's life will be at the front line in future cancer research.

The treatment of cancer used to entail surgery, chemotherapy or radiation, or any combination of the three. Although these types of procedures have altered the medical landscape for treating cancer over the past 100 years, new methods have emerged that bypass invasive or problematic surgeries. Researchers have begun to understand, for example, how the body fights cancer on its own through the immune system. Many of the developments in fighting cancer have come through the harnessing of the immune system's ability to produce antigens to combat cancerous cells. Therapy in the form of cancer vaccines has been largely experimental. Recently, however, the FDA approved a major breakthrough in cancer prevention using vaccines.

The development of a vaccine against the human papillomavirus (HPV) marked the first vaccine to gain approval in the fight against cancer since the hepatitis B vaccine. HPV is a leading cause of cervical cancer and, to a lesser degree, other types of cancer. The vaccine, which has gained FDA approval, was shown to be 100 percent effective against two of the leading types of HPV virus. These two strains account for 70 percent of all cervical cancers worldwide.

Vaccines for cancer can either prevent cancer directly (therapeutic vaccines) or prevent the development of cancer (prophylactic vaccines). Therapeutic vaccines are used to strengthen the body against existing cancers to prevent the recurrence of cancerous cells. Prophylactic vaccines, like the one for HPV, prevent viruses that ultimately cause cancer. The HPV vaccine represents a significant breakthrough in cancer research. There are no officially licensed therapeutic vaccines to date, though numerous prophylactic vaccines are being tested by the National Cancer Institute.

Vaccines are part of a growing area of treatment known as biological therapy or immunotherapy. Biological therapy uses the body's immune system to fight cancer or lessen certain side effects of other cancer treatments. The immune system acts as the body's defense system, though it does not always recognize cancerous cells in the body and often lets them go undetected. Furthermore, the immune system itself may not function properly, allowing cancerous cells to recur in a process called metastasis, wherein the cancerous cells spread to other parts of the body. Biological therapy seeks to step in to enhance or stimulate the body's immune system processes.

One of the new dimensions of cancer research has been the revolution of personalized, or molecular, medicine in the fight against cancer. Personalized medicine takes into account knowledge of a patient's genotype for the purpose of identifying the right preventive or treatment option. With the success of the Human Genome Project, new approaches have emerged in the field of cancer research. Approaching cancer from the perspective of "disease management" will lead to more customized medical treatments.

The successful implementation of such a revolutionary way of handling the disease will require that a vast amount of genetic data be classified, analyzed, and made accessible to doctors and researchers to determine the treatments for individual patients. In 2004 cancer centers across the United States took part in the implementation of the National Cancer Institute's caBIG (cancer Biomedical Informatics Grid), a virtual community that seeks to accelerate new approaches to cancer research. The caBIG community aims to establish an open-access database that provides researchers the necessary infrastructure for the exchange of genetic data.

New methods for detecting cancer have also been making headlines. One such method has been gene expression profiling, a process that is capable of identifying specific strains of cancer using DNA microarrays. These microarrays identify the activity of thousands of genes at once, providing a molecular profile of each strain. Research has demonstrated two important guidelines in cancer identification and treatment. Even though certain types of cancer look similar on a microscopic level, they can differ greatly on a molecular level and may require vastly different types of therapy.

The most notable example of this type of process has been used to identify two different strains of non-Hodgkin's lymphoma (NHL), a cancer of the white blood cells. Two common but very different strains of NHL have radically differing treatments such that the ability to easily diagnose which strain is active has been a great boon for therapy. As a result of misdiagnosis of the different strains, there were errors in determining the appropriate therapy that unnecessarily led to a lower survival rate.

Another innovation in cancer detection involves the field of proteomics. Proteomics—the study of all the proteins in an organism over its lifetime—entered into the discussion about cancer detection when it was discovered that tumors leak proteins into certain bodily fluids, such as blood or urine. Because tumors leak specific types of proteins, it is possible to identify the proteins as "cancer biomarkers." If such proteins can be linked to cancers, then examining bodily fluids could greatly increase the ability to screen potentially harmful cancers early.

Certain proteins have already been implemented as cancer biomarkers. Levels of certain antigens—types of protein found in the immune system—can indicate cancer of the prostate (in men) or of the ovaries (in women). This method of detection has not yet proved to be 100 percent effective. It may give false negatives in which the test may not detect cancer when it is actually present or even false positives where it may detect cancer in cancer-free patients.

As processes for detecting cancer improve, the number of cancer diagnoses is likely to increase; this would increase the overall rate of cancers but decrease their lethal consequences.

Traditional forms of cancer treatment—surgery, chemotherapy, and radiation therapy—are also undergoing significant breakthroughs. Developments in traditional cancer treatments involve refining existing procedures to yield better outcomes and reducing the side effects typically associated with such treatments. For example, chemotherapy regimens for head and neck cancers, typically difficult to treat, have improved through recombination of chemotherapy treatments with radiation, the first such major improvement for that type of cancer in 45 years.

Chemotherapy solutions are also being affected by the genetic revolution. A burgeoning field called pharmacogenomics seeks to tailor pharmaceutical offerings to a patient's genetic makeup, abandoning the one-size-fits-all or "blockbuster" drug of previous years. Drugs will now be matched using knowledge of a patient's gene profile, avoiding the trial-and-error method that is often practiced in trying to find the correct treatment program for a patient. Patients will be able to avoid unwanted side effects from unnecessary drugs, as well as lower the cost of health care and reduce repeat medical visits.

Much ground must still be covered before a pharmacogenomics revolution can take place. Drug alternatives must be found for numerous genotypes to avoid leaving patients without any options if their genotypes do not match the drugs available. Drug companies must also have incentives to make specialized drugs, given the exorbitant cost of offering one single drug on the market.

The effects of cancer and cancer treatments will continue to be studied as more information becomes available on the long-term effects of certain diseases. New examples of long-term complications with cancer have emerged recently in both breast cancer survivors and childhood cancer survivors. Breast cancer survivors have reported fatigue 5 and even 10 years after their therapies. Similarly, long-term research into childhood cancer survivors has shown that children who survive cancer are much more likely to have other frequent health problems, five times more than their healthy siblings. A large percentage of childhood survivors often developed other cancers, heart disease, and scarring of the lungs by age 45. Such evidence underscores the complicated nature of cancer survival and how long-term studies will continue to play an important role.

There are now more than 10 million cancer survivors in the United States alone. The cancer survival rate between 1995 and 2001 was 65 percent, compared to just 50 percent from 1974 to 1976. As more is known about cancer itself, more must also be known about the effects of cancer after remission. Studies examining post-cancer patients 5 to 10 years after surgery are revealing that the effects of cancer and cancer treatment can extend beyond the time of treatment.

Not all research into cancer has been positive: certain types of cancer—namely skin cancer, myeloma (cancer of plasma cells in the immune system), and cancers of the thyroid and kidney—are on the rise. The reasons for the increase in cancers are wide-ranging and require further research to be fully understood.

With the fight against cancer continuing to evolve, new advances continue to converge from different fronts—in the use of human bio-specimens, in nanotechnology, and in proteomics. Each of these fields individually has contributed to the efforts at detecting, preventing, and treating cancer, but if their efforts can be streamlined and pooled, the fight against cancer will have won a major battle.

As the fight has taken on a more global character, developments in knowledge sharing and community support have provided cancer researchers, patients, and survivors with new means of battling the life-threatening disease. As the technologies and infrastructures change, however, public policy will also need to change the way advancements in medical science are linked with their accessibility to patients, so that financial means are not a prerequisite for these new treatments.

See also Genetic Engineering; Human Genome Project; Immunology.

Further Reading: Khoury, M. J., and J. Morris. *Pharmacogenomics and Public Health: The Promise of Targeted Disease Prevention.* Atlanta: Centers for Disease Control and Prevention, 2001; Nass, S., and H. L. Moses, eds. *Cancer Biomarkers: The Promises and Challenges of Improving Detection and Treatment.* Washington, DC: National Academies Press, 2007; National Cancer Institute. "Cancer Vaccine Fact Sheet." http://www.cancer.gov/cancertopics/factsheet/cancervaccine; Ozols, R., et al. "Clinical Cancer Advances 2006: Major Research Advances in Cancer Treatment, Prevention, and Screening—A Report From the American Society of Clinical Oncology." *Journal of Clinical Oncology* 25, no .1 (2007): 46–162; Sanders, C. "Genomic Medicine and the Future of Health Care." *Science* 287, no. 5460 (2000): 1977–78.

Michael Prentice

CENSORSHIP

Censorship refers to blocking distribution of, or access to, certain information, art, or dissent. Censorship is imposed by ruling authorities to benefit, in their view, society as a whole. This of course is where the debate begins. On one side are those who argue that all censorship should be avoided and in fact is damaging to society. On the other side are those who argue that rulers have a mandate to protect society from falsehood, offense, or corrupting influences, and therefore censorship is a power for good. In between are those who regard censorship as undesirable but who accept that it is required even in a society committed to freedom of expression.

Technology always has some role to play in censorship. New technology seems to invite censorship of its perceived dangers, and in turn it always offers new ways to avoid censorship and new ways to impose it.

The opposite of censorship is freedom of expression, or freedom of speech as it is known in the First Amendment to the U.S. Constitution. Most Western democracies have similar provisions to protect information, art, and dissent from censorship (for example, Article 19 of the Universal Declaration of Human Rights and Article 10 of the European Convention on Human Rights). The purpose of laws guaranteeing freedom of expression is first to ensure the free flow of ideas essential to democratic institutions and responsive government. Many nations have histories in which censorship was used to block political reform. They therefore have adopted constitutions to make it difficult to restrict minorities or their opinions. The central argument against censorship is that any censorship, however well chosen in the short term, will in the long term undermine the freedom of expression required for a just society. The price of such freedom is the defense of freedom of expression even for those whose viewpoints are despised. The classic example is the American Civil Liberties Union (ACLU) fighting to protect the right of Nazi demonstrators to parade through Skokie, a Jewish suburb in Illinois, in 1978. Were attempts to block the march on the grounds of potential violence justified or simply the censorship of unpopular views?

Second, freedom of expression and the avoidance of censorship promote the pursuit of truth. Censorship has a shameful record of suppressing scientific progress and picking the wrong side in the advance of truth. The oft-cited example is Galileo and the Church. Although the details of this conflict as understood by historians of science are not what they seem to be in the public imagination, Galileo did argue that the sun, not Earth, was the center of the solar system at a time when the Church (and supposedly the Bible) claimed otherwise. Thousands of scientific advances have at first been censored by the authorities "for the good of society" and have then turned out to be true.

Freedom of expression is not absolute, however, and there are exceptions to justify censorship. For example, one may not shout "fire" in a crowded theater or incite a crowd to riot. All modern countries prohibit libel (publishing untrue or deliberately demeaning content) and hate literature (content meant to incite hatred). Likewise, one may not publish military secrets, distribute private financial and medical data, or lie in court. There are many other forms of common-sense censorship that pass without comment today. The debate is about less obvious situations concerning obscenity, sexually oriented content, political dissent, and literature offending ethnic or religious sensibilities. All have been legally censored at one time or another, and some censorships remain in place today. For example, in some European countries it is illegal to distribute Nazi literature or deny the Holocaust. In some Muslim countries no one may publish criticisms of Islam. Since 9/11, any expression remotely promoting terrorism has been strictly censored in many countries.

In a free society all questions of censorship end up in the courts. Laws vary, but in general, Western democracies approach the question this way: government and authorities such as school boards may censor information, art, or dissent within strict limitations. The censorship may not be justified simply by offense caused to a few or even to a majority of people. Though much censorship is requested by religious groups, religion cannot be used as a justification for it. Reasons for censorship may not be arbitrary or irrational.

For censorship to be justified, there must be a danger that is pressing and substantial and that only censorship may address. This accounts for the inflamed rhetoric accompanying many requests for censorship. It is common to claim that without censorship of certain television content, literature, or graphic imagery, children will be irreparably harmed and the future of civilization put in peril. Opponents of censorship tend to make the same argument, in both cases making decisions about censorship difficult to reach.

A common pattern is for government to impose censorship and find it challenged later and denied, often because the perceived dangers have turned out to be baseless. Examples here are nudity in art, profanity in literature, and representation of nontraditional families on television. None has corrupted society even though censors once claimed they would. In other cases, the court upholds censorship. Examples are hate literature, child pornography, and a surprising number of restrictions to free speech in the school system. Neither teachers nor students are free from censorship. School newspapers are not protected by freedom of the press, and teachers may not express some opinions in areas of religion, sexuality, and politics.

There are two general ways censorship is applied: prepublication and post-publication. Prepublication censorship means that approval is required before something is distributed or expressed. Censors approve soldiers' personal mail so that it does not give away military secrets. In the former Soviet Union, books and magazine articles required an okay from state censors before they were printed. In many totalitarian countries, daily newspapers print only what has been checked by government censors. Less and less of this prepublication censorship goes on in modern societies in part because it is so hard to achieve. Digital distribution through the Internet also makes content quite hard to control.

Nevertheless, there is considerable indirect prepublication censorship going on in even the most open of societies. For example, arts groups funded by the government know well that offensive art is unlikely to be funded. Criticism by large corporations will bring a storm of lawsuits against which few individuals can afford to defend themselves. It is easier to keep quiet. Political dissenters know that their tax returns will be audited or private information released if they criticize the government. This is called the "chill effect" and is simply a form of censorship in which people censor themselves.

Post-publication censorship is harder now because of technological advances. The ability to distribute information, art, or dissent has never been greater. Suppressing them has never been more difficult. Mass media increasingly are in the hands of individuals. Publication on the World Wide Web (WWW or Web) is cheap, easy, and relatively unfettered. There are 70 million blogs on the Web expressing every imaginable opinion. Discussion forums, picture galleries, and social sites seem to operate without restriction. WWW pioneers thought their new technology was a dream come true for democracy. They spoke of the "dictator's dilemma." Would repressive regimes grant access to the Web with all the freedom it brought—or block access to the Web and all the economic benefits that come with it?

In fact, the introduction of Internet technologies has given repressive regimes even more tools to suppress dissent. Dictators simply switch off the Internet in their countries if it distributes embarrassing political commentary. In turn, internet technology itself makes it easier to track dissidents who then face their own dilemma—either use the Internet to distribute information and risk being tracked by authorities or forgo the Internet and risk being unable to publish anything at all. It is ironic that along with the opportunity to distribute information, art, and dissent worldwide has come technology to track it. It may prove easier to block a single web site than to retrieve a thousand flyers. The dictator's dilemma in fact may be the dictator's dream. In a decade of explosive Internet growth, there has been no decrease in repressive regimes or the abuse of human rights that come with them.

How did a technology designed for endless distribution of information become a tool for censorship? In a supreme irony, high-tech companies from so-called free nations provide advanced technologies to filter, block, and track Internet use in repressive regimes. Nowhere is this truer than in China, where Internet use was once considered the key to political reform. Its 200 million–plus Internet users are surrounded by what is popularly called "the Great Firewall of China." Access to certain sites, such as the BBC, among hundreds of news sites, is blocked outright. The Chinese version of Google blocks certain keywords such as Tiananmen Square (the 1989 political protest), Falun Gong (a banned religion), Taiwan (a rebel province), and other political topics deemed inappropriate by the authorities. Furthermore, users are warned that such searches are blocked, leading many to assume their actions are logged and may lead authorities back to them. With such tools of censorship in place, it simply is not worth browsing topics not approved by the government. Users censor themselves out of fear of government retribution. Censorship, which has always been strictly applied in China, operates successfully on the Web and in any repressive nation that wants it.

Though the Internet is the newest technology for distribution of information, art, and dissent, it follows the same pattern as the technologies before it. Each new technology inspires a burst of free expression and seems to promise freedom from old censorship methods. In turn, somebody always calls for regulation of the dangerous new medium. Some form of censorship is imposed and then is protested and finally bypassed by the next new technology.

For example, when books were written by hand, information, art, and dissent were easy to control. The new printing press in 1450 upset the rules, and soon Church and state censors began to ban books. Soon opponents of censorship took up calls for a freedom of the press. The same patterns can be seen with the introduction of photography in the nineteenth century. Along with the new art form came calls to restrict the dangers to society posed by pictures of naked people, something that had been a subject of art for thousands of years. Cinema and the twentieth century, and particularly the addition of sound, inspired the laws to keep films from corrupting the values of the young. The American Hays code inspired by religious groups went beyond prohibiting overt sexuality, violence, and swearing in films; it imposed an entire fabric of principles on moviemakers. It was replaced in 1966 by voluntary censorship in the form of the ratings system in use today. This system does not regulate what is in films, but rather only who is permitted to see them, based on age and parental guidance. Though this is still a form of censorship, it has withstood court challenges and the changing norms of society.

Radio and television inspired the same burst of free expression followed by attempts to censor it. It is possible still to see older programs where a married couple's bedroom has twin beds to preclude any suggestion that they might be sleeping together. Restrictions have lessened through the years, but religious and conservative political groups still lobby to keep what they consider offensive off the air. In 2004 a fleeting glimpse of a pop singer's breast in the Super Bowl halftime show brought down a $550 thousand fine on the network that broadcast it.

Self-censorship still oversees television and radio content with a rating system similar to that used by cinema. Instead of age group, practical censorship uses broadcast times to shift programs with adult themes to times after children have gone to bed. With satellite distribution, time shifting, and recording devices, this method of censorship is acknowledged by everyone to be unworkable in practice. In its place the American government promotes the use of the V-Chip, an embedded device parents use to block certain types of television content. This removes censorship headaches from the government and transfers them to parents who as yet have not been hauled before the Supreme Court to defend their choices.

The current frontier for censorship debate is the Internet. Once again the pattern repeats. Early adopters filled the WWW with formerly censored information, art, and dissent. They trumpeted the end of censorship. By the mid-1990s reports about the dangers of the Internet appeared everywhere, and soon groups were demanding controls on it. The 1996 Communications Decency Act took a traditional censorship approach to keep pornography away from young users. Publishers were to be fined. The act suffered various constitutional setbacks, as did the Child Online Protection Act (COPA) of 1998, which imposed fines for the collection of information from minors. The Children's Internet Protection Act (CIPA) of 2000 took a different route, using federal funding to require libraries and other public institutions to install filtering software. Though adults could request access to any site, the available Internet was filtered of objectionable content.

The question, of course, is what is objectionable content? It turns out that medical sites, sex-education sites, and alternate lifestyle sites were blocked. No one was entirely sure who created the standards. The law, however, has withstood several challenges and appears to have become a model for other nations. Australia recently introduced default filtering of pornography at a national level. Though theoretically all Internet sites are available on request, privacy and free-speech advocates worry that requesting exceptions is still a form of censorship and a violation of privacy.

Censorship continues to be a polarizing issue. Technology will not decide the outcome for either side, but it will continue to make the debate even more important as each new technology arrives.

See also Computers; Information Technology; Internet; Privacy.

Further Reading: Burns, Kate. *Censorship*. Chicago: Greenhaven Press, 2006; Heins, Marjorie. *Sex, Sin, and Blasphemy: A Guide to America's Censorship Wars*. 2nd ed. New York: New Press, 1998; Herumin, Wendy. *Censorship on the Internet: From Filters to Freedom of Speech*. Berkeley Heights, NJ: Enslow, 2004.

Michael H. Farris

CHAOS THEORY

Chaos theory demonstrates, on the one hand, the ability of scientists to grasp the increasing complexities they encounter as they explore systems of scale from galaxies and galaxy clusters to the ever-burgeoning world of elementary particles. On the other hand, chaos theory demonstrates how easy it is to take scientific ideas, especially ideas stripped of their own complexities, and apply them to the world of everyday life or of particular professional and occupational circles of practice (such as theology or the self-help movement).

Chaos theory is part of a network of ideas, ranging from autopoiesis and self-reference to dissipative structures that fall under the conceptual umbrella of self-organization. The pedigree for these ideas can be variously traced to late nineteenth-century thermodynamics, to the works of Turing and von Neumann in the 1940s, and even to Nietzsche's criticism of Darwinian theory. The standard history traces the invention of chaos theory to a paper by Edward Lorenz published in 1972 whose title has given us an "everyman"/everyday summary of the theory: "Predictability: Does the Flap of a Butterfly's Wings in Brazil Set Off a Tornado in Texas?"

The modern fascination, in particular, with complexity and chaos (these two ideas are practically a paired concept) has a blinding effect on sciences that do not share a paradigmatic home with the physical and natural sciences. The social sciences have, however, inevitably been drawn into the love affair with complexity and chaos. These ideas can lead to strange bedfellows indeed. It is not unusual for New Age references to suddenly bubble up in a discussion of complexity and chaos. One minute we are reading about Turing and von Neumann, and next minute we find ourselves in the company of children of the Age of Aquarius writing about Taoism and physics, Buddhism as a philosophy, and the virtues of free love and communal living. In something like the way the term "relativity" in "relativity theory" tends to draw our attention away from the fact that relativity theory is a theory of invariance (originally called in German "invariantheorie"), chaos theory's title draws attention away from the fact that it is a deterministic theory. The battleground here is the affinity of chaos theory with New Age thinking versus a sober scientific concern for the dynamic properties of nonlinear systems.

Chaos theory may be as much a reflection of late twentieth-century popular culture as it is of the impact of studying increasingly complex systems from the weather to plate tectonics and from electrical networks to planetary magnetic fields and population ecology. It would be interesting to ask the average man and woman on the street where they get their knowledge about chaos. It would not be surprising if that knowledge came from seeing movies such as *Jurassic Park* and *The Butterfly Effect* or from reading books by Michael Crichton (*Jurassic Park* and *The Lost World*) and Ray Bradbury (*A Sound of Thunder*). Of course, James Gleick's book *Chaos: Making a New Science* (1987) was a best seller and introduced a wide public to some of the basics of chaos theory without the mathematics. Efforts to engage the public in issues of science are important and should be encouraged. It is not clear, however, that such efforts can do much more than create an impression in the reader that he or she is actually learning science. (It is not unusual for reviewers of popular science books to write that they know nothing about the particular subject matter of the book they are reviewing but that the author explains things really well!)

It is important to consider why we choose certain words to express ideas. For example, it would have caused less confusion in the public mind if relativity had been simply referred to as "invariant theory" or "a theory of invariance." The term *chaos* is more likely to raise images of randomness than of deterministic systems. "Chaos," like "relativity," reflects the disorders of wars, worldwide threats of extinction, and environmental catastrophes that characterized the twentieth century. "Nonlinear systems dynamics" does not have the same cachet in a world that often seems to randomly (and yes, chaotically) surprise us with earthquakes, volcanoes, tsunamis, and deadly viruses.

The scientific battleground here has to do with whether chaos theory tells us something new about determinism and predictability and whether it encourages or discourages thinking about ourselves as agents with free will. A distinguished scientist, James Lighthill, publicly apologized on behalf of the scientific community in a paper presented to the Royal Society in 1987. The title of his paper tells the story: "The Recently Recognized Failure of Predictability in Newtonian Dynamics." There is a resonance here with the poet W. B. Yeats's theologically inspired observation that "things fall apart; the center cannot hold; Mere anarchy is loosed upon the world." This has its secular echo in the philosopher Jacques Derrida's famous speech on decentering: "In the absence of a center, everything becomes discourse." These are not easy ideas to grasp or explain in this moment. They are suggestive, however; they provoke us to think of chaos as one of a set of ideas emerging across the spectrum of disciplines that tell us more about the contemporary human condition than about purely scientific developments.

We can see this more clearly when we consider how chaos theory is used to promote certain human agendas. The theologian James Jefferson Davis, for example, has linked Lighthill's remarks to the priest-physicist William Pollard's theological answer to Einstein's query concerning whether God plays dice with the universe. Einstein's answer was no; Pollard's answer is yes. This is just one example of the efforts of theologians to tie their horses to the latest model of the science cart. Many schools of theology (and a variety of religious leaders, defenders of the faith, lay believers, and so on) are engaging science on a variety of levels. Some are dedicated to making their ideas, beliefs, and faith fit into the scientific worldview that is dominating more and more of our educational and intellectual life at the expense of religion. Others react to the cultural power of science by arguing for a separation of domains so that the issue of making religion compatible with science does not arise. And there are other strategies too. One of the main issues turns on this point: if chance and uncertainty have supposedly entered science, some in the religious camp feel they must demonstrate how this not only fails to derail God's providential plan in any way but also in fact supports it. In the wake of relativity theory, quantum probabilities, Gödel's incompleteness theorems, Heisenberg's uncertainty principle, and chaos theory, we are given a God of Chance; we are given a scientific basis for agency and "free will" compatible with the theology or religious sentiment of the moment. There are then religious and theological battlegrounds within the science and religion arenas of dialogue, engagement, and conflict. The problem here and in science is that chance, probability, statistics, complexity, and chaos are all compatible with science's concern for controlling and predicting outcomes. The controversies here are due in great part to a failure to distinguish the lawful behavior of systems from the deterministic behavior of systems. It is more important to notice that the evidence for complexity, chaos, randomness, and uncertainty has no impact on the scientific community's efforts to control and predict outcomes.

Suppose we started to speak of complexity and eliminated the term "chaos" from our vocabulary? Complexity is certainly a significant focus of research and theory across the disciplines today, and problems in complexity are addressed by researchers at universities and think thanks (notably, for example, the Santa Fe Institute in New Mexico). Or suppose we used the term "deterministic chaos," which accurately describes systems sensitive to initial conditions and fully determined by them? What if mathematicians did a better job of communicating the idea that randomness is an ordered phenomenon? And what if sociologists of science got their message across, that science is a method for creating order out of disorder or perhaps, in the extreme, out of chaos?

Consider further the idea that Newtonian mechanics brings absolute order and predictability to the universe and is thus at odds with chaos theory. Contrary to a prevalent view in and outside of science, Newton's God did not set the clockwork in motion and then retreat. Newton's God is an ever-active agent, ready and able to intervene at any time to reset the clockwork or adjust it in some way or other. Newton's universe may have needed an occasional tune-up to sustain its orderliness and predictability, and God was the ultimate omnipresent mechanic. There was more innate chaos in Newton's universe than the earthly experience of predictability in mechanical systems conveyed.

When a field such as theology has to deal with the dynamics—the chaotic behavior—of science, constantly trying to adjust belief and faith to the latest discoveries, the latest science may contradict or complexify the science last used to explain or justify a theological position. This places theology in a wearying game of catch-up. We see here another piece of the science and religion puzzle discussed in other entries in this volume. This is a significant piece of the nonlinear dynamics of the contemporary intellectual landscape that helps to explain the ubiquitous presence of questions of meaning, represented in terms of science and religion, in the discourse these volumes depict.

See also Quarks; Space.

Further Reading: Davis, John J. "Theological Reflections on Chaos Theory." *Perspectives on Science and Christian Faith* 49 (June 1997): 75–84; Gleick, James. *Chaos: Making a New Science.* New York: Penguin, 1988; Prigogine, Ilya, and Isabelle Stengers. *Order Out of Chaos.* New York: Bantam 1984; Stewart, Ian. *Does God Play Dice? The Mathematics of Chaos.* Oxford: Blackwell Publishers, 1990.

Sal Restivo

CHEMICAL AND BIOLOGICAL WARFARE

Chemical and biological warfare (CBW) uses harmful or deadly chemical or biological agents to kill or incapacitate an enemy. These agents have the potential to kill thousands and should be considered Weapons of Mass Destruction (WMD). Chemical and biological weapons are particularly dangerous because of the ease with which they can be procured and used.

Mention chemical weapons, and most people think of the Great War (World War I). But chemical warfare (CW) existed long before that. The first recorded use of poison gas occurred during the wars between Athens and Sparta (431–404 B.C.E). Arabs made use of burning naphtha to hold the armies of the Second Crusade (1147–49 C.E.) at bay during sieges.

The nature of CW changed around 1850. Europeans began experimenting in the field that became known as organic chemistry and in so doing made many fascinating discoveries. By 1900, Germany was the world leader in making dyes for clothes; many of these compounds were used to produce CW agents.

The first large-scale CW attack occurred near the Belgian village of Ypres in April of 1915. The Germans stockpiled over 5,000 cylinders of chlorine gas (from dye production) in their trenches. Once the wind conditions were right, they opened the valves, and the chlorine gas flowed over the French and British lines. French colonial troops quickly broke and ran, but the Canadians, serving as part of the British Imperial Forces, held their position throughout the day.

By 1916, the Germans had developed another chlorine-based irritant called mustard gas. In the first month of use, the British suffered more casualties from mustard gas than from all other CW gases combined. All told, the various combatants used 124,000 tons of CW munitions, mostly in 65 million artillery shells, between 1914 and 1918. With an almost 20 percent dud rate, areas of France and Belgium are still dangerous today, and as recently as 1990, a CW shell claimed another victim.

The European powers learned from their experiences in the Great War and did not use CWs during World War II despite their availability. The only recorded post-1918 use of chemical weapons was during the Sino-Japanese War (1937–45). Record keeping and postwar secrecy cloud the numbers, but most authorities accept the figure of 100,000 Chinese casualties from CW. The last known use of chemical weapons occurred during the Iran-Iraq war (1980–88). In March of 1984, the United Nations confirmed Iraqi use of mustard gas as well as the nerve agent GA.

Biological warfare (BW) agents, like their chemical counterparts, have a long history. Some of the earliest attempts came about in 400 B.C.E., when Scythian archers dipped their arrows in decomposing bodies. In siege warfare, poisoning a city's water supply by dumping animal carcasses in wells was common. Several authors attribute the outbreak of the Black Plague in Europe (1346) to a BW attack during the siege of Kaffa, when bodies of people infected by the plague were catapulted over the walls into the city. Biological agents are WMDs; once unleashed, they can be impossible to control.

Much like their chemical counterparts, biological weapons received a boost in the nineteenth century. While scientists experimented with dyes, medical practitioners studied microbial cultures created by Louis Pasteur and others. By 1878, people understood how to grow, isolate, and use these substances for war.

In 1914 most nations had biological weapons programs; however, all agreed these weapons should not be used against humans. Animals were another story. Both German and French agents infected enemy horses used for hauling artillery. The Versailles Peace Treaty highlighted the main difference between biological and chemical weapons. Although the Germans were forbidden from developing chemical agents, the same restriction was not applied to biological weapons, partly because these weapons were not used against humans.

Biological weapons in World War II mirrored chemical weapons. The major European powers restrained themselves from using biological weapons, whereas the Japanese made use of them in their war with China. Although reports vary, Japanese and Chinese researchers agree that approximately 270,000 Chinese soldiers died in BW attacks.

Because of Japanese use of biological weapons against the Chinese, the United States began to rethink its policies on BW. By 1942, the United States had adopted a proposal to conduct research into BW and to develop BW agents. By 1943 a BW program was in existence, although full-scale production did not begin until 1945. After the war, the United States scaled back its efforts in BW, mostly to laboratory-style research, where, for the most part, it remains today.

Unlike nuclear weapons, chemical and biological weapons are easily obtainable under the guise of legitimate uses. CW agents can be synthesized from easily obtainable chemicals. Biological weapons can be grown from cultures used in any medical lab. The difficulty lies in designing and delivering the agents on the battlefield, a process known as weaponization.

Whether delivered by aircraft or artillery shell, the agent must be able to survive the production process, encapsulation in a bomb, temperature extremes that can reach −30 degrees Fahrenheit or colder at altitude, the heat buildup during free fall, and finally, the explosion when the weapon strikes the ground. All of these factors challenge the weapons designer and limit the agents that can be used.

Sometimes these agents are dispersed into the air, a procedure the Japanese used against the Chinese. Low-flying aircraft, working much like modern-day crop-dusters, do this quite effectively, although the aircraft are very vulnerable. Regardless of delivery technique, weather is always a factor. High wind speeds and the accompanying turbulence disperse CBW agents, particularly those delivered by spraying, to the point of ineffectiveness, and rain cleanses agents that are carried through the air.

There are five main classes of chemical weapons: choking gases; blister agents; blood agents; nerve agents; and incapacitants. Choking gases were used as early as the Great War and consist of agents such as chlorine (Ypres, 1915), phosgene (Somme, 1915), diphosgene (Verdun, 1916), chloropicrin (or chlorpicrin) (Eastern Front, 1916), and perfluoroisobutylene (PFIB) (never used). All cause irritation of the nose, throat, and lungs. In the extreme, the victims' lungs fill with liquid, and they eventually drown. Because all of these agents possess a strong odor and dissipate very quickly, they are rarely used today.

Blister agents, also called vesicants, are some of the most widely stockpiled and used agents. The list includes mustard, nitrogen mustard, lewisite, phosgene oxide (or "nettle gas"), and phenyldichlorasine (PD). Mustard gas, first used near Ypres in 1917, was known for almost 90 years at that time. Highly toxic, exposure to less than 1 gram for 30 minutes will likely cause death. Lewisite, first synthesized in 1904, and rediscovered by W. Lee Lewis in 1918 and used by the Japanese against the Chinese, is similar in its toxicity and effects to mustard gas. However, it is easier to produce and is a greater threat in times of war.

As the name implies, blister agents cause blisters on the surface of the skin. They penetrate the skin, dissolve in the body's fat, and move inside the body to attack other organs. The results are usually fatal.

Hydrogen cyanide, cyanogen chloride, arsine, carbon monoxide, and hydrogen sulfide are examples of blood agents. These agents block oxygen transfer to and from the blood, effectively asphyxiating the victim. They tend to be unstable and highly volatile, so their utility as weapons is limited. Their volatility gives them one advantage: relatively little time is needed after an attack before the area is safe to occupy.

Nerve agents are toxic versions of organophosphates used in insecticides. The two main agents stockpiled or used today are sarin, also known as GB, and VX. Each is representative of a series of agents known as the G-series and the V-series. Both kill by paralyzing the respiratory musculature, causing death within minutes. The G-series persists for hours, whereas the V-series lingers for two weeks or more

Finally, there are the incapacitants. Perhaps the most well-known example is the street drug LSD. Although there have been unconfirmed reports of the use of incapacitants, to date no one has proven their use in combat.

BW agents are organized into three main classes: bacteria, viruses, and biological toxins. Bacteria and viruses are infectious microbial pathogens, and biological toxins are poisons extracted from biological sources.

Bacteria are single-cell, free-living organisms that reproduce by division. Bacteria cause disease in humans by invading tissues or producing toxins. Common BW agents include anthrax, plague, tularaemia, glanders, Q-fever, and cholera.

Anthrax can enter the body either by absorption through the skin or by inhalation of the spores. Once inside the body, the spores travel to and attack the lymph nodes. Victims experience fever, fatigue, a general malaise, and within 36 hours, death. Anthrax spores live in soil for decades and can be absorbed by animals and then passed on to humans. The spores can easily survive weaponization and delivery. Anthrax is easily available and easy to grow, making it attractive to terrorist groups.

Plague comes in two varieties: bubonic plague, transmitted by flea bites; and pneumonic plague, spread through the air. Bubonic plague kills about 50 percent of its victims, whereas pneumonic plague is almost 100 percent fatal, usually within one week. Plague bacteria are not as hardy as anthrax and usually die after several hours.

Tularemia, glanders, Q-fever, and cholera have all been made and in some instances used. All can cause death but are usually controlled with antibiotics. None are hardy, and problems with weaponization limit their uses in war.

Viruses consist of smallpox, hemorrhagic fever viruses, Venezuelan equine encephalitis, and foot-and-mouth disease. Unlike bacteria, viruses cannot reproduce on their own; they require a host cell for replication. Although all of these diseases can be fatal, the mortality rate is below 30 percent, and Venezuelan equine encephalitis and foot-and-mouth disease rarely cause death. The difficulty of delivery and likelihood of survival makes this class of BW agents less than ideal for warfare.

Biological toxins cannot be grown in large scale at this time but instead are harvested from living organisms. They include mycotoxins, fungi, botulinum toxin, staphylococcal enterotoxin type B, ricin and saxitoxin, and trichothecene mycotoxins (T2). Almost all the agents in this class are useful for medical research. Most can cause death, but again the mortality rate is generally low. Delivery remains a problem, so their utility on the battlefield is limited.

The threat of CBW agents has forced countries to address production and use of these weapons in war. The first agreement in this area came shortly after World War I. In 1925 the major powers banned the use of chemical weapons. Given the wide use of chemical weapons during the Great War, it is understandable that the 1925 Geneva Protocols focused on CW agents, though there was also a provision for the banning of bacteriological weapons of war. This convention came into effect in 1928, and although it handled chemical weapons, its weaknesses on biological weapons eventually led to another series of talks in Geneva in 1959. The talks dragged on, and in 1969 President Nixon publicly renounced the use of chemical and biological weapons by the United States. This provided the impetus to complete the Biological Weapons Convention (sometimes called the Biological and Toxin Weapons Convention or BTWC). Signed in 1972, it came into force in 1975. This convention bans the use, development, production, and stockpiling of biological weapons. Finally, in 1993 the Chemical Weapons Convention provided the same restrictions on chemical weapons. The main difference between the 1972 Biological Convention and the 1993 Chemical Convention lies in their verification provisions. Under the 1972 Convention, there are no verification requirements, whereas the 1993 agreement lays out the requirements to allow inspectors to verify regulatory compliance.

From the first gas attacks in Ypres in 1915 to the modern agreements in place today, chemical and biological weapons hold a particularly fearsome place in the minds of the people who use them or are victimized by them. As WMDs, they can kill thousands, but the deaths are rarely quick and painless. Most involve extensive suffering, illness, or worse. For these reasons, most countries restrict their development and use. The major fear for Western countries lies in the indiscriminate use of these weapons by terrorist groups, however, which are not bound by international conventions.

See also Asymmetric Warfare; Epidemics and Pandemics; Warfare.

Further Reading: Croddy, Eric. *Chemical and Biological Warfare: A Comprehensive Survey for the Concerned Citizen.* New York: Copernicus Books, 2002; Taylor, Eric R. *Lethal Mists: An Introduction to the Natural and Military Sciences of Chemical, Biological Warfare and Terrorism.* New York: Nova Science, 1999.

Steven T. Nagy

CLONING

To clone is simply to produce an identical copy of something. In the field of biotechnology, however, *cloning* is a complex term referring to one of three different processes. DNA cloning is used to produce large quantities of a specific genetic sequence and is common practice in molecular biology labs. The other two processes, therapeutic cloning and reproductive cloning, involve the creation of an embryo for research or reproductive purposes, respectively, and have raised concerns about when life begins and who should be able to procure it.

DNA cloning, often referred to as recombinant DNA technology or gene cloning, is the process by which many copies of a specific genetic sequence are produced. By creating many identical copies of a genetic sequence through a process known as amplification, researchers can study genetic codes. This technology is used to map genomes and produce large quantities of proteins and has the potential to be used in gene therapy.

The first step in DNA cloning involves the isolation of a targeted genetic sequence from a chromosome. This is done using restriction enzymes that recognize where the desired sequence is and "cut" it out. When this sequence is incubated with a self-replicating genetic element, known as a cloning vector, it is ligated into the vector. Inside host cells such as viruses or bacteria, these cloning vectors can reproduce the desired genetic sequence and the proteins associated with it. With the right genetic sequence, the host cell can produce mass quantities of protein, such as insulin, or can be used to infect an individual with an inherited genetic disorder to give that person a good copy of the faulty gene.

Because DNA cloning does not attempt to reproduce an entire organism, there are few ethical concerns about the technology itself. Gene therapy, however, which is currently at an experimental stage because of safety concerns, has raised ethical debates about where the line falls between what is normal genetic variation and what is a disease.

Somatic cell nuclear transfer (SCNT) is the technique used in both therapeutic cloning and reproductive cloning to produce an embryo that has nuclear genetic information identical to an already-existing or previously existing individual.

During sexual reproduction, a germ cell (the type capable of reproducing) from one individual fertilizes the germ cell of another individual. The genetic information in these germ cells' nuclei combine, the cell begins to divide, and a genetically unique offspring is produced. In SCNT, the nucleus of a somatic cell (the type that makes up adult body tissues) is removed and inserted into a donor germ cell that has had its own nucleus removed. Using electrical current or chemical signals, this germ cell can be induced to begin dividing and will give rise to an embryo that is nearly identical to the individual from which the nucleus came, rather than a combination of two parent cells. This "clone" will not be completely identical to the parent. A small number of genes that reside within mitochondria (small organelles within a cell that convert energy) will have come from the germ cell donor. Therefore, the embryo will have nuclear genetic information identical to the parent somatic cell, but mitochondrial genetic information that is identical to the germ cell donor.

SCNT is controversial because it involves the artificial creation of an embryo. Many people who feel that life begins at conception take issue with the technology because a germ cell is induced to divide without first being fertilized.

Similar ethical concerns are raised about therapeutic cloning, also referred to as embryo cloning, which is the production of embryos for the purpose of research or medical treatment. The goal of this procedure is to harvest stem cells from an embryo produced by SCNT.

Stem cells are useful because they are not yet differentiated. Not all cells in the human body are the same; a muscle cell, a bone cell, and a nerve cell have different structures and serve different functions. They all originally arise from stem cells, however, which can be used to generate almost any type of cell in the body. With further research, stem cells may be used to generate replacements cells that can treat conditions such as heart disease, Alzheimer's, cancer, and other diseases where a person has damaged tissues. This technology might provide an alternative to organ transplants, after which the donated organs are frequently rejected by the receiver's body because the cells are recognized as not being the person's own. With stem cells generated from a person's own somatic cells, rejection would not be an issue.

Because the extraction of stem cells destroys the embryo, people who feel that life begins with the very first division of a cell have ethical concerns about this type of research. Before this technology progresses, it will be important for society to define the rights of an embryo (if rights can be defined) and decide whether embryos can be manipulated for the treatment of other people.

Reproductive cloning is the process by which a nearly identical copy of an individual is created. In one sense, this type of cloning already occurs in the natural world. Although sexual reproduction of plants and animals involves the genetic information of two individuals combining to create a unique hybrid, asexual reproduction occurring in plants does not involve the combination of genetic information. In this case, an identical copy of the plant is naturally produced. Artificial reproductive cloning has enabled the cloning of animals as well. In this procedure, SCNT is used to create an embryo that has identical nuclear DNA to another individual. This embryo is then cultivated until it is ready to be inserted into the womb of a surrogate parent. The embryo is gestated, and eventually a clone is born. The first mammal to be successfully cloned and raised to adulthood was Dolly, a sheep, in 1997.

Since Dolly, many other animals have been cloned, including goats, cows, mice, pigs, cats, horses, and rabbits. Nevertheless, cloning animals remains very difficult and inefficient; it may take over 100 tries to successfully produce a clone. Previous attempts have also shown that clones have an unusually high number of health concerns, including compromised immune function and early death.

The inefficiency of current cloning technology, along with the compromised health of clones, raises further ethical concerns about the artificial creation of life and the manipulation of individuals for the benefit of others.

The American Medical Association (AMA) has issued a formal public statement advising against human reproductive cloning. The AMA maintains that this technology is inhumane because of both the inefficiency of the procedure and the health issues of clones. The President's Council on Bioethics worries that cloning-to-produce-children creates problems surrounding the nature of individual identity, as well as the difference between natural and artificial conception.

Although some individuals and groups have claimed to have successfully cloned a human, these claims have not been substantiated. In the United States, federal funding for human cloning research is prohibited, and some states have banned both reproductive and therapeutic cloning.

See also Eugenics; Genetic Engineering; Reproductive Technology; Research Ethics; Stem Cell Research.

Further Reading: American Medical Association Web site. http://www.ama-assn.org; Fritz, Sandy, ed. *Understanding Cloning*. New York: Warner Books, 2002; The President's Council on Bioethics Web site. http://www.bioethics.gov/reports; Shmaefsky, Brian. *Biotechnology 101*. Westport, CT: Greenwood Press, 2006; Wilmut, Ian, et al. "Viable Offspring Derived from Fetal and Adult Mammalian Cells." *Nature* 385, no. 6619 (1997): 810–13.

Heather Bell

COAL

Coal is essentially a kind of "compacted sunlight." It is a combustible material derived from leafy biomass that has absorbed energy from the sun and has been compressed in the earth over geologic time. It is usually found in seams associated with other sedimentary rock. Historically, Earth went through the Carboniferous age about 350 to 290 million years ago. During this period, Earth was like a hothouse with a higher average temperature than today and a steamy atmosphere that caused plants to grow rapidly. Using sunlight and moving through their life cycle, layer upon layer of plants accumulated on the surface of the earth. These plant materials gradually developed into peat bogs, and many of the bogs became covered with other material and were subjected to pressure over geologic time. The result today is that we find an abundance of coal, often associated with sedimentary rock such as limestone, sandstone, and shale.

From a human perspective, coal is a nonrenewable resource. From a geological perspective, coal could be renewed from sunlight and plants over eons, but it would require another carboniferous (hothouse) era of the world, which would not be very congenial to humans.

Peat is the first stage of the development of coal. It has very high water content and is not a good fuel if actual coal is available. When peat is compressed, it first becomes lignite or "brown coal." With further compression, brown coal becomes bituminous coal (soft coal). Finally, with both heat and high compression, we get anthracite or "hard coal," which has the least moisture content and the highest heat value.

Coal mining directly impacts the environment. Surface mining produces waste materials, including destroyed trees and plants, but also substantial amounts of waste rock. When a small mountain is stripped for coal, waste rock is often dumped in valleys, and this can generate acid contamination of water. Surface mining also generates considerable dust (the technical name for this is "fugitive dust emissions"). Underground mining occurs largely out of sight but can result in large areas of subsidence. The generation of methane (and other gases) and acid mine drainage into local aquifers can also occur. After coal is mined, the next step is called coal beneficiation. In this step, coal is cleaned of some of the impurities that have interpenetrated it because of surrounding rock formations and geologic activity over several million years. This generates waste streams, including coal slurry and solid wastes that must go somewhere. Then, the cleaned coal has to be stored, handled, and transported. Handing and transportation produce more fugitive dust emissions.

There are examples of both surface and underground mining in which great care has been taken to mitigate these and other environmental effects. However, the effects on local environment can be severe, as shown in many other cases.

Coal combustion byproducts (CCBs) are the waste material left over from burning coal. CCBs include fly ash, bottom ash, boiler slag, and flue gas desulfurization (FGD) material. Between 30 and 84 percent of this material can be recycled into other products such as concrete, road construction material, wallboard, fillers, and extenders. The rest is waste that may include toxic elements that can cause human health problems if they are inhaled (as dust in the wind) or if they get into groundwater.

Emissions from coal combustion include water vapor (steam), carbon dioxide, nitrogen, sulfur, nitrogen oxides, particulate matter, trace elements, and organic compounds. The sulfur dioxide released may transform into sulfur trioxide (sulfuric acid). Nitrogen oxides contribute to the formation of acid rain. Particulate matter causes lessened visibility and can have serious health consequences if the particles are breathed, including asthma, decreased lung function, and death. Carbon dioxide is a major component of greenhouse gases. A certain balance of greenhouse gases is necessary to keep the planet habitable, but too much greenhouse gas contributes strongly to global warming. Carbon sequestration is the term for capturing carbon dioxide and putting it somewhere.

Carbon sequestration is the attempt to mitigate the buildup of carbon dioxide in the atmosphere by providing means of long-term storage, for example by capturing carbon dioxide where coal is burned and injecting the carbon dioxide into the earth, injecting it into the oceans, or attempting to absorb it into growing biomass. The questions to ask about proposed methods of carbon sequestration are the following: How long will it stay sequestered before it is released back to the atmosphere? And will there be any unintended side effects of the carbon dioxide in the place in which it is to be put? We also need to be aware of what is sometimes called "silo thinking," that is, trying to solve an important problem without being aware of interactions and linkages. Right now, fish stocks are declining, and ocean coral is dissolving because the oceans are becoming more acidic. Putting huge additional amounts of carbon dioxide in the oceans might help to make power plants "cleaner," but it would more quickly kill off the existing forms of aquatic life.

Despite some of these effects, however, coal will continue to be the dominant fuel used to produce electricity because of its availability and lower price compared with other forms of electricity generation. At the same time, carbon dioxide released in the burning of coal is a large contributor to rapid global warming. This is a contradiction without an easy solution. If efficiency, widespread availability, and lowest cost are the relevant criteria, then coal is the best fuel. If we choose in terms of these standard market criteria, we will also move quickly into global warming and climate change. The physical root of the problem is primarily one of scale: a small planet with a small atmosphere relative to the size of the human population and its demand for the use of coal.

It is a simple fact today that the use of electricity is increasing all over the planet. The intensity of electricity use is growing gradually, year by year, throughout the economically developed portions of the planet, particularly because of the ubiquitous use of computers and the placing of increasing machine intelligence into other business and consumer devices. The poor and so-called backward regions of the planet continue to electrify, largely in response to their penetration by multinational corporations as an aspect of globalization. At the same time, intermediately developed countries with rapidly growing economies, such as India and China, are experiencing the emergence of strong consumer economies and rapid industrial development. For the near and intermediate future, these (and other) major countries will require substantial numbers of new central generating stations. Meaningfully lowering the demand for electricity would require major changes in our patterns of life, such as moving away from a consumer society and business system and a reorientation of housing and cities to maximize the use of passive solar energy and a transition to local DC power systems in homes.

Historically, the high-quality heat developed from good quality coal is responsible for much of the success of the Industrial Revolution in the Western economies. The transition from the stink of agricultural life and the stench and illnesses of early industrial cities to clean, modern living, characterized by mass production of consumer goods, is highly dependent on clean electricity. Coal kept us warm, permitted the manufacture of steel products, and gave us much of our electricity over the last century. With only a little coal, natural gas, and oil, the human population of the planet would have been limited largely to the possibilities of wind and sun power; history would have developed very differently, and the human population of the planet would be only a small percentage of its size today. It is important to know that doing without coal, gas, and oil would have the reverse implication for the carrying capacity of the planet. At root, it is not only the historic and continuing advancement of civilization but also the size and quality of life of populations of nations that are dependent on coal, natural gas, and oil. That is why securing these resources is so integral to the trade and military policies of nations.

At the same time that coal has been a wonderful resource for human development and the multiplication of the human population, there is a paradox—electricity, which is so clean at point of use, if generated from coal, is associated with extreme carbon loading of the atmosphere. This contradiction originally existed only at a local level. As an illustration, Pittsburgh, a major industrial center in America, was long known as a dirty coal and steel town, with unhealthy air caused by the huge steel plants, the use of coal for electricity generation, and the general use of coal for home and business heating in a climate with long cold winters. The air was often dirty and the sky burdened with smoke and dust.

This was initially taken as a sign of economic vigor and prosperity. Pittsburgh's air was cleaned up in the early 1950s by the requirement of very high smoke stacks and a shifting away from nonindustrial uses of coal for public health and civic betterment reasons. The tall smoke stacks, however, though they provided a local solution, simply transferred the problem to places downwind. This is a reality of pollutants: they do not go away; they go somewhere else. Places downwind of the Midwestern power plants (such as New York City) experienced more unhealthy air days, and lakes in the mountains downwind began to die because of acid rain. This is the local level of the paradox—clean electricity and efficient large-scale industry produce local or regional pollution problems because of the use of coal.

Similarly, the global level of the paradox is that the use of coal is responsible for significantly fouling the planet, leading to a common future filled with the multiple disasters associated with global warming. Just a few of these experiences we have to look forward to include submergence of coastal areas, loss of ice at the poles, loss of snowpack on mountains, invasions of species from other areas against weakened natural species, dramatic food shortages, and an increasing number of riots in poor areas where the rising cost of food cannot be met within the local structure of wages—not a war of "all against all," but of increasing numbers of persons increasingly shut out of the economic system against those still protected by remaining institutional arrangements or by wealth. As resources contract, in addition to the problems of food shortages and new outbreaks of disease, the resulting income gap likely signals a return to the social inequalities of the Victorian era.

An underlying variable, of course, is the size of the human population. If we were facing a few new power plants and limited industrial production, the myth of unlimited resources that underlies conventional economics would be approximately true. It would not matter much if we fouled a few localities if the human population was one-hundredth or one-thousandth of its current size, and the planet was covered with vibrant meadows and ancient forests. With a much smaller human population, the fouling of the planet would be less of an immediate problem. But given the size of the human population, the need is for several hundred new power plants. The demand through market forces for consumer goods, industrial goods, and electricity, particularly from the portion of the human population engaged in unsustainable modern market economies, drives the need for hundreds of new central power plants in the immediate to intermediate future. Industry in India and China, in particular, is taking off along a huge growth curve, different from, but in many ways similar to, that of the Industrial Revolution in the West. In our current situation, coal is, on the one hand, the preferred market solution because it is relatively inexpensive, is a widespread and still abundant resource (in contrast to gas and oil), and can provide power through electricity generation that is clean at point of use. The problem at the global level is the size of the planet and the limited atmosphere in relation to the size of human population. The scale of what is required will generate far too much pollution for the planet to handle in ways that keep the planetary environment congenial to humankind.

It is possible, however, to talk about "clean" coal. "Clean coal" has two meanings. First, some types of coal are cleaner than others, and some deposits of coal contain much less foreign material than others. Cleaner coal is more expensive than more dirty coal. Second, the phrase is a slogan of the coal industry pointing toward the concept of capturing gas emissions from coal burning. As a slogan, it serves the purpose of conveying the image of a future in which commercial-scale coal-burning power plants would emit no carbon dioxide. Research on this problem is ongoing, but there are no such plants at the present time. The U.S. FutureGen project is on hold after federal funding from the Department of Energy was pulled. The questions to ask about the promised clean coal future are these: What is the scale of transfer of carbon dioxide that would be required (if it could be captured)? What would be done with the massive quantities that would have to be sequestered, and would this have any unintended consequences?

Coal is less expensive than other fuels, but this is due in part to the free market system in which the social and environmental costs of coal are treated as what economists like to call "externalities." That is, these costs are left for other people—for regional victims of pollution—and for global society to bear. Several systems have been proposed to transfer all or part of these costs to companies that burn massive amounts of coal, such as electric utilities. In fact, a sector of the electric utility industry is currently campaigning to have some form of carbon trading or carbon tax imposed. It is generally expected that this will occur in the not-too-distant future, given that many industry leaders would like to resolve the ambiguity and uncertainty of what form these costs will take and to speed the new system into place. This may substantially increase the cost of coal as an energy resource.

Coal has had and continues to have a major role in the advancement of civilization. It is currently more abundant and more easily available than other major fuels. Its concentrated energy (high heat content) permits us to create steel products. Without coal, natural gas, and oil, the human carrying capacity of the planet would be a small percentage of the current human population. Yet there is a contradiction inherent in the massive use of coal and in the building of hundreds of new coal generating stations because carbon release will hasten global warming and also produce other environment effects not helpful to human life. This is a contradiction without an easy solution.

See also Fossil Fuels; Global Warming.

Further Reading: McKeown, Alice. "The Dirty Truth about Coal." Sierra Club monograph, http://www.sierraclub.org/coal/dirtytruth/coalreport.pdf; Miller, Bruce G., *Coal Energy Systems.* San Diego, CA: Elsevier Academic Press, 2005.

Hugh Peach

COLD FUSION

Cold fusion is the popular term for low-energy nuclear reactions occurring at room temperature and pressure. In a fusion reaction, two atom nuclei are forced together to form one nucleus. The products of this reaction are energy, neutrons, and other subatomic particles. The major application of hot nuclear fusion has been military in nature in the form of thermonuclear (fusion) weapons, the first of which was detonated by the United States in 1952. The major technical barrier to nuclear fusion as a nonmilitary technology has been the extremely high temperatures, similar to those on the surface of the sun, that seem to be required. Any possibility of fusion at temperatures closer to those found on the surface of the Earth—that is, cold fusion—would constitute a scientific or technological breakthrough with world historical implications.

A highly contentious debate arose in March 1989 when two electrochemists at the University of Utah, Martin Fleischmann and Stanley Pons, made the claim via a press conference that they had successfully conducted cold fusion experiments. Fleischmann and Pons used a surprisingly simple electrochemical cell with heavy water, which has more neutrons than regular water, to produce a cold fusion reaction. Their primary evidence for fusion was excess heat produced by the cell that they argued could have come only from a nuclear fusion reaction. Other scientists immediately began trying to replicate the Fleischmann-Pons results, some with claimed success but most with failure. This failure by others to replicate the Fleischmann-Pons results has caused many to doubt the validity of the original claims.

Fleischmann and Pons were electrochemists and not physicists, which may have contributed to the controversy. Cold fusion has traditionally been the domain of condensed-matter nuclear physics, and physicists wanted evidence of cold fusion in the form of neutrons released from the fusion reaction. Fleischmann and Pons were not experts at measuring subatomic particles, so part of their experiment was attacked by physicists. They nevertheless continued to assert their success using excess heat as the primary evidence, and subsequent discussion in the scientific community has focused on whether the amount of excess heat was enough to prove the existence of fusion and enough to make cold fusion viable commercially.

There have also been extensive debates about the methods Fleischmann and Pons used to announce their results. Scientists are expected to make their results public and usually do so by publishing in a peer-reviewed journal. The publishing process can take several months, and a scientist's research is reviewed by others in their field. Cold fusion is such a revolutionary possibility that anyone who proves the existence of the phenomenon first has much to gain. Some speculate that Fleischmann and Pons, and the University of Utah's lawyers and public relations people, did not want to risk losing priority and patent protection. Thus they chose to announce their results via a press conference rather than publishing in a journal. Additionally, it is known that Steven Jones, a physicist from Brigham Young University also working on cold fusion, was planning to announce his results in May 1989. Amid rivalry and miscommunication between the Jones group and their own, Fleischmann and Pons decided to go ahead with a press conference before they had written up their results. This led to a circus atmosphere of many scientific groups proclaiming via the press either their ability to replicate the Fleischmann-Pons results or their disproof of such results. Many scientists who witnessed this media frenzy believe the original results should have been carefully reviewed by the scientific community and, once their importance was verified, then simply published in a scientific journal.

One alternative to cold fusion that is currently being pursued is commonly known as plasma fusion. This is a form of hot fusion in which the extremely hot reacting particles are controlled by a magnetic field so that they do not come in contact with their container, which would be destroyed by the high temperatures. The energy required to control such large magnets and heat the particles to such high temperatures limits the economic viability of plasma fusion. As of now, the fusion energy output is not greater than the energy input.

Aside from the scientific difficulties of cold fusion, there are other factors that make cold fusion currently impractical as a large-scale energy producer. Any large-scale production of energy would need to be more reliable than the Fleischmann-Pons experiments. Also, scientists who have tried to produce cold fusion using electrochemical cells have had problems maintaining the integrity of their experiment vessels. Both types of fusion still have rather large logistical hurdles, in addition to the scientific issues, to overcome before they can become commercial options for energy consumption.

Current economic and political concerns involving energy independence make cold fusion attractive. Some private companies working on cold fusion have been granted patents, but no one has yet produced a commercial application. Anyone successful in producing such a method will gain a great deal of power and money.

The U.S. Department of Energy convened two panels, one in 1989 and one in 2004, to assess the promise of experimental claims as well as the excess heat phenomenon. The earlier panel concluded that cold fusion was not a practical source of energy, but the second panel was evenly split as to whether cold fusion should be pursued. It seems that the current concern over energy prices and independence has lead some to rethink the possibility of cold fusion. The Japanese, for instance, are continuing to invest in cold fusion research.

The first step undoubtedly is to successfully prove the existence of cold fusion. Once that is accomplished, cold fusion could be commercially developed. These two steps will require a great deal of time and resources if they are possible, so the likelihood of using energy produced by cold fusion in the near future is low, but in a few generations humans may be provided with an unlimited source of clean energy in the form of fusion.

See also: Nuclear Energy; Unified Field Theory.

Further Reading: Collins, Harry, and Trevor Pinch. "The Sun in a Test Tube: The Story of Cold Fusion." In *The Golem: What You Should Know about Science,* pp. 57–77. Cambridge: Cambridge University Press, 1998; Taubes, Gary. *Bad Science: The Short Life and Weird Times of Cold Fusion.* New York: Random House, 1993.

Ursula K. Rick

COMPUTERS

Computers have become ubiquitous throughout Western society. Modern business completely depends on them to conduct daily affairs. They have become instrumental in virtually all the sciences, medicine, engineering, and manufacturing. Record keeping has been revolutionized by modern databases, as censuses, tax information, birth and death certificates, and other essential government records are now all digitized. Communication takes place instantly, effectively, and (perhaps most important) cheaply through contemporary information structures.

It may seem counterintuitive that there are any debates about computers, which at face value seem so beneficial, but perhaps the most problematic feature of the information revolution is the digital divide debate. Digital divide opponents acknowledge the strength of contemporary computers but argue that only the privileged have access to what has become an essential infrastructure in developed societies. Other criticisms of computing focus on the ease of engaging in anonymous criminal activity. Such activities include "phishing," copyright violation, aggressive lending, and endangering the welfare of children and women. Finally, some opponents of computing note the ease with which racist, sexist, and other derogatory materials such as pornography travel easily through the Internet and that there are no overarching institutions or common standards that monitor such materials.

It is important to note that computers have been present in society for much longer than many people think, if one accepts the literal definition of the device as something that aids in calculation. The word *computer* itself is derived from the eighteenth- and nineteenth-century title given to human workers, often women, who professionally kept sums of accounts or other lengthy calculations.

The earliest known computer is in fact the abacus, known most often as a rack of wires on which beads are suspended, though other forms exist. It is likely the most widely used computer in history, used in various forms throughout ancient Babylonia, Mesopotamia, Persia, India, China, Egypt, and Africa. It had many traditional uses, though it was almost certainly most often used for trade and inventory accounting. Most abacuses utilized a biquinary notation system, where one set of numbers represented base 2 numerals, and the other base 5. With this relatively simple system, addition and subtraction could be performed with ease, though higher-order operations were usually out of the question. As societies progressed, however, more advanced forms of the abacus emerged, such as the suanpan of China. The suanpan was able to perform decimal and hexadecimal operations and allowed for more complex operations such as multiplication, division, and even square root calculation with relative speed and ease.

The abacus was the computer of choice for much of the history of civilization, with a variety of incarnations in virtually every society. Interestingly, numbers—their operations and their significance—were a constant fascination for scholars in the ancient and medieval world. In addition to practical computing, numerology developed alongside mathematics in Ancient Greece, Rome, Babylonia, Egypt, and India and within Jewish communities. Numerology supposed that numbers were entities in their own right, possibly with mystical or magical powers. With the proper system and operations, the future could be told with numbers in a process known as numerological divination. Complex codes with deep-seated significance began to merge and be sought out. Numerology and mathematics were not separate disciplines as they are now. Numerological concepts became embedded in religious texts, a feature that is often played up in Hollywood depictions of numerological codes.

The abacus and numerology cross in an interesting fashion at one point in history. Numerology's association with divination called for more practical tools for telling the future, given that previously, priests needed to be consulted in the temple to tell the future through complex auguries. This was often done through the use of divining rods, or arrows, which were tossed by the priest in response to a specific question. The rods pointed to specific allegorical symbols on the temple wall, which corresponded with various celestial forces. In this way, the future could be read, but it required a trip to the temple. Because the abacus was such a convenient, portable means of counting and arithmetic, it was reasoned that at some point temples themselves could benefit from a portable version of themselves. Throughout Ancient Egypt, Rome, and Greece, a religious cult was formed around a mythical figure known as Hermes Trismegistus, meaning literally "three-fold great Hermes." Hermes in this case was a figure who was Hermes to the Greeks, Mercury to the Romans, and Thoth to the Egyptians, whom all regarded as a god among a pantheon of gods (some even speculate that this same figure was Moses to the Hebrews).

We may never know whether Hermes Trismegistus existed or not, but he did have followers who crafted a "Book of Thoth," a series of unbound leaves that depicted the allegorical images on the temple walls erected by his followers. The book and its illustrations are believed by many historians, after passing through many hands, to eventually have formed the *Tarocci,* or Tarot deck. The illustrations were known as *atouts,* what we commonly think of as face cards, and *pips,* which match standard numerical counterparts, were later added. Tarot cards were originally an instrument of divination, but as they circulated throughout the ancient world and Europe, they developed into what we think of as standard playing cards.

The usefulness of playing cards increased exponentially, however, when xylography, or wood engraving, was invented. The same technology that Gutenberg used to print copies of the Bible was used to create copies of decks of playing cards, and whereas the abacus was commonly available, playing cards were even more commonly available, their pips able to be used in the same manner as an abacus was used for basic addition and subtraction. Additionally, playing cards were standardized into a form of almanac: 4 suits for 4 seasons, 13 cards to a suit for a 13-month lunar calendar, 52 cards total for 52 weeks in a year, and 365 pips total for the days in a year, a convention that stands to this day. It is even rumored that major events in the Bible could be tied to cards in a deck, made famous through various folk tales. In any case, the deck of playing cards was a sort of "personal computer" in the Middle Ages and Renaissance, in that it could be used for mundane accounting, assisted in keeping records, and of course, could be used to play games.

Mathematics flourished in the Renaissance, and complex operations began to surface that required intensive computing. This drove two trends in mathematics research: the ability to develop complex notation, algebraic reduction, and other "elegant" solutions of complex operations to save on actual elaborate calculation and the search for a more efficient, mechanical means of calculation. Blaise Pascal, a mathematical prodigy who made enormous contributions to the field of mathematics, especially probability, also developed one of the earliest forms of mechanical computer to assist his father with tax accounting. Developed over a period of years in the mid-1600s, the "pascaline," as it came to be called, was a clockwork series of gears turned by a crank that provided mechanical computation in the decimal system. Fifty of the devices were made in all, but they did not come to replace their human equivalents.

Mechanical computation was a tinkerer's delight for many years, but the next major innovation in mechanical calculators did not come until Charles Babbage conceived of the difference engine in 1822 and, later, the analytical engine. His life's work, though never fully completed by Babbage himself, the difference engine is credited with being the first programmable mechanism intended for computation. The design layout of Babbage's engines strongly reflects the same layout of today's modern personal computers, with separate routines for data storage, operations, and input and output. The London Science Museum engineered a version of Babbage's difference engine based on his designs, and it successfully performed complex operations while being built from methods that were indeed available in Babbage's day. Babbage's associate, Ada Lovelace, was one of a few contemporaries who fully grasped the impact of what Babbage was proposing. Based on his designs for the difference and analytical engine, she published a method for its use in computing Bernoulli numbers, though some historians debate how much of a role Babbage himself had in its design. The modern computing language Ada is named in her honor.

An alternative point of view holds that the honor of being the first successful programmer and hardware operations designer belongs to Joseph Jacquard. Working in the early 1800s, and as a precursor to Babbage's work, Jacquard devised a series of punch cards that fed instructions into a textile loom. Complex patterns were represented on the cards, and effectively transferred to the loom such that changing the pattern produced was as simple as swapping out the punch cards, like a cartridge or memory card. Though Jacquard was not a mathematician or aspiring computer hardware designer, the concept of his interchangeable processing is the heart and soul of a software–hardware interface.

Representing information on punch cards was deemed so useful that Herman Hollerith relied on them to develop a tabulating machine for the 1890 U.S. census. Immigration into the United States had produced an overwhelming need to process census data, vital to taxation and representation, at a rate that could keep pace with the influx of new citizens. Hollerith's system relied on holes in the punch card being able to complete a circuit, allowing for census records in the punch card format to be quickly tabulated. The census, which was expected to take up to 13 years, was completed instead in 18 months. Bolstered by his success, Hollerith founded the Computing Tabulating Recording Corporation, which eventually developed into International Business Machines, or IBM.

It is important to note that tabulating machines did not perform any computations per se; their usefulness was in maintaining and efficiently operating on vast quantities of information. It is taken for granted today that personal computers and large servers are able to perform both of these functions, but advanced computation and record keeping were originally separate tasks.

Relays are essentially a kind of switch, like a light switch, but they are controlled by another circuit. As electric current passes through the relay, it activates a small mechanical arm, which toggles a larger, more powerful circuit. Relays were developed in the nineteenth century and found some of their most useful application in the telephone switching industry. The essential formula of the relay as either on or off allows them to operate in a binary fashion, albeit clumsily. Konrad Zuse, a German pioneer of computer engineering, used discarded telephone relays during World War II to create the Z3, a programmable electromechanical computer that had a 64-bit floating point, sophisticated by even today's standards. The actual operation of the Z3 was likely something to behold, as its hundreds of relays had to physically toggle on and off to perform operations, like the keystrokes of a vintage typewriter.

The basic design of relay-based computing could be shrunk down, and the postwar era saw an explosion of miniature electromechanical calculators useful for home and office accounting. Calculators such as the Monroe Company's line of "monromatics" weighed in at a relatively light 20 to 25 pounds, ran on regular AC wall current, and could fit on a single desk. They performed addition, subtraction, multiplication, and division with ease and were able to store and recall previous results. Along with the slide rule, they were a major computing staple for calculation-intensive applications such as engineering and accounting for much of the twentieth century.

Of course, the vacuum tube changed everything. Vacuum tubes, which looked something like light bulbs, relied on thermionic emission, the heated discharge of electrons, to toggle another circuit on and off, conceptually almost identical to how a relay performed in a computing context. The advantage to the vacuum tube, however, was that there were no mechanically moving parts that actually toggled, and its operation was thus much quicker. The disadvantage of vacuum tubes was that they were fragile, were very energy-inefficient, required significant cooling, and were prone to failure. Nonetheless, vacuum tubes are a reliable technology still in use today, most notably as cathode ray tubes (CRTs) in television and computer monitors, though it appears that LCD, plasma, and other flat-screen technologies are the coming thing. The first fully electronic computer to rely on the use of vacuum tubes was ENIAC, the Electrical Numerical Integrator and Computer, designed principally by John Mauchly and J. Presper Eckert at the University of Pennsylvania's Moore School of Electrical Engineering between 1943 and 1946.

MOORE'S LAW

Gordon Moore, a chemist and physicist by training, is also the cofounder of the Integrated Electronics company (Intel), one of two dominant computer microprocessor firms (the other being Advanced Micro Devices, or AMD). Writing in 1965, Moore observed that the capacity of embedded transistors in microprocessor packages was doubling roughly every two years, such that the overall capacity was increasing exponentially. Pundits noticed that in addition to microprocessors, Moore's Law, as it came to be known, also held for digital electronics in general as memory capacity, speed, and the ability to manufacture components also rose (or fell, in the case of price) exponentially.

Taken at face value, Moore's Law appears to be true; the capacity of digital electronics has increased exponentially. The reasons for this, however, are not straightforward. Some technological enthusiasts see Moore's Law as an expression of nature in general, as a sort of evolutionary path that draws parallels with the human genome and human progress in general. Along these lines of thinking, many artificial intelligence pundits, such as Ray Kurzweil and Hans Moravec, point to Moore's Law as one reason why artificial intelligence is inevitable.

Another perspective, however, is that Moore's Law is a self-fulfilling prophecy of the digital electronics industry. Industry designers such as Intel, AMD, Fairchild, and Texas Instruments were all aware of the exponential nature of their achievements, and a high bar was set for these companies to remain competitive with one another. Though Moore's Law still holds today, some cracks are beginning to show in the process. CPUs, for example, benefit from the process most, whereas hard disk drives and RAM do not obey Moore's Law. Although the engineering challenges of sustaining the law are mounting, ultimately transistor design would have to take place at the atomic level to keep up, the ultimate wall. Gordon Moore himself, speaking in 2005, admitted that his law would ultimately have to be broken.

Subsequent development of computers relied heavily on the lessons learned from the ENIAC project; most notable, of course, was the lesson that electric circuits are vastly superior to electromechanical systems for computation. ENIAC was crucial in the calculation of ballistic tables for military application, and one of its major contributors, John Von Neumann, improved the original design in many ways. Von Neumann was instrumental, along with Eckert and Mauchly, among others, in designing ENIAC's successor, EDVAC (Electronic Discrete Variable Automatic Computer). EDVAC switched from a decimal base to a binary base, the common standard in contemporary computers, and was also a much larger step toward a general computer, easily reprogrammable for solving a specific problem.

With the enormous success of electric computers such as ENIAC, EDVAC, UNIVAC, and other similar systems, computers began to look more like what one would recognize as a computer today. They also began to enter the popular imagination and were often referred to as "electric brains" or other such popular representations in science fiction and the media. Alongside the computer, however, the seemingly innocuous transistor, also a sort of switch, like the relay and the vacuum tube, was also developing. It is not apparent to whom the credit belongs for inventing the transistor. Julius Lilienfield, an Austrian physicist, took out a patent on a transistor-type device in 1925, as did German physicist Oskar Heil, though neither appears to have built the devices or published papers regarding their application. Recent controversy on the matter, however, points to the fact that Bell Labs researchers based much of their work on these original patents without acknowledgment.

The transistor remained an obscure idea until 1947, when researchers at Bell Labs demonstrated a working transistor prototype. The name itself captures the peculiar characteristics and electrical properties of the transistor as a semiconductor, thwarting electric conduction under some conditions (known as resistance) and introducing gain in others. Conceptually, a transistor can be thought of as a plumbing faucet, where a small turn of the handle can unleash a torrent of water, the water in this case being electric current flow. Bell began rapidly developing the transistor with phenomenally successful applications in radios, televisions, and ultimately, computers.

Transistors in computer circuits are strung together in sequences to form logic gates. Collectively, such layouts are referred to as integrated circuits, or solid-state devices. They have all the advantages that vacuum tubes have with respect to being nonmoving, or solid-state, but transistor integrated circuits are exponentially smaller, cheaper, and more energy-efficient. With decades of refinement in their design methods and application, solid-state integrated circuits have culminated in the microprocessor, the keystone to contemporary computing. As of this writing, Intel's most recent quad-core CPU microprocessor, the QX9650, contains 820 million transistors in a roughly one-square-inch (2.2 square cm) package.

Computers have a long history of military application, but aside from the original impetus of designing ENIAC to compute ballistics tables, perhaps no other military project has affected computing as much as ARPANET. Conceived of and implemented at the Defense Advanced Research Projects Agency (DARPA), ARPANET's initial development was led by computer scientist Joseph Licklider. His concept of an "intergalactic computer network" was the basis for the possibility of social interactions over a computer network. After Licklider's initial development of the idea, his successors—Ivan Sutherland, Bob Taylor, and MIT researcher Lawrence G. Roberts—became convinced of its feasibility, and ARPA decided to implement the system. The initial stumbling block the ARPA team ran into was the total inadequacy of telephone circuit switches for allowing computers in different parts of the country to communicate together. Leonard Kleinrock, also at MIT, convinced Roberts that packet switching, which allowed for the transmission of data between computers over standard communication lines without relying on the telephone switching system, was the solution.

After the initial success of packet switching enabled ARPANET, the project quickly expanded and developed into the backbone of the current Internet, initially being used primarily for e-mail. A controversy in the history of ARPANET is the commonly held belief that its original formulation was meant to safeguard military communications in the face of nuclear war, which historians and researchers at the Internet Society (www.isoc.org) claim is a misconception. Although it is true that ARPANET is resilient if large portions of the network are destroyed, members of the Internet Society (including Kleinrock) claim that this resiliency was added as a feature only because the telephone switching system was so brittle, and the network itself was constantly riddled with errors.

The personal computer, or PC, is what most people think of today when the word *computer* is mentioned. The first incarnation of the PC was arguably the Altair 8800. Sold through mail-order electronics magazines, the Altair was a microcomputer intended for home hobbyists with basic electronics knowledge. Several important ingredients fueled the Altair 8800's popularity. The first was that it was based around an Intel 8080 microprocessor but otherwise used a motherboard to allow other, modular components to be installed and later swapped out. This modular, upgradeable design feature is a hallmark of contemporary personal computers. Additionally, the 8800 was able to be programmed with a high-level programming language, Altair BASIC (Beginner's All-Purpose Symbolic Instruction Code), sold to Altair for distribution by Paul Allen and Bill Gates, who later formed "Micro-Soft."

The basic design was modified by many, but perhaps the next most successful iteration was IBM's PC XT and AT. These models were the first PCs to come standard with a processor, RAM, a hard disk drive, a floppy disk drive, and ISA slots for additional components. Monitors and printers were standard components as well, and virtually all personal computer architecture to date follows the basic architectural archetype of IBM's version of the PC. The XT and AT also came standard with PC-DOS, an operating system developed in cooperation with Microsoft, and standard applications were the BASIC programming language and compiler and a word processor. The IBM PC and Microsoft operating system became the standardized tool for the business computing environment.

Many other companies were also successfully designing PCs, including Commodore, but the largest rival to IBM and Microsoft was Apple Inc. Founded by Steve Jobs, Steve Wozniak, and Ronald Wayne, the company was essentially a commercial enterprise made out of "homebrew" computing ideals and a community orientation. Their first personal computers, the Apple I and Apple II, were not as commercially successful as Commodore or IBM kits until Apple entered the game of software design. Their first highly successful application was VisiCalc, a spreadsheet program that was the basis for Microsoft's Excel and Lotus 1-2-3. Apple's true claim to fame began with the design of Lisa and the Macintosh, the first application of a mouse-driven GUI (Graphical User Interface). This was clearly the basis for all future GUIs such as Microsoft's Windows operating system. A bitter conflict erupted between Apple and Microsoft over intellectual property rights, though nothing ever came of Apple's lawsuit against Microsoft.

Countercultural movements making use of digital electronics have existed for quite some time. One of the earliest examples of this phenomenon was known as "phreaking," wherein users would manipulate the telephone switching system to receive free long-distance phone calls. Josef Carl Engressia Jr. was perhaps the first well-known phreaker. Born blind in 1949, Engressia was also gifted with perfect pitch. Through serendipity, Engressia, who later changed his name to Joybubbles, discovered that a perfectly whistled 2600-hertz tone would send phone circuits into a form of "debug" mode. In this mode, the user could take advantage of the recently installed automated switching system to place calls anywhere in the country. This discovery led to the creation of "blue boxes," homemade electronic instruments that produced the 2600 hertz specifically for this purpose; *2600* magazine, a well-known electronics counterculture publication, is so named in honor of the blue box tone. Apple's cofounder, Steve Wozniak, owned and operated a blue box currently on display in the Computer History Museum.

The advent of widely available computer modems allowed users to connect their computers over standard phone lines, and modems commonly also became used for the practice of "war-dialing." Illicit computer users would program software to dial every telephone number across a certain range, usually the local area because the calls were free. Occasionally, the war-dialer would find another modem-enabled computer, sometimes owned and operated by a major corporation or government agency. This allowed the dialer to then play a sort of game, in which he or she would attempt to log on to the unknown system and infiltrate its file system. This was one of the first known, among many, kinds of computer hacking. The practice itself was made famous in the 1983 film *Wargames,* in which a war-dialer unwittingly hacks into a Department of Defense mainframe and triggers potential nuclear war.

With further refinements in PCs and modems, a number of bulletin board services, or BBSs, sprang up during the 1980s and 1990s. A BBS system operator, known as a "sysop," would knowingly leave a usually high-end computer running to which other users would dial in with their own PCs. A BBS would typically serve as a message center, common to contemporary forums, and counterculture files would be distributed across the relatively anonymous network (though telephone logs could always be called upon to see activity). Such files were an "anarchist cookbook," in various incarnations, a guerilla field manual for pranks, vandalism, and "social engineering," the manipulation of basic services for gain. BBSs also were one of the first methods of "warez" distribution, wherein computer software was willfully distributed in violation of copyright law. Serial numbers and generators were distributed with the software needed for users to "crack" it. Not all BBSs were associated with warez, anarchy, or other counterculture elements, however, though their users usually self-identified as part of a definitive subculture.

The basic services of the Internet were also a tremendous boon to computer counterculture, and eventually BBSs were phased out because of the exploding popularity of widespread Internet access. Warez distribution skyrocketed, initially spreading out through the "scene," power-users who often connected to one another via a "darknet," a connection similar to ARPANET's standards, but in which users are not responsive to outside network queries in the usual fashion. USENET, also a major component in the downfall of the traditional BBS, is a message posting protocol of the Internet that has been in use since 1980.

Originally intended just for message communication, USENET has exploded in popularity because of its ability to transfer extremely large files such as movies, games, and music as text-encoded binary files. This is done with a powerful level of anonymity with little risk to the warez distributor, usually but not always a group in the "scene."

With widespread broadband Internet a feature of today's computer picture, peer-to-peer file sharing has become commonplace. This is achieved by point-to-point ad hoc connections between users, usually managed by a software client such as a Gnutella network–based client or, more recently, a torrent client. File sharing is prolific, with hundreds of thousands of users swapping digital copies of music, films, software, and other digital products with extreme ease. Ad hoc file transfer protocol, or FTP, connections (also a backbone of the Internet dating back to the original ARPANET) are also common in the distribution of illicit software. Examples of such software are files that circumvent security in other digital devices, such as video game consoles and digital satellite systems (DSS) that allow for access to free games and commercial programming. In other ways, however, the Internet has weakened counterculture movements by making them widely accessible and thus easily monitored by outside agencies in law enforcement and groups such as the Recording Industry Association of America (RIAA), who aggressively sue users who violate copyright laws.

The digital divide refers broadly to the disparity between those who have access to computer technology and those who do not. This disparity can take a variety of forms. Within the United States, many rural and urban public schools do not have the resources to offer instruction in computer technology. Given how vital computers are to contemporary business, education, and manufacturing, these students are at a serious disadvantage. Even if computers themselves are available, they may be out of date and, more importantly, may not have access to

OPEN SOURCE SOFTWARE

There is an active subculture of computer users and programmers who believe that software should be freely distributed, with elegant designs programmed on a voluntary basis for recognition within the community. Collectively, this is known as the "open source" movement, and it usually revolves around the GNU (Not Unix) public license for "free" software. Although the software itself may not be free to acquire, the GNU public license gives end users the freedom to copy the software and distribute it as much as they wish (without modifying it and not mentioning the modification, the freedom to tweak the source code programming of the software if they wish, and to redistribute their own version of the software to the community).

Such acts of "copylefting" have a fervent and active community surrounding them, ranging from free distributions of operating systems such as Linux to alternatives to commercial software, such as the Open Office word processor. Open source software thus challenges and subverts copyright laws and digital property without actually breaking laws, as warez hackers do when they crack software and redistribute it illegally. One major criticism of open source software, however, is the high level of expertise usually required to use it.

the Internet. There was a time when the Internet was not a vital aspect of learning to use computers, but its role is becoming increasingly central to computer use.

These same critiques hold in the economy in general. Broadband Internet has not penetrated some areas because they are too distant from existing infrastructure or they are not priority markets. This is especially true in many developing countries, for whom computers do not yet have as much value because their infrastructure is not geared toward the "global," industrial political economy.

Many philanthropic programs and foundations take this problem seriously and are attempting to make computer technology accessible and affordable for these areas. Critics of such programs make two important points. First, as wonderful as computers are, they should not be a priority over sustainable agriculture and food production, adequate health care, sanitary drinking water, and an end to violence. Computers may be a tool to help with these issues, but they are not as vital as the issues themselves; technological fixes do not often work for problems that have social and political underpinnings. Second, there is often an assumption that computers are an end in and of themselves, meaning that the requisite training in their use and maintenance is not included in social programs that attempt to cross the digital divide.

See also Internet; Information Technology; Search Engines; Software.

Further Reading: Dreyfus, Hubert. *What Computers Still Can't Do: A Critique of Artificial Reason.* Cambridge, MA: MIT Press, 1992; Moore, Gordon E. "Cramming More Components onto Integrated Circuits." *Electronics Magazine,* 38, no. 8 (1965); Moravec, Hans. *Robot: From Mere Machine to Transcendent Mind.* Oxford: Oxford University Press, 1999; Raymond, Eric S. *The Cathedral and the Bazaar: Musings on Linux and Open Source by an Accidental Revolutionary.* Sebastopol, CA: O'Reilly Media, 2001; Shurkin, Joel N. *Engines of the Mind: The Evolution of the Computer from Mainframe to Microprocessor.* New York: Norton, 1996; Van Rensselaer, Mrs. John King. *Prophetical, Educational, and Playing Cards.* Philadelphia, PA: George W. Jacobs, 1912.

Colin Beech

CREATIONISM AND EVOLUTIONISM

Creation stories explain not only the origin of the world but also how people ought to live and worship. One creation story, common in various ways to the Jewish, Christian, and Islamic traditions, is of a divinity with absolute transcendence on whom humans and the world are wholly dependent. Thus, when science, within the ambit of the Christian tradition, became committed to a theory of the world's formation by means of evolutionary change, there arose a major confrontation between science and religion, particularly in the United States, that has yet to be fully resolved. Attempts at resolution have taken at least three forms: fundamentalist affirmation of creationism, a religion-rejecting affirmation of evolution, and efforts to adjudicate some intermediate position.

"How did the world come to be?" is a very old question. Aristotle (384–322 B.C.E.) thought of everything in the world as having an end goal or purpose—for example, trees grow leaves to capture sunlight, sharks have fins to swim, and so on—and that there must be some first cause that produces this effect. For Aristotle this cause or "unmoved mover" was conceived as part of the cosmos. By contrast, the Judeo-Christian-Islamic creation story tells of a wholly transcendent God who gave intelligence to the world but is beyond complete human comprehension. The medieval Christian theologian Thomas Aquinas (1224–74), for instance, used Aristotelian-like arguments to demonstrate the existence of a first mover God who functions throughout time to make the world as it is, but he did not think reason alone could prove God created the world at some point in the past. Divine revelation, as found in the Bible, was the basis for such a belief. At the same time, following a tradition of interpretation that goes back to the use of metaphors in the Bible, especially as developed by biblical commentators such as Philo Judaeus of Alexandria (20 B.C.E.–50 C.E.) and Augustine of Hippo (354–430), Aquinas adopted a nonliteral interpretation of the creation story. Six days need not have been six 24-hour periods, nor need there have been two individuals named Adam (which in Hebrew simply means "man") and Eve.

In the wake of the work of natural philosophers such as Nicolaus Copernicus (1473–1543) and Isaac Newton (1642–1727), new theories about the cosmos began to explain physical phenomena in ways that did not require a Prime Mover, divine or otherwise. Although some later interpreters of the punishment by the Roman Catholic Church of Galileo Galilei (1564–1642) made it into a conflict between religion and science (and there is evidence to the contrary), over the next two hundred years God was out of favor in European intellectual circles as an explanation for terrestrial events. As part of the Enlightenment promotion of science, philosophers defended the possibility of explaining *all phenomena* in natural scientific terms. Theorists such as Pierre Laplace (1749–1827) sought to understand the world through physical causes alone. When Napoleon asked about his work, noting how he never once mentioned God, Laplace reportedly replied that he had "no need of that hypothesis." This approach became known as naturalism. Half a century later, Charles Darwin (1809–82) and the theory of evolution extended naturalistic explanation from the physical to the biological realm.

It was Darwinian evolution, much more than Galilean astronomy or Laplacian physics, that some nineteenth- and early twentieth-century American thinkers felt presented a deep challenge to Christian beliefs. If human beings are at most indirect creations of a God who sets in motion evolutionary processes, it becomes increasingly difficult to give much meaning to the idea of humans as created "in the image" of God (Genesis 1:27), and God himself becomes an increasingly remote or abstract reality.

Darwin's theory, which proposes that humans as well as all other species evolved over millions of years from simpler life forms, created a challenge for some Christians, who were already heatedly engaged in debates over the literal interpretations of the Bible and the separation of Church and State. In 1860, after publication of Darwin's *Origin,* a famous debate on evolution took place in London between Bishop Samuel Wilberforce (opposing) and Thomas H. Huxley (supporting). The effect was to draw a line between Christian belief and scientific theory. In England and Europe this opposition became less and less severe as liberal Christian theologians worked to reconcile the two views, often simply by saying that God could use evolution or any way he chose as a means of creation. In the United States, however, political differences compounded arguments over the separation of Church and State and the literal interpretation of the Bible and led to a much more heated and extended debate, with "creation" and "evolution" represented as the key protagonists.

These North American difficulties can be traced back to what is called the fundamentalist movement to reaffirm the literal truth of the Bible. This movement, with which 45 percent of North Americans identified in 1991, has given rise to three major creationist oppositions to biological evolution: Scopes trial–era creationism, creation science, and intelligent design. The emergence of this movement and its different viewpoints provide the basic framework within which the creationism versus evolution debate has developed.

Within a few decades after publication of *On the Origin of Species* (1859), Darwin's ideas were becoming widely accepted in both the scientific and the public arenas. Among the greatest official opposition to evolution was the fundamentalist movement in early twentieth-century America. This movement was in part a reaction to the German theological movement toward a cultural, historical, and literary interpretation of the Bible, using creation as a test case. The rise of science and technology for these first fundamentalists seemed to bring with it a deterioration of traditional human values.

Christian fundamentalists wished to preserve the "fundamentals" of Christianity as defined by widely distributed booklets called *The Fundamentals,* published between 1910 and 1915. Not all of the booklets were antievolutionary; some maintained that a divine creator and evolution could coexist. Yet it was the fundamentalist movement that broke open the evolution and creationism debate and caused it to spill into the realms of politics, law, and education.

A challenge to some recently passed Tennessee legislation against the teaching of evolution, the 1925 Scopes trial was the first clash of creationism and evolution in the courtroom. William Jennings Bryan, a former presidential candidate and liberal politician known for supporting workers' rights, prosecuted a high school biology teacher, John T. Scopes, for teaching evolution. Bryan, a Christian fundamentalist and creationist, won the case and effectively inhibited the teaching of evolution in U.S. high schools; by 1930 evolution was not taught in 70 percent of American classrooms. Yet the newspaper reporting of events during the Scopes trial left a very different impression on the American public. The evolutionists were set against the fundamentalists, who were portrayed as foolish religious zealots. Decades later, the popular play *Inherit the Wind* (1955) satirically presented the ruling of the trial as being in violation of free speech and freedom of conscience. Despite the negative image of the fundamentalists, evolution was infrequently taught in high school classrooms.

The launch in 1957 of Sputnik—the world's first satellite—by the Soviet Union stimulated a desire to intensify science education in the United States. In the years leading up to Sputnik, more and more evidence had been gathered by scientists around the world to support evolutionary theory, and James Watson and Francis Crick had identified and explained the existence of DNA.

Consequently, the teaching of evolution was determined by the National Science Foundation (NSF) to be integral to the best science education. School boards across the country were pressured by their communities to choose new, up-to-date textbooks with the NSF stamp of approval. Evolution once again became a prominent theme in biology textbooks for political reasons.

It was during this time that a second major form of the creationism-versus-evolution debate arose. In the decade before American scientific education was refocused by the space race, Henry M. Morris, a trained engineer, began defending creationism with science. His most famous work, *The Genesis Flood* (1961), cemented the modern creation-science movement. In 1970 Morris established the Institute for Creation Research (ICR), which continues to be active today. The institute "equips believers with evidences of the Bible's accuracy and authority through scientific research." The most decisive blow to creation science being taught in the science classroom, however, came in 1981 when a law that required it to be taught in Arkansas public schools alongside evolution was struck down as a violation of the separation of church and state. The trial, *McLean v. Arkansas Board of Education,* was dubbed "Scopes II."

Since the 1980s the creationism-versus-evolution debate has taken on a number of new dimensions. Of particular importance are theories of intelligent design (ID), which claim to be based in science, and a religion-rejecting affirmation of evolution that is also said by its defenders to be based in science.

A central concept for ID is the *inference to design.* ID is different from creation science in that proponents of ID do not claim outright that the designer is the God of Genesis. They do claim that the religious belief that gives impetus to their work is justified because it is equivalent to the evolutionists' belief in naturalism. For ID proponents, searching for origins by way of purely physical causes (naturalism) is no more objective than doing so by way of a combination of physical causes and design.

Given the current scientific evidence, ID proponents argue, to infer that a species must have been designed is more reasonable than to say it evolved. University of California at Berkeley law professor Phillip E. Johnson attempts to shoot holes in evolutionary theory in *Darwin on Trial* (1991). Supporting Johnson's work, biologist Michael Behe in *Darwin's Black Box* (1996) explains in detail his notion of the "irreducible complexity" of particular physical bodies from which we can infer a designer. Guillermo Gonzales and Jay Richards alert us to the finely tuned conditions necessary for life in *The Privileged Planet: How Our Place in the Cosmos Is Designed for Discovery* (2004). The coordinating and funding agency for research in ID is the Discovery Institute.

On the other side of the debate are evolutionists who believe evolutionary theory proves the nonexistence of God. Perhaps the most famous current religion-rejecting evolutionists are Richard Dawkins, author of the best-selling *The God Delusion* (2006), and Daniel C. Dennett, author of the best seller *Breaking the Spell: Religion as a Natural Phenomenon* (2007). They argue that the theory of evolution is strong enough to explain how species and complex bodies evolved. Given a world that can be explained using naturalist processes, they see a conflict between religion and science that is insurmountable. How can these creation stories be true, when we know how the world creates itself?

Although there is no certainty as to the origin of species, mainstream scientists are convinced that evolutionary theory is the key to understanding such origins. If evolutionary theory is correct, and scientists are someday able to explain the origins of life using it, does this rule out God? There are possible syntheses between the two positions, as religious thinkers and scientists outside North America have more readily explored, because one does not necessarily rule out the other. Considering the evidence for both theories unfortunately seems far less popular in some circles than the desire to generate heat (rather than light) on the subject, and so the debate continues.

See also Culture and Science; Religion and Science.

Further Reading: Barbour, Ian G. *When Science Meets Religion.* San Francisco: Harper Collins, 2000; Discovery Institute. http://www.discovery.org; Institute for Creation Research. http://www.icr.org; Pennock, Robert T., ed. *Intelligent Design Creationism and Its Critics: Philosophical, Theological, and Scientific Perspectives.* Cambridge, MA: MIT Press, 2001; Scott, Eugenie C. *Evolution vs. Creationism: An Introduction.* Westport, CT: Greenwood Press, 2004; Woodward, Thomas. *Darwin Strikes Back: Defending the Science of Intelligent Design.* Grand Rapids, MI: Baker Books, 2006.

Michael J. Bendewald

Creationism and Evolutionism: Editors' Comments

In the sociology and anthropology of religion, the claim is that creation myths are not about the creation of the universe but about the creation of new societies, new nations. Thus, the Jehovah of Genesis represents a new vision of what a human being is, in contrast to the Mesopotamian view of "man" as a slave. The opening lines of Genesis follow the creation story associated with Marduk killing the Goddess of traditional Mesopotamia, the concordance being one of number and structure as outlined in the first video of John Romer's *Testament 7* video series. The movement from the Mesopotamian creation story to that of Genesis reflects large-scale civilizational changes that replaced traditional agricultural societies symbolized by the Goddess with more settled urban nations (e.g., Egypt) symbolized by God. The fundamental idea in the sociology of religion that gods and religions symbolize societies was crystallized in the work of Emile Durkheim (*The Elementary Forms of Religious Life,* 1912). Contemporary new atheists continue to mislead themselves and the public by ignoring the social functions of religion as one form of organizing the moral order all societies require—the order that defines right and wrong, good and bad behavior. New atheists and other antireligious critics tend to try to explain religion and the gods in terms of physical and natural theories, when the most plausible theory is one that recognizes the symbolic, allegorical nature of religious texts.

Further Reading: Collins, Randall. "The Sociology of God." In *Sociological Insight,* pp. 30–59. New York: Oxford University Press, 1992; Restivo, Sal "The Social Construction of Religion." In *The Sociological Worldview,* pp. 149–59. Boston: Blackwell, 1991.

CULTURE AND SCIENCE

Over the centuries science has done battle with itself, as theories and facts are updated and changed to reflect new experimental and observational findings. Of course, science also does battle with forces outside itself. Given that science can be defined as a body of knowledge, a method, a process, and a set of ideals, occasionally a bit of knowledge is bound to challenge an ideal, or an ideal for science's potential will come into conflict with a particular method. With the money and power attached to science within Western (and now global) culture, it is not surprising that science is a source of contention as its claims are played out in the culture it has helped to create.

Although all of science's battles in the past two hundred years have at least tacitly questioned both what science is good for (its value) and how science works (its rules), each of the incidences here focuses explicitly on either science's values or its rules with respect to culture. Only 50-odd years ago, science was trumpeted as a way to save the world from poverty and war, only to be shot down as being too menial and material for such lofty pursuits. That particular encounter, between C. P. Snow and F. R. Leavis, marked the cusp of science and technology's primacy and total integration into culture and began an ongoing public conversation about whether education should be merely practical (i.e., skills based) or whether it should continue to embrace humanistic values reflected, for example, in literature and history. The second battle might more aptly be called a war over what science should value and how, in turn, science is valued by culture. Since it first appeared in the mid-nineteenth century, the theory of Darwinian evolution has caused cultural conflict beyond its value as an explanation of origins or inheritance in ways that cast doubt on the scientific approach to knowledge and its relationship to social or cultural meta-theory.

Science faced its first high-profile battle of the industrialized age in 1959, after novelist and cultural commentator Charles Percy (C. P.) Snow delivered a speech for an annual public lecture at Cambridge known as the Rede lecture. Snow's lecture, titled "The Two Cultures and the Scientific Revolution," posited a wide gulf between scientists and literary intellectuals. Snow did not elaborate a great deal on the makeup of these groups, but the substance of his lecture implied that "literary intellectuals" included both the professoriate and those who wrote and reviewed literature more popular than what is typically included in a formal canon of literature. "Scientists," of course, included both academic scientists and those who worked in government and industrial laboratories and workshops. Snow characterized literary intellectuals as arbiters of a traditional culture institutionalized by the British education system and concerned with the expression of the human soul. Their culture, Snow made plain, was stalled and stultifying because it did not concern itself with changing material realities. In the era of Snow's lecture, the role of the academic literary critic—the English professor—was almost strictly evaluative; that is, scholars of literary criticism in the early and mid-twentieth century decided what counted as "good" and "bad" literature and what would could count as a classic or merely popular. Regardless of Snow's broad groupings on this particular occasion, typically literary intellectuals disdained the popular and busied themselves mainly with interpreting and promoting what they believed qualified as literature.

On the other hand, Snow characterized the culture of scientists as forward thinking and responsive to the world around them. Rather than render judgments, as the literary intellectuals seemed to, scientists solved problems. Because scientists were concerned with physical well-being, Snow characterized them as politically progressive, whereas literary intellectuals, when they took an interest in politics at all, were more likely to side with authoritarian regimes. (Here Snow was thinking of some modernist authors who notoriously defended fascism during the 1920s and 1930s.) Despite these words, Snow still supported the overall goals of literary intellectuals: to preserve the best thoughts and words of eras and to examine and characterize the human condition. Snow used the Rede lecture, however, to propose that these two cultures be brought together, in order to bring scientists' problem-solving abilities to bear on literary intellectuals' concern with existential inward conditions.

In his lecture, Snow claimed that nothing less than the fate of the world and solutions to global problems, poverty in particular, depended on these two cultures being able to communicate with one another. Snow acknowledged that his proposal was not an easy one to implement and that it was made more complicated by the strict tracking within the British educational system. This system educated students exclusively according to their interests and talents from the beginning of their education. British students understood the literary and the scientific to be utterly separate enterprises. Students who showed interest in the humanities were all but disallowed from learning any science from adolescence on. Similarly, students who showed aptitude and inclination to work with their hands and solve problems were practically prevented from a very early age even from reading a novel. Moreover, in the deeply embedded and stratified British class system, a distinction was attached to these disciplinary boundaries. For example, when technical institutes opened in northern England in the mid-nineteenth century in order to educate those who would be working in the growing industries in the region, there emerged a socioeconomic division between those who worked in industry and those who worked with or on literature. The class divide followed historical lines: those who worked in industry worked with their hands and therefore were "lower"; those who worked in culture worked with their minds and therefore were "higher." In other words, Snow was up against a great deal in making what seemed at face value to be a simple proposition; he was indicting not only the wide divide between science and the humanities but also the institutions that produced and sustained it.

What seemed like a straightforward argument—that practical problems needed practical solutions—touched several sensitive nerves that Snow may or may not have anticipated. For instance, his claim that literary intellectuals were divorced from material reality rested on his identification of literary intellectuals as Luddites. In the early nineteenth century, Luddites were a group of Englishmen who destroyed machinery that they saw as replacing their work and therefore putting them out of jobs (think of robots today that work on automotive assembly lines where people used to work). By the time of Snow's Rede lecture, in the middle of the twentieth century, the term Luddites had lost its working-class roots and had come to designate anyone who was suspicious of or hostile to technological change. From Snow's point of view, because literary intellectuals saw themselves as the guardians of high culture, which was strictly understood to be the realm of the humanities, they were naturally disinclined to trust technological or scientific change, let alone to embrace it. Snow argued that getting over their hostility to technology was crucial if literary intellectuals were to involve themselves in alleviating global poverty and hunger.

Although his wording might have been somewhat dramatic, Snow was not entirely wrong about literary intellectuals' attitudes toward technology and toward the Industrial Revolution. The decades prior to Snow's Rede lecture included huge changes in the applications of science to everyday life, from the widespread integration of electricity into homes and businesses to the invention of the automobile and so on. Because literary intellectuals came from an embedded tradition that emphasized ideas over the material or, put another way, abstraction over the concrete, they saw themselves as divorced from technological change—after all, so the reasoning went, they were concerned with the human soul and how it might best be ennobled, and such a project was sullied by dealing with machinery. Snow's suggestion that literary intellectuals should embrace technology drew unwanted attention to a vaunted tradition of esoteric concerns that saw itself as existing above the mundane concerns of everyday life, let alone something as common as *eating*.

So it was that Snow's suggestion, that science could save the world, was hardly greeted with open arms by the literary elite. Snow had implied that literary intellectuals were immoral because they seemed unconcerned with global poverty, and the intellectuals balked. In the public (and widely publicized) retort to Snow, literary intellectuals were represented by Cambridge literature professor Frank Raymond (F. R.) Leavis. Leavis launched attacks on Snow's argument, not the least of which was against Snow himself and his lack of qualifications to make pronouncements about either science or culture. Leavis portrayed Snow as a second-rate novelist and failed scientist whose overall intellectual mediocrity denigrated the occasion of the prominent Rede lecture. Moreover, Leavis claimed, Snow's role as Rede lecturer in 1959 evinced the decline of civilization into tasteless moralisms. As an arbiter of high literary culture and advocate for the careers of modernist authors T. S. Eliot and D. H. Lawrence himself, Leavis also objected to Snow's argument on the grounds of the necessity of literature and therefore also of the role of the literary critic. The literary intellectual, according to Leavis, should not concern himself with the lives of the public but instead should devote his time to interpreting the refined expressions of humanity.

Additionally, Leavis reasonably claimed that Snow had mischaracterized literary intellectuals, mistakenly including popular critics and those who wrote in widely read magazines among their number. Rather, Leavis claimed, the real literary intellectuals were those who, like Leavis, occupied academic posts and held positions historically and institutionally recognized as arbitrating the most esoteric forms of culture. For Leavis, Snow's literary intellectuals included mere poseurs, rendering Snow's claims groundless. Leavis also thought Snow had no business accusing the literati of Luddism because the scientific revolution had brought with it a degeneration of the mind, and therefore their theoretical rejection of technology (though not practical given that they used electricity, indoor plumbing, telephones, and the like) was a rejection of the automation that shaped everyday life. Indeed, Leavis accused Snow himself of automatism, by reason of Snow's repeating easy and empty slogans which showed his "intellectual nullity." In short, Leavis felt that his life's work and his most deeply held values were under attack, and he struck back hard with a vengeance not blunted by time (Leavis's lecture in retort was delivered three years later). By not directly addressing the economic issues cited by Snow, Leavis obliquely, but no less powerfully, defended the social structures, in particular the class-producing educational system, that Snow implied were responsible for such issues. Simply stated, Leavis wanted to shore up the very divisions Snow wanted to eliminate.

Snow's and Leavis's back-and-forth, and that of their respective supporters, went on for some years, though it was each man's initial lecture, rather than subsequent commentary, that garnered the most attention. Whether or not the scientific method itself was under attack is not clear. What remains clear is that the very fast pace of scientific and technical development over the decades encompassing industrialization and the world wars presented challenges to long-held class-based and enshrined values in Great Britain. Swift and widespread technological changes also demanded a response by virtue of the dramatic material shifts in the lives of people in industrialized and industrializing nations. Although the rigidly codified British class system contributed to the Snow-Leavis controversy, other countries, notably the United States, at the same time had to deal with similar issues regarding the structure of an educational system that reflected and propagated cultural values both ethical and timely. The Snow-Leavis controversy was perhaps the first modern high-profile challenge to the relationship between science and literature, and although some aspects of the controversy are outdated, the basic concerns it presented, in particular those surrounding the ethical duties of the sciences and of the humanities, remain relevant today in the debates about the relationship between science and culture. In this instance, it was science that took the humanities to task with the help of the broad ambiguity of widely used terms. Rather than succumbing to Leavis's withering disdain, however, science and technology's great practical power and huge earning potential gave it resiliency against those who would criticize its materialistic focus.

The other key chapter in the controversies between science and culture is the concern, particularly in North America, over the relationship between evolutionism and creationism. During the "Two Cultures" debate, given the recent widespread changes in lifestyles, people were becoming increasingly apt to link science to an improved material world. Prior to the 1950s, however, science was frequently the tool of oppression, rather than a means of liberation. Because Darwinian evolution asks those who take it seriously to substantively reconsider the shape and content of what could be a deeply ingrained worldview, it has provoked (and continues to provoke) conflict because of the implications it bears for religion and politics and of course for science.

When Darwin's *The Origin of Species* was first published in 1859, it made a lot of waves for several reasons. Although Darwin was not the first to suggest that a process such as evolution was responsible for producing separate species of plants and animals, his was the most thorough, public, and eloquent explanation of evolution's principles and processes. Central to Darwin's treatise was the idea that plant and animal species had, contrary to received scientific wisdom of the era, changed slowly and by way of infinitesimal changes over time rather than having remained fixed and distinct from the moment they appeared. Moreover, he claimed that all species had developed from a common root organism. Darwin's book had (and still has) several scientific and religious implications. For example, it challenged the idea that man—that is, humanity—was intrinsically different from other kinds of animals: according to this new idea, humans were just another species of animal. Regarding religion, Darwin's ideas challenged the notion that humans were created in God's image and were therefore among the most perfect of God's creations. Even those scientists who were not religious thought that humans were the best or most finely and complicatedly developed creature. Darwin's treatise challenged this assumption by asserting that humans were an accident of change over time in exactly the same way that every other species was an accident of change over time. Darwin made it difficult to think of a species or organism's intrinsic "improvement" because the species was always a product of its long-term environment.

Darwin's hypothesis was undergirded by geological and political economic theories developed prior to the *Origin*'s publication. In the late eighteenth century, geologists had determined that Earth was far older than anyone had thought previously, radically expanding the timeline in which changes within species could take place. Political economics, in the form of *An Essay on the Principle of Population* by Thomas Malthus, explained the growth of populations—of considerable interest in the nineteenth century as the Western human population exploded—in terms of geometric, rather than arithmetic, progression. That is, if population growth were graphed on an xy grid, populations would be inclined to grow on an exponential (and somewhat more vertical) curve, rather than along a true diagonal. Malthus observed that with such growth, there must be external factors that limit populations because populations could not maintain such a growth rate indefinitely. Malthus posited that limits were the exhaustion of resources needed to support the population.

With these ideas in mind—eons upon eons of geologic time and organisms' tendencies to reproduce themselves at faster rates than there were resources to support—Darwin returned to his field notes and samples from a trip he had taken as a young man to the Galapagos Islands off the coast of South America. Darwin noticed that many animals had very similar corresponding species on a different continent. For example, each of the continents he visited seemed to have some version of a finch (a kind of bird). Using these notes, along with samples and organisms brought back to him by dozens of colleagues who went on field expeditions, Darwin documented change over time among organisms. Because this was well before the advent of genetics, Darwin did not have the concepts of genes or mutations as we know them, and so he could only surmise that changes in organisms happened in response to their environment. He argued that those traits that best suited an organism to its environment enabled the animal to live long enough to reproduce, thus preserving or passing on the particular trait that suited it to its environment. This process Darwin called "natural selection" to designate the organic way in which some organisms flourished and others died out, all in response to the environment. It would not be until the 1930s, with the modern synthesis of evolutionary theory with genetic theory, that people would begin to understand that changes within the organisms were in fact random mutations, some of which were suited to the environment and some of which were not, and those that were could thus be passed down through generations in both sexual and asexual reproduction. For his part, Darwin thought that these changes took place according to a process suggested by one of his predecessors in the field, Jean-Baptiste Lamarck. Lamarck thought that traits that were peculiar to certain animals, which aided them in survival, had developed in response to their environments. For example, according to a Lamarckian framework, giraffes' necks lengthened as they stretched them to reach leaves high in trees, and each subsequent generation of giraffes had necks the length of their parents.

Although current scientific thinking rejects some aspects of Darwin's theory (such as his own Lamarckianism), and although Darwin worked without benefit of the knowledge of genes, the contours of his treatise remain meaningful for people working in the biological and genetic sciences. Because the field of genetics contributed knowledge of how exactly traits are passed along from parents to offspring, evolution received the piece Darwin was missing, which was an explanation of how his theory worked. Put simply, Darwin provided an account of the process, and geneticists of the 1930s and beyond provided an account of the substance. Before this modern synthesis could be accomplished, however, ideas and events inspired by Darwin's ideas put evolution to the test and developed its political, religious, and social stakes.

Even within scientific discourses, evolution is something of an anomaly. It diverges from the scientific method because it does not have a falsifiable hypothesis whose validity can be proven through repetitive tests. Because Darwinian evolution is a theory about what happened in the distant past, it can be neither definitively refuted nor finally proven. Therefore, over the years it has been alternately revered and reviled, as well as shaped to the instrumental use of powerful people and institutions, with sometimes deeply deleterious effects on society. For example, so-called social Darwinism was a movement that produced such effects, not only on the lives of particular individuals but also in the parallel institutions of education and law. Despite its name, social Darwinism was a far cry from anything Darwin had anticipated when he originally published his theory.

Darwin did not intend for his theory of biological development of individuals and of species to be used to understand contemporary society, although others have adopted it for such purposes. In the late nineteenth century, the nascent field of sociology developed against the backdrop of exploding urban populations and accompanying increase in poverty, and in this atmosphere, prominent people interested in the new science of society made their own sense of Darwin's theory; the term *social Darwinism* was coined. Social Darwinists saw natural selection at work among humans and at a visible rate: in a competitive society left to follow "natural" laws, those who were fit flourished, and those who were unfit suffered. The late nineteenth century was marked by almost unlimited faith in anything that seemed even vaguely scientific, and so Darwinism struck many people in positions of political and financial power as a promising and novel way to improve society. Advocates of this new pseudoscience of social management were quick to capitalize on rhetorical associations with science, although the so-called science actually used under this rubric was, even at the time and ever since, thoroughly debunked. A British man named Francis Galton with broad and deep interest in the developing social sciences invented the term *eugenics* from the ancient Greek roots meaning "well-born." Based on the newly but only partially understood principles of inheritance discovered by the Czech monk Gregor Mendel, Francis Galton's idea behind eugenics was to encourage the propagation of what he saw as the good genes in the general population. Society, so the thinking went, could be improved by improving its genetic stock.

Although a far cry from anything Darwin had proposed and from anything biologists and geneticists recognized as the principle of evolution, then or now, social Darwinism was used by its advocates to maximize what they saw as desirable traits in humanity in much the same way that farmers and animal breeders over the centuries had sought to maximize desirable traits in livestock. For example, by controlling the breeding of animals, people could cultivate cows that produced more milk or chickens that laid eggs faster. Similarly, eugenicists, as people working in this field were known, sought to breed excellent humans. The human traits eugenicists were interested in breeding for included Nordic features, good manners, and employability. Conditions such as epilepsy, stuttering, poverty, and illiteracy were seen as social scourges, and therefore eugenicists sought to eliminate them. Eugenicists sought to make a science of developing what they saw as positive traits in the population and eliminating those traits they saw as negative.

Eugenics could take different forms, but all eugenicists were interested in creating policies that would impel those processes of natural selection they determined to be best for society. "Positive" eugenics involved encouraging people with desirable qualities to reproduce with one another. The British, led by Francis Galton, tended to emphasize positive eugenics. "Negative" eugenics, on the other hand, used most broadly in the United States, meant cutting off defective germ-plasms (the substance taken to be the culprit for passing on undesirable traits) before they could develop further through more generations. Negative eugenics involved forcibly sequestering and sterilizing people deemed unfit to reproduce in American society, such as those with odd or unsociable mannerisms and the "feebleminded." Many American institutions all but abandoned positive eugenics and embraced the negative version, with as many as 35 states at one time having enacted legislation that would house the unfit and sterilize them, all against their will. Although, during the very early years of the twentieth century, eugenics gained a great deal of momentum in the United States, backed by large amounts of resources and national credibility, it nevertheless had its detractors. One of eugenics' major critics was William Jennings Bryan, a well-known politician and orator who was dedicated to the rights of the common man and to the principles of majority rule. On principle, Bryan objected to eugenicists' violation of the rights of so many people who could not defend themselves. His passion and principles would lead him to be a major figure in the Scopes trial, where the controversy between creationism and evolutionism, and thus between science and culture, was crystallized.

Although the trial occasionally and derisively called the "Scopes monkey trial" is frequently invoked as evidence of the backward thinking of the American conservative religious South, the case was brought to trial in part because William Jennings Bryan, a major figure in the trial, saw evolution brandished as a tool of oppression. Bryan saw evolution as wielded by social Darwinists and eugenicists as dangerous not only to religious belief but also to efforts to bring rights to the disenfranchised. Bryan was concerned about the extent to which *The Origin of Species* was cited in eugenics policy advocacy, which said that there was something in the nature of people that made them deserve to be poor and meager—that society was only working how it was supposed to, by rewarding the "fittest," or the most worthy of rewards. Because of social Darwinism's high-profile status, this would have been the most meaningful version of evolution to Bryan.

After the passage of the Butler Act, which prohibited the teaching of evolution in the public schools, the American Civil Liberties Union (ACLU) had advertised all over Tennessee for someone willing to participate in a test case, testing the Butler Act in court. The ACLU saw the Butler Act as infringement on a citizen's right to free speech. Dayton, Tennessee, had recently suffered financial and population setbacks, and civic leaders saw the ACLU advertisement as an opportunity to put their town on the map by attracting national interest and businesses. Dayton public prosecutors approached John Scopes, the math teacher and football coach at a public high school in Dayton, with their idea, and Scopes agreed to it. In the school year after the implementation of the statewide Butler Act, Scopes had been called on to substitute teach biology. He used a textbook written by advocates of eugenics, titled *Civic Biology*. The textbook taught, among other things, that man had descended from "lower" forms of life. He was "arrested," which amounted to filing paperwork before Scopes went to play his regularly scheduled tennis match, and the paperwork was sent to the ACLU. From the beginning, the Scopes trial was managed and staged for purposes of civic pride. However, those who took interest in it found their stakes elsewhere, and the event did indeed receive a great deal of national attention.

Far from being simply a stage production, however, the trial garnered wide attention because of the nerve it touched in a nation rife with controversy over whether the Bible should be literally or metaphorically understood and over the government and religion's role in public education. Although much about the Scopes trial was political theater, the crowds and press that it drew indicated that the question of how science could live in the same world as religion was far from settled. Each seemed to require a certain amount of faith: religion because it did not have empirical evidence and science because few people had access to its specialized knowledge. In the years preceding the Scopes trial, the American religious fundamentalist movement, based on taking the Bible as the word of God at face value, had gained much ground and many followers. At the same time, and in a different segment of the population, secularism also had taken strong hold, in a nation grown more cynical after World War I and more educated. The result of this growing opposition was that opinions about religion and science strengthened, grew apart, and gained mass. Also emergent around the time of the Scopes trial was a "modernist" approach, which allowed religious people to accept the doctrine of evolution by loosening the mandate for literal interpretation of the Bible. In other words, religious people also dedicated to science could see the Genesis story of creation as being metaphorical rather than as an exact account of what happened. The modernist approach eased the stark choice between either belief in God necessarily accompanied by a belief in the Bible as literal or atheism or agnosticism necessitated by belief in evolution, and it appealed to a great many people. Rather than bring a peaceful resolution to an intractable problem, however, the modernist approach only caused people at either end of the debate to dig their heels in further. John Scopes was at the center of a maelstrom of controversy surrounding humanity's most bedrock values and conceptions of itself, characterized in terms of how science was shaping Western culture in the twentieth century.

Battle lines were drawn on the field of evolution and public policy in the decades before the Scopes trial even started. William Jennings Bryan was devoted to the principle of populism, or of the majority rule. He was at least equally devoted to his Christian values, which for him included social progress in the form of equality of the races and sexes, with that equality being reflected in social institutions such as schools. Clarence Darrow, another famous political personality, was selected to be the lawyer for the defense. Darrow supported the rights of individuals in the minority, just as passionately as William Jennings Bryan supported the principle of majority rule. Darrow was also vehemently antireligious and saw organized monotheistic religion as detrimental to society's ethical obligations to its members. Thus, the Scopes trial was as much about the will of the majority versus the rights of the minority as it was about religion versus science, but these stakes are not as different as they seem. Both are about the status of religious and governmental authority, the status of the written word, and the status of dissenting opinion. The Scopes trial was much bigger than the sum of its parts.

The trial ultimately was decided against Scopes, appealed, and then dismissed on a legal technicality, leaving more of a legacy of legal celebrity than of actual direct policy. It did, however, introduce a conversation—continuing today—regarding public schools, religion, science, and the separation of church and state. In more iconic terms, the trial continues to emblematize conflict between science and religion as opposing social doctrines.

Over the course of recent decades, more conflicts involving science and religion or politics have emerged, and many of them follow the contours of the debate over evolutionism. Science frequently challenges the status quo, and although little has challenged the status quo as deeply as Darwinian evolution, other scientific ideas have similarly made suggestions that are far from neutral.

The debate over global warming, also known as climate change, is one such non-neutral issue because it affects governments and corporations. For many years, some have denied that anything called "global warming" exists, chalking up evidence cited by scientists who said global warming was an urgent problem to alarmist hearsay or as incomplete. Far from being simply a matter of establishing true facts, the great deal of investment on either side of the debate has made it difficult for policy makers to sort through the evidence and claims. For instance, corporations that have pollution-emitting factories would incur a great cost if they were shown to be contributing to climate change. Similarly, scientists, laboratories, and foundations dedicated to protecting Earth's air and water have poured a great deal of time and money into coming to conclusions about the state of things, and they stand to gain more if their predictions and advice are taken. Parallel to the "two cultures" controversy, questions about global warming ask the public to make choices about priorities and the ethical status of science; parallel to the various controversies surrounding evolution, global warming provides opportunity for ideological division and occasionally pits the government or other powerful institutions against the public they ostensibly serve. Proponents and opponents of the veracity of global warming are somewhat less apt to stick to only the facts than they are to concentrate on the other ideological issues. Besides global warming, other scientific controversies are with us and unlikely to be resolved any time soon, including stem cell research and species extinctions, to name two current issues. As controversies develop, no doubt both new and very familiar arguments will be expressed that attempt to define and therefore direct the relationship between science and culture.

See also Creationism and Evolutionism; Religion and Science; Science Wars; Scientific Method.

Further Reading: Black, Edwin. *War against the Weak: Eugenics and America's Campaign to Create a Master Race*. New York: Thunder's Mouth Press, 2004; Darwin, Charles. *The Origin of Species*. Introduction by Julian Huxley. New York: Signet Classics, 2003; Larson, Edward J. *Summer for the Gods: The Scopes Trial and America's Continuing Debate over Science and Religion*. New York: Basic Books, 2006; Leavis, F. R. *Two Cultures? The Significance of C. P. Snow*. New York: Pantheon Books, 1963; McKibben, Bill. *The End of Nature*. 10th anniversary ed. New York: Anchor, 1997; Snow, C. P. *The Two Cultures*. Introduction by Stefan Collini. London: Cambridge University Press, 1993.

Elizabeth Mazzolini

D

DEATH AND DYING

Death is the ultimate battleground, in science and technology, in our lives, on our planet, and in the universe. Every other battleground is eventually transformed into or leads to death. On the level of our lives, it might seem that whatever the death and dying battlegrounds, there would not be a conflict over definitions. And yet, definitions are where the conflict crystallizes. The medical community has given us at least three definitions of death: you are dead when you suffer heart-lung failure; you are dead when you suffer whole-brain death; or you are dead when you suffer higher-brain death. The technical details of these medical definitions can be skipped here without losing sight of the fact that the definition we choose will have implications for how people are treated in hospitals, how we deal with organ donations and transplants, and what we do about abortion, stem cell research, and scientific research on corpses. If there was ever a time when death seemed to be a rather simple fact of life ending, that has all changed thanks to mechanical hearts, breathing machines, intravenous technologies, and other technologies that have given rise to the notion that people can be kept technically alive in a "vegetative state," or in short- and long-term comas. In more general philosophical terms, consider the different senses of death as a state, a process of extinction, or the climax or end point of that process. We can furthermore distinguish these senses of death from events that cause death (e.g., automobile accidents and gunshots).

If your heart loses the capacity to pump blood effectively, doctors can attach a left-ventricle assist device (LVAD) to one of the heart's main chambers and the aorta. The aorta is the main artery supplying blood to the body. LVADs were originally designed as temporary "hold" devices for use while heart patients were waiting for a transplant or surgery. Increasingly, they are being used to prolong life independently of surgical or transplant options.

This raises the question of the role of physicians on this battleground. In some ways, they seem to be the ultimate arbiters of the life-and-death boundary. Many of them now use the electroencephalogram (EEG) to answer the question "Is this person dead?" The EEG shows brain wave activity. It is at this point that we enter a new battleground. Is it the heart that defines who we are, or is it the brain? If it is the heart, then we would expect the heart to be at the center of our definition of death; if it is the brain, we would expect death to be defined in terms of the brain. Currently, the stopped heart defines death. This seems to be a clear-cut scientific matter. If our heart is beating, we are alive; if it stops beating we are dead. Issues arise when we consider whether the brain can be used in the same way as the heart to define life and death. The brain would appear to be a good candidate for this purpose because there is brain activity (which means we are alive in at least some sense) or there is no brain activity (which should mean we are dead). Why this a battleground can be demonstrated by considering the so-called 24-hour rule used at the Massachusetts General Hospital in Boston. If the EEG is flat for 24 hours and stays flat even in the presence of outside stimuli (a loud noise, for example), the patient can be declared dead. Any demonstration of muscular or papillary reflexes is sufficient to delay the declaration that the patient has died, even if the EEG is flat. The other condition is that there must be no heartbeat or respiration other than that provided mechanically. But it gets even more complicated. If the patient has suffered barbiturate poisoning or has been exposed to extreme cold for a long time, he or she might have a flat EEG for hours and still recover fully. We can begin to see why physicians are the ultimate arbiters and why their judgments are so powerful in the face of other considerations.

This is not the end of the story, though, because some agents want to take into account the mind and the soul. It should be clear that we are once again close to entering the science and religion battleground.

For those who believe in "the human spirit," the soul, or the mind as a nonphysical feature of life that does not end with death, there is more to death and dying than stopped hearts and dead brains. These advocates of a spiritual dimension in life believe that the soul, the mind, or the spirit of a person does not cease to exist when the heart stops or the brain dies. Some of them give the brain priority over the heart because they believe that the soul, mind, or spirit is a product—even if nonmaterial—of the brain. We can boil all this down, then, to a battleground with scientific materialists on the one side and religious or perhaps spiritual advocates on the other. The extreme scientific materialists would argue that when you die, you die; life ends, you end, and your body becomes a lifeless biophysical mass added to the earth by way of burial or cremation. Spiritual advocates believe to different degrees that some essential part of our self, our personhood, continues to exist after we die and may in fact be immortal. Some debate whether we retain consciousness of our earthly selves in this post-life state. In any case, death and dying bring into play a wide range of players—physicians, lawyers, the dying, the relatives of the dying, religious and spiritual leaders, biomedical technologists, advocates of hospice, euthanasia supporters, and ethicists. Here as elsewhere, we see the importance of the sociocultural idea that context is everything. There are as many definitions of death as there are contexts of dying, and death carries different symbolic meanings and values in different cultures and in different times and places. The body may be the ultimate boundary object, one entity subject to a multiplicity of perspectives that can leave the individual a rather weak player in the game of life and death.

Until recently, suicide was our only recourse when it came to deciding the time, place, and method of our death. The emergence of complicated life-support technologies has led to a situation in which we can find ourselves debating when to plug people into life and when to pull the plug. It might seem simple at first—that everyone has the right to decide if, when, and how to end his or her life. The question then is why the state, religious institutions, the family, and the community enter into this decision-making process with more power than the individual.

Death is in fact a social, cultural and community matter. We are born into societies with existing views about the nature and symbolic value of death. The loss of a member upsets the solidarity of a family and a community at least temporarily. Suicide has been tied to social solidarity by sociologists and anthropologists. Too much or too little solidarity can provoke suicide, as can rapid changes (whether positive or negative; a rapid upturn in the stock market will provoke suicides in the same way that a rapid downturn will). Suicide and community norms, values, and beliefs are tightly knit together.

The ceremonies we have created to deal with death have two functions. One is to reestablish the solidarity upset by the loss of a member. Funerals, then, are more about us—those still living—than they are about the person who has died. The other is to ensure that the dead person stays dead and in a sense respects the boundary that separates the living from the dead. Religious beliefs can complicate this process by creating a tension between the real loss we feel when someone close to us dies and the belief that the person goes on to live in a "better" place, in heaven for example. Some people believe that this transcendent life begins immediately on death and that departed loved ones can now look down on them from "above." One rarely if ever hears of the dead looking up at us from the netherworld. In any case, the new technologies of life and death and new levels of awareness about the process of dying have given individuals new powers over their own lives during the final moments, months, or years of their lives. Many individuals want the right to decide for themselves when to pull the plug.

The question of who owns your body might seem strange at first. Ownership implies the idea of property. We do not normally think of our bodies in this way. And yet, your body is a piece of property. The state owns it and can put it to work in a war or take it out of circulation if you do something of which the state does not approve—if you break the law. This idea of the body as property, and even as a commodity (even body parts and organs are commodities in today's biomedical market places), affects your ability—and even your desire—to control the conditions under which you die.

Philosophers, of course, have had a lot to say about death and dying. Because philosophy is a discipline of logic and analysis, they have sorted out the main issues and arguments that help to define this battleground. We have already discussed one of their major concerns: what life is and when it ends. They have also identified what they refer to as the symmetry argument and the immunity argument. These arguments are both designed to deal with the "harm thesis." The symmetry argument is basically that we have been "dead" before and indeed dead for as long as we are going to be dead after we die. Our first period of nonexistence was not bad or harmful in any way, so in a sense we have been through it all before. A simple refutation of this argument might be that during our first period of nonexistence, our eventual birth was going to come about; our second period offers no such promise. Time is an issue here. If we argue—as philosophers, not as scientists—that time makes sense only in the world of the living, time stops when you die. So you are not dead forever, or for all time; you are not dead for any time at all. These sorts of ideas are not likely to mitigate the fear of dying and of not existing ever again. Is it possible to embrace death, to accept it fully as a condition of life the way we embrace and accept breathing and the need for nourishment? This seems to be possible in different degrees for different people and in different cultures. How we answer this question depends at least in part on if and how we value life. In "The Problem of Socrates," the philosopher Friedrich Nietzsche (1844–1900) claims that the sages in all ages have viewed life as "worthless"; Socrates, he reminds us, said that life means being sick a long time. It is easy to see how unappealing this view would be to most people who value life (certainly their own lives) without the sort of philosophical reflection that can lead to the problem of Socrates.

In addition to the symmetry problem, philosophers have identified the so-called timing puzzle, which deals with the possible harm that might befall you as a subject in death and after death. The answer to this puzzle is fairly simple—no subject, no harm. You can be a subject only within time and space. So at death and after death, no harm can befall you. Once again, this—like other philosophical problems posed in the Western logical and analytical traditions—would be viewed quite differently if you believed (1) that you possessed an immortal soul or spirit; (2) that you would be resurrected in body, in soul, or in body and soul by one god or another at some end-of-time cosmic juncture; or (3) that you would be reincarnated. The idea of reincarnation is widespread across cultures and comes in various forms. Although there is not space here to review the variety of views on reincarnation, two things should be noted. First, the Buddhist idea of rebirth is different from the idea of reincarnation in the Hindu traditions; there is no self or soul in Buddhism that could be reincarnated. Second, reincarnation has taken hold in our own time amongst neo-pagans and New Agers of various stripes. For every position on death, whether in the personal, popular, scientific, religious, or philosophical imagination, there is one or more opposing position. This entry provides a few hints on the shape of those overlapping battlegrounds, but the reader will have to visit and negotiate them on his or her own.

Increasingly sophisticated research, development, and applications in the field of robotics and social robotics have raised new questions about what it means to be alive and what it means to die. It is already clear that the more human qualities we are able to build into machines, the more they will provoke us to label them "alive." Once we have attributed (legislated, one might say) life in the case of a machine, we are immediately faced with the question of whether a machine can die. We are just beginning to negotiate this threshold that (barely) separates the literal from the metaphoric.

One of the most, if not the most, important death and dying battleground involves the conflict, controversies, and debates over euthanasia. In the Hippocratic Oath, Hippocrates (400–300 B.C.E.) wrote that under no circumstances would he prescribe a drug or give advice that might cause a patient to die. The ancient Greeks and Romans were not intolerant of suicide, however, if no relief was available to a dying person. The Stoics and Epicureans were more radical and supported personal suicide decisions. Suicide and euthanasia are generally considered criminal acts now and have been so considered since at least the 1300s and through the 1900s in the West, following English common law. The debates over euthanasia in the modern context were fueled by the development of anesthesia.

Stripped down to the basics, the controversy over euthanasia concerns whether it is a method of merciful death and death with dignity or whether it is murder or at best a potentially harmful method in the hands of abusive physicians or others, especially in respect to certain vulnerable persons and populations. The contemporary debate in the United States was stimulated by the case of Karen Ann Quinlan (1954–85). At the age of 21, Quinlan lapsed into a coma after returning from a party. She was kept alive using a ventilator, and after some months her parents requested that all mechanical means being used to keep her alive be turned off. The hospital's refusal led to a national debate and numerous legal decisions concerning euthanasia. Quinlan was taken off life support in 1975 but lived for another 10 years in a comatose state. The Quinlans were Catholics, and Catholic theological principles were prominent in the various legal battles surrounding this case and others that followed around the world. The case is credited with promoting the concept of living wills and the development of the field of bioethics.

Dr. Jack Kevorkian is perhaps the most visible advocate of assisted suicide, and his actions prompted Michigan to pass a law against the practice. In 1999 Kevorkian was tried and convicted of murder after one of his assisted suicides appeared on television. Earlier, in 1990, the Supreme Court had approved non-aggressive euthanasia, that is, simply turning off life-support systems. Passive euthanasia, by contrast, is widely accepted and a common practice within the informal organization of hospitals. In this method, common medications and treatments are simply withheld, or a medication to relieve pain may be used with full recognition that it may result in the death of the patient. Passive euthanasia is considered legitimate during the final stages of a terminal illness. Aggressive or active euthanasia uses lethal substances or other direct methods in assisting suicide or in mercy killing.

Oregon legalized assisted deaths in 1994 for terminal patients with no more than six months to live. The law was approved by the Supreme Court in 1997.

In *Gonzales v. Oregon* (2001), the Bush administration tried to overturn the Oregon legislation by using drug laws. The effort failed, and in 1999 Texas legalized nonaggressive euthanasia. The Netherlands decriminalized physician-assisted suicide (PAS) in 1993 and reduced restrictions further in 2002. Belgium approved PAS in 2002, but in 1997 Australia's Federal Parliament overturned a euthanasia bill passed by the Northern Territory in 1995. One of the major organized efforts to promote and define the legalization of euthanasia has been conducted by the Hemlock Society, which was founded in 1980 by Derek Humphry of Santa Monica, California. The society grew into the largest organization in the United States supporting the legal right-to-die, voluntary euthanasia, and PAS. The term *hemlock,* referring to the root of the weed used as a poison in ancient Greece and Rome, is associated in educated circles with the death of Socrates. The name "Hemlock" did not appeal to a lot a people and may be one of the reasons for the declining membership in the society during the 1990s. The Hemlock Society per se is no more. It survives, however, in End-of-Life Choices (Denver) and Compassion in Dying (Portland), now merged in Compassion and Choices. Other related entities include the Death with Dignity National Center and the Euthanasia Research & Guidance Organization. Hemlock Society supporters dissatisfied with the board decisions that led to these changes have formed the Final Exit Network.

This entry has covered a lot of territory, so it might be valuable to conclude by summarizing the basic arguments for and against euthanasia. These arguments in general apply to suicide too. Somewhere on this battleground there may be some way to deal with the issues raised for the individual by the certainty of death. Arguments in favor of euthanasia and suicide include the right to choose (a general liberal, democratic assumption that becomes problematic at the boundary that separates the body and the community, thus the issues raised by abortion and suicide); quality of life (here there will be tensions between individual, community, medical, legal, and ethical standards); economic costs and resources; and avoiding illegal and uncontrolled methods of suicide (here the arguments are similar to those in favor of legal abortions; legalizing abortions eliminates the need for back alley and other dangerous methods of abortion). Arguments against euthanasia range from statements found in the original Hippocratic Oath and its modern variants to moral and theological principles. The need for euthanasia can almost always be disputed; families may resist euthanasia in order not to lose a loved one, and pressure might be put on the patient to "pull the plug" in order to save his loved ones, the hospital and physicians, and society at large money. Perhaps this battleground will lead to a better appreciation of death in the evolutionary and cultural scheme of things, and greater education and reflection will eventually allow more and more people to embrace death and dying.

One of the issues that surfaces on a regular basis in discussions about death and dying is the way in which attitudes toward death in Western societies have undergone radical shifts. At the extreme, commentators on the social dimensions of religion have observed that Western culture is, in effect, death-denying. They point to the decay of traditional rites of passage, from a community-based structure of meaning within which the life of the deceased was remembered to a random set of funeral home liturgies and burial practices. From the custom of families washing the body, dressing it for burial, and laying out the deceased relative in the front parlor, there came, first, ever more stylized embalming and open casket ceremonies in funeral chapels and second, the quick shuffle off to the crematorium, so that most people do not see the body of the deceased at all—just an urn, unopened, full of ashes. The obsession with youth—or at least the look of youth—behind the various types of plastic surgery and Botox injections is equally viewed as the result of the psychological inability to age gracefully and the need to deny the reality and inevitability of death. At the extreme end of this denial, one might argue, is the effort to preserve bodies in cryogenic storage until some future date when all illness, and death itself, might be healed.

One wonders if the denial of death is another manifestation of the mechanical metaphor for life, in which what is broken should be able to be fixed (with new parts, if necessary); the concept of death, as a natural and inevitable destination, is resisted seemingly at all costs. This refusal to die gracefully—or an unwillingness to die at all—has led to resistance to palliative care and end-of-life decision making, only one small part of which relates to euthanasia in any of its forms. Resources are spent on finding ways of keeping people alive—by whatever mechanical means possible—rather than on easing their departure, preferably in the comfort and familiarity of their own homes. The relatively minimal expenditures of Western health care systems on palliative care as opposed to the provision of heroic measures, and the entanglements that result when care should be withdrawn, create a host of moral and ethical dilemmas other generations or other cultures simply have never needed to face.

The response can be made that these dilemmas and how they are handled are situational, depending on the situation and the people who are participants in it. Such a response, however, does not fairly depict the institutional character of this denial of death, this inability to accept death as the inevitable outcome of life and to see it as part of the natural order instead of a failure of some mechanical system.

See also Health and Medicine.

Further Reading: "Death." *Stanford Encyclopedia of Philosophy.* http://plato.stanford.edu/entries/death; Humphry, Derek. *Final Exit.* 3rd. ed. New York: Dell, 2002; Nicol, Neal. *Between the Dying and the Dead: Dr. Jack Kevorkian's Life and Battle to Legalize Euthanasia.* Madison: University of Wisconsin Press, 2006; Nietzsche, Friedrich. *Twilight of the Idols.* 1888. New York: Penguin Classics, 1990.

Sal Restivo

DRUG TESTING

When does the greater social good achieved by drug testing outweigh a person's reasonable expectation of privacy? High school students and parents have been on the front lines protesting against the testing of bodily fluids for illegal drugs, which typically occurs in drug treatment programs, workplaces, schools, athletic programs, and the military. Urine testing for marijuana, cocaine, methamphetamine, PCP, and opiates has been upheld as a constitutional search under the Fourth Amendment of the U.S. Constitution. Recently, alternatives such as sweat patches, saliva testing, or hair testing came onto the market, and results found with these methods are now being upheld by the courts. Although drug testing is assumed to deter illegal drug use, there is little evidence that it does.

Decisions about who to test are supposed to be based on reasonable suspicion that a person has used an illegal drug. Urine tests are a relatively accurate way to establish the presence of drug metabolites. By the 1980s, EMIT, an enzyme-based drug assay still used today, had become accurate enough to hold up in court. Accurate testing first requires a preliminary screening for samples that may be positive and then another more sensitive step to confirm or disconfirm that impression. There is always the possibility of environmental contamination, tampering, or human error because the process leaves room for interpretation. Sometimes false positives and inaccurate results are obtained. Most importantly, drug tests cannot document habitual drug use or actual impairment. Drug tests can detect only discrete episodes of drug use. The window during which use will remain detectable varies by drug and testing technology. Because the technology itself has become less cumbersome, testing kits are now relatively easy to use (although they still require laboratory processing).

Today drug-testing kits are marketed to educators and even suspicious parents so that they can administer tests without involving police. The Drug Policy Alliance, a policy reform organization, opposes this surveillance practice on the grounds that it is an incursion into civil liberties and privacy rights. They argue that testing does not deter drug use but instead humiliates youth and creates distrust between parents and children. It is very important for students to recognize what rights they do and do not have when it comes to urine-testing programs.

Mass urine testing was first widely used by the U.S. military as a result of heroin addiction during the Vietnam conflict. Veterans' organizations and unions challenged testing programs in a series of mid-1970s court cases that led to urine testing being understood as a legal form of "search and seizure" under the Fourth Amendment. Although individuals who are not under any suspicion of drug use routinely find themselves tested, such programs have been upheld.

Transportation accidents have also been the basis for broader testing of workers. In *Skinner v. Railway Labor Executives Association,* 489 U.S. 602 (1989), the courts suggested a "balancing test" to determine whether testing was conducted for purposes of public safety or criminal prosecution (which would contradict the presumption of innocence). Balancing tests attempt to weigh individual expectations of privacy against government interests. In *National Treasury Employees Union v. Von Raab,* 489 U.S. 656, 665 (1989), government interests outweighed individual interests in a case involving U.S. customs agents. Random drug testing now occurs mainly where public safety is concerned.

The urine-testing industry itself is a $6 billion industry. Individual tests currently cost $25 to $35, making the cost of urine testing prohibitive for many school districts and sports programs. The federal government is the industry's largest customer because of drug-free workplace policies. Once court challenges to military testing programs proved unsuccessful, the Reagan administration extended such programs to federal workplaces, transportation, health care delivery, and educational settings. The goal of a "drug-free federal workplace" made urine testing commonplace in the United States—but it remained uncommon elsewhere. By 1996 over 80 percent of major U.S. firms had testing programs. School districts followed suit once educational testing programs survived court challenges from students and parents. However, a 1997 case clarified that school testing must be based on reasonable suspicion of individuals and cannot subject an entire "suspect class" (such as student athletes) to blanket testing. Drug testing is expensive for school districts, despite government subsidies.

Many are now convinced that drug-testing programs are ineffective: they identify few drug users, and their effects on workplace safety and productivity are unimpressive. Some argue that the effects of testing programs can even be negative because they produce a climate of suspicion and antagonism.

Hiring is one of the main points at which testing takes place. Pre-employment screening is used to deter anyone who has used illegal drugs from even applying for jobs. Once an individual is on the job, there are both random testing programs and those provoked by accidents. Another locus for testing is the criminal justice system, where testing occurs especially in community corrections and drug court programs. Hospitals have extended routine urine testing beyond diagnosis and monitoring infection to testing for illegal drugs in the case of pregnant women. The U.S. Supreme Court has ruled that urine-testing pregnant women without their knowledge or consent is an unlawful violation of the Fourth Amendment (*Ferguson v. City of Charleston*, 532 U.S. 67 [2001]). This case suggests that there are times when individual privacy interests are understood to outweigh state interests. The Fourth Amendment, however, safeguards only U.S. citizens from government intrusion—it does not guard against nongovernmental intrusions. For instance, the NCAA drug-testing program for college athletes held the organizations' interest in the health and safety of athletes over and above the individual right to privacy.

See also Drugs; Drugs and Direct-to-Consumer Advertising; Medical Marijuana; Off-Label Drug Use.

Further Reading: Alderman, Ellen, and Caroline Kennedy. *The Right to Privacy.* New York: Vintage Books, 1997; American Civil Liberties Union. *Drug Testing: A Bad Investment.* New York: ACLU, 1999; Hoffman, Abbie. *Steal This Urine Test: Fighting Drug Hysteria in America.* New York: Penguin, 1987; Kern, Jennifer, Fatema Gunja, Alexandra Cox, Marsha Rosenbaum, Judith Appel, and Anjuli Verma. *Making Sense of Student Drug Testing: Why Educators Are Saying No.* 2nd ed. Oakland, CA: Drug Policy Alliance, 2006.

Nancy D. Campbell

DRUGS

Drugs enjoy a social significance different from other commodities, technologies, or artifacts. Celebrated by artists and visionaries from the nineteenth-century Romantics to the twentieth-century Beats to twenty-first-century hip-hop, drugs have been seen to shape minds and bodies in socially positive and problematic ways. Prescription drugs are credited with improving health, productivity, and well-being, whereas nonprescription drugs are blamed for destroying minds and bodies. How society views drugs depends on who produces them, how they are distributed and marketed, who consumes them and how. There are many controversies surrounding the cultural work of these fascinating, functional, and sometimes dangerous technologies.

History reveals a remarkable parade of "wonder drugs"—such as heroin, introduced in 1898 by the German pharmaceutical company Bayer as a nonaddicting painkiller useful for treating tuberculosis and other respiratory diseases. Bayer introduced aspirin a few years later as a treatment for rheumatoid arthritis but promoted it aggressively for relief of headache and everyday aches and pains. Today, aspirin is the world's most widely available drug, but there was a time when pharmacists smuggled it across the U.S.–Canadian border because it was so much more expensive in the United States than elsewhere. Cocaine, distributed to miners in the Southwest as an energizing tonic, was used much as amphetamines and caffeine are used in postindustrial society. Barbiturates; sedative-hypnotics such as thalidomide, Seconal, or Rohypnol; major and minor tranquilizers; benzodiazepines such as Valium; and so-called painkillers or analgesics have all been promoted as wonder drugs before turning out to have significant potential for addiction or abuse and are also important for medical uses—for instance, cocaine is used as an oral anesthetic.

Wonder drugs are produced by pharmacological optimism—the myth that a drug will free human societies from pain and suffering, sadness, anxiety, boredom, fatigue, mental illness, or aging. Today "lifestyle drugs" are used to cope with everything from impotence to obesity to shyness to short attention spans. Yet adverse prescription drug reactions are the fourth leading cause of preventable death among adults in the United States. Some drugs, we think, cause social problems; we think others will solve them. Drugs become social problems when important interest groups define them as such. Recreational use of illegal drugs by adolescents has been considered a public health problem since the early 1950s, when the U.S. public attributed a wave of "juvenile delinquency" to teenage heroin addiction. Since our grandparents' generation, adolescence has been understood as a time when many experiment with drugs. Today a pattern of mixed legal, illegal, and prescription drug use has emerged among the first generation prescribed legal amphetamines and antidepressants. Many legal pharmaceuticals have been inadequately tested in children, and the short-term effects and long-term consequences of these drugs are unknown.

Portrayed as double-edged swords, drugs do not lend themselves to simple pros and cons. Drug controversies can best be mapped by asking which interest groups benefit from current policies, whose interests are at stake in changing them, and how "drugs" are defined differently by each group of producers, distributors, and consumers.

The basic terms through which drug debates are framed are not "natural" and do not reflect pharmacological properties. The meaning of drug use is best thought of as socially constructed because it is assigned meaning within social and historical contexts. Varied meanings were attributed to the major subcultural groups of opiate addicts in the early twentieth-century United States. Opium smoking by nineteenth-century Chinese laborers in the United States was tolerated until the labor shortage that attracted them became a labor surplus. Although laborers have long used drugs to relieve pain, stress, and monotony, the larger population of nineteenth-century opiate addicts was white women, born in the United States, who did not work outside the home. Pharmacy records indicate that rates of morphine addiction were high among rural Southern women from the upper and middle classes—and almost nonexistent among African Americans. Male morphine addiction was concentrated among physicians, dentists, and pharmacists—professions with access to the drug.

Why did so many native-born white people rely on opiates through the early twentieth century? Prior to World War II, when antibiotics were found useful for fighting infection, doctors and patients had few effective treatments. Opiates were used to treat tuberculosis because they slow respiration and suppress cough, for diarrhea because they constipate, and for pain (their most common use today). Physicians and patients noticed that opiate drugs such as morphine and heroin were habit-forming, however. They used the term "addict" to refer to someone who was physiologically or psychologically dependent on these drugs. With entry into the twentieth century, physicians began to refrain from prescribing opiates except in cases of dire need. Improved public health and sanitation further reduced the need, and per capita opium consumption fell. Despite this, the United States could still be termed a "drugged nation."

Since the criminalization of narcotics with the Harrison Act (1914), U.S. drug policy has been based on the idea of abstinence. There was a brief period in the early 1920s when over 40 U.S. cities started clinics to maintain addicts on opiates. This experiment in legal maintenance was short-lived. Physicians, once the progenitors of addiction, were prosecuted, and they began to refuse to prescribe opiates to their upper- and middle-class patients. By the 1920s the opiate-addicted population was composed of persons from the lower or "sporting" classes. Drug users' median age did not fall, however, until post–World War II. The epidemiology, or population-wide incidence, of opiate use in the United States reveals that groups with the greatest exposure to opiates have the highest rates of addiction.

Exposure mattered, especially in urban settings where illegal drug markets took root. Urban subcultures existed in the nineteenth century among Chinese and white opium smokers, but as users switched to heroin injection or "aged out" of smoking opium, the Chinese began to disappear from the ranks of addicts. Older "dope fiend" subcultures gave way to injection heroin users, who developed rituals, "argots" or languages, and standards of moral and ethical behavior of their own. Jazz musicians, Hollywood celebrities, and those who frequented social scenes where they were likely to encounter drugs such as heroin, cocaine, and marijuana were no longer considered members of the respectable classes. The older pattern of rural drug use subsided, and the new urban subcultures trended away from whites after World War II. African Americans who had migrated to northern cities began to enjoy increased access to illicit drugs that had once been unavailable to them. So did younger people.

Social conflict between the so-called respectable classes and those categorized as less respectable often takes place around drugs. Debates over how specific drugs should be handled and how users of these drugs should be treated by society mark conflicts between dominant social groups, who construct their drug use as "normal," and subordinate social groups whose drug use is labeled as "abnormal," "deviant," or "pathological." As historian David Courtwright points out, "What we think about addiction very much depends on who is addicted." How drugs are viewed depends on the social contexts in which they are used, the groups involved, and the symbolic meanings assigned to them.

Recent medical marijuana campaigns have sought to reframe marijuana's definition as a nonmedical drug by showing its legitimate medical uses and backing up that assertion with clinical testimonials from chronic pain patients, glaucoma sufferers, and the terminally ill. Who are the dominant interest groups involved in keeping marijuana defined as nonmedical? The voices most often heard defending marijuana's status as an illegal drug are those of drug law enforcement. On the other hand, the drug policy reform movement portrays hemp production as an industry and marijuana use as a minor pleasure that should be decriminalized, if not legalized altogether. The range of views on drug policy range from those who want to regulate drugs entirely as medicines to those who are proponents of criminalization. A credible third alternative has emerged called "harm reduction," "risk reduction," or "reality-based drug policy." Asking oneself the question "Whose voices are most often heard as authoritative in a drug debate, and whose voices are less often heard or heard as less credible?" can be a method for mapping the social relations and economic interests involved in drug policy. Who was marginalized when the dominant policy perspective was adopted? Who lost out? Who profited? Although the frames active in the social construction of drugs change constantly, some remain perennial favorites.

Not all psychoactive substances used as recreational drugs are currently illegal. Alcohol and tobacco have been commonly available for centuries despite attempts to prohibit them. Both typically remain legal except where age-of-purchase or religious bans are enforced. Alcohol prohibition in the United States lasted from 1919 to 1933. Although Prohibition reduced per-capita consumption of alcohol, it encouraged organized crime and bootlegging, and repeal efforts led to increased drinking and smoking among the respectable classes. Prohibition opened more segments of the U.S. population to the recreational use of drugs such as the opiates (morphine and heroin), cannabis, and cocaine. Although cannabis, or marijuana, was not included in the 1914 legislation, Congress passed the Marijuana Tax Act (1937) during a period when the drug was associated with, for example, Mexican laborers in the southwestern United States and criminal elements throughout the country. Cocaine was relatively underused and was not considered addictive until the 1970s. Although cocaine was present in opiate-using subcultures, it was expensive and not preferred.

Social conflicts led legal suppliers to strongly differentiate themselves from illegal drug traffickers. The early twentieth-century experience with opiates—morphine, heroin, and other painkillers—was the real basis for U.S. and global drug control policy. The Harrison Act was a tax law that criminalized possession and sale of narcotic drugs. It effectively extended law enforcement powers to the Treasury Department responsible for enforcing alcohol prohibition. After repeal of Prohibition, this unit became the Federal Bureau of Narcotics (FBN), the forerunner of today's Drug Enforcement Agency (DEA).

Pharmaceutical manufacturing firms began to use the term "ethical" to distance themselves from patent medicine makers. Pharmaceutical firms rejected the use of patents on the grounds that they created unethical monopolies. Unlike the patent medicine makers with their secret recipes, ethical firms avoided branding and identified ingredients by generic chemical names drawn from the U.S. Pharmacopeia (which standardized drug nomenclature). Ethical houses did not advertise directly to the public like pharmaceutical companies do today. They limited their business to pharmacists and physicians whom they reached through the professional press. Around the turn of the twentieth century, however, even ethical firms began to act in questionable ways, sponsoring lavish banquets for physicians and publishing advertisements as if they were legitimate, scientifically proven theories. Manufacturing facilities were not always clean, so the drug industry was a prime target of Progressive campaigns that followed publication of Upton Sinclair's muckraking book *The Jungle,* which was about the meatpacking industry. The Pure Food and Drug Act (1905) created a Bureau of Chemistry to assess fraudulent claims by drug makers. After more than one hundred deaths were attributed to a drug marketed as "elixir of sulfanilamide," which contained antifreeze, in 1935, the U.S. Congress passed the Food, Drug, and Cosmetic Act (FDCA) in 1938. The FDCA created the Food and Drug Administration (FDA), the government agency responsible for determining the safety and efficacy of drugs and approving them for the market. Relying on clinical trials performed by pharmaceutical companies themselves, the FDA determines the level of control to which a drug should be subjected. In 1962 the FDCA was amended in the wake of the thalidomide disaster, and the FDA was charged not only with ensuring the safety and effectiveness of drugs on the market but also with approving drugs for specific conditions. Companies must determine in advance whether a drug has "abuse potential" or is in any way dangerous to consumers. Despite attempts to predict accurately which "wonder drugs" will go awry, newly released drugs are tested on only a small segment of potential users. For instance, OxyContin, developed by Purdue Pharma as a prolonged-release painkiller, was considered impossible to tamper with and hence not "abusable." Soon known as "hillbilly heroin," the drug became central in the drug panic.

Drug panics are commonly recognized as amplifying extravagant claims: the substance at the center of the panic is portrayed in mainstream media as the "most addictive" or "most dangerous" drug ever known. Wonder drugs turn to "demon drugs" as their availability is widened and prices fall. This pattern applies to both legal and illegal drugs. Another major social frame through which drugs are constructed, however, is the assumption that medical and nonmedical use are mutually exclusive.

Medical use versus nonmedical use is a major social category through which drugs have been classified since the criminalization of narcotics. If you are prescribed a drug by a medical professional and you use it as prescribed, you are a medical user. The old divisions between medical and nonmedical use break down when we think about something like cough medicine—once available over-the-counter (OTC) with little restriction despite containing small amounts of controlled substances. Today retail policies and laws restrict the amount of cough medicine that can be bought at one time, and purchasing-age limits are enforced. Availability of cough suppressants in home medicine cabinets led to experimentation by high school students with "chugging" or "robo-tripping" with Robitussin and DM-based cough suppressants.

Practices of self-medication blur the medical-versus-nonmedical category. In some places illegal drug markets have made these substances more widely available than the tightly controlled legal market. Many people who use heroin, cocaine, or marijuana are medicating themselves for depression, anxiety, or disease conditions. They lack health insurance and turn to drugs close at hand. Legal pharmaceuticals are also diverted to illegal markets leading to dangerous intermixing, as in the illegal use of legal benzodiazepines as "xani-boosters" to extend the high of an illegal drug. The social construction of legal drugs as a social good has been crucial to the expansion of pharmaceutical markets. The industry has distanced itself from the construction of illegal drugs as a serious "social bad," but this has become difficult in the face of a culture that has literally adopted "a pill for every ill."

Drug issues would look different if other interest groups had the cultural capital to define their shape. Some substances are considered to be essential medicines, whereas others are controlled or prohibited altogether. When drugs are not used in prescribed ways, they are considered unnecessary or recreational. Like the other frames discussed, this distinction has long been controversial.

The history of medicine reveals sectarian battles over which drugs to use or not use, when to prescribe for what conditions, and how to prescribe dosages. The main historical rivals were "regular" or allopathic physicians, who relied heavily on "heroic" doses of opiates and purgatives, and "irregular" or homeopathic physicians, who gave tiny doses and operated out of different philosophies regarding the mind–body relation. Christian scientists and chiropractors avoided drugs, and other practitioners relied primarily on herbal remedies. As organized medicine emerged as a profession, allopathic physicians became dominant. After World War II, physicians were granted prescribing power during a period of affluence and optimism about the capacity of technological progress to solve social problems. By the mid- to late 1950s, popular attitudes against using "a pill for every ill" turned around thanks to the first blockbuster drug, the minor tranquilizer Miltown, which was mass-marketed to middle-class Americans for handling the stresses of everyday life. Miltown was displaced first by the benzodiazepine Valium and then by the antidepressants Prozac and Zoloft and the antianxiety drugs Xanax and Paxil. A very high proportion of U.S. adults are prescribed these drugs, which illustrates the social process of "medicalization."

Medicalization is the process by which a social problem comes to be seen as a medical disorder to be treated by medical professionals and prescription drugs. Many of today's diseases were once defined as criminal or deviant acts, vices, or moral problems. Some disorders have been brought into existence only after a pharmacological fix has become available. During "Depression Awareness Week," you will find self-tests aimed at young people, especially at young men. Typically, women medicalize their problems at higher rates, but the male market is now being tapped. Health care is a large share of the U.S. gross national product, and pharmaceutical companies maintain the highest profit margins in the industry, so there are huge economic stakes involved in getting you to go to your doctor and ask for a particular drug. Judging from the high proportion of the U.S. population on antidepressant prescriptions at any given time, these tactics have convinced people to treat even mild depression. Antidepressants are now used as tools to enhance productivity and the capacity to "balance" many activities, bringing up another active frame in the social construction of drugs: the difference between drugs said to enhance work or sports performance and drugs said to detract from performance.

Performance enhancement drugs first arose as a public controversy in relation to steroid use in professional sports and bodybuilding. However, this frame is also present in the discussion of Ritalin, the use of which has expanded beyond children diagnosed with attention deficit and hyperactivity-related disorders. Amphetamines, as early as the late 1940s, were known to have the paradoxical effect of settling down hyperactive children and allowing them to focus, but today the numbers of children and adolescents diagnosed with ADD and ADHD is extremely high in the United States. Stimulants such as cocaine, amphetamines, and caffeine are performance-enhancing drugs in those who are fatigued. Caffeine is associated with productivity in Western cultures but with leisure and relaxation in southern and eastern Europe, Turkey, and the Middle East, where it is consumed just before bedtime. Different cultural constructions lead people to interpret pharmacological effects differently. Today caffeine and amphetamines are globally the most widely used legal and illegal drugs—the scope of global trading of caffeine exceeds even that of another substance on which Western societies depend: oil.

Performance detriments are typically associated with "addictive" drugs, a concept that draws on older concepts of disease, compulsion, and habituation. With opiates, delight became necessity as individuals built up tolerance to the drug and became physically and psychologically dependent on it. Addiction was studied scientifically in response to what reformers called "the opium problem" evident on the streets of New York City by the early 1920s. The U.S. Congress created a research laboratory through the Public Health Service in the mid-1930s where alcohol, barbiturates, and opiates were shown to cause a physiological "withdrawal syndrome" when individuals suddenly stopped using them. The Addiction Research Center of Lexington, Kentucky, supplied data on the addictiveness of many drugs in popular use from the 1930s to the mid-1960s. During that decade, the World Health Organization changed the name of what they studied to "drug dependence" in an attempt to destigmatize addiction. They promoted the view that as a matter of public health, drug dependence should be treatable by medical professionals, whose treatment practices were based on science. This view brought them into political conflict with the expanding drug law enforcement apparatus, which saw the problem as one to be solved by interrupting the international trafficking. Public health proponents lost out during the 1950s when the first mandatory minimum sentences were put into place by the 1951 Boggs Act. These were strengthened in 1956. By the end of the decade, law enforcement authorities believed that punishment-oriented drug policies had gotten "criminals" under control. They were proven wrong in the next decade.

Patterns of popular drug use often follow the contours of social change. Several factors tipped the scale toward constructing drug addiction as a disease in the 1960s. The U.S. Supreme Court interpreted addiction as an illness, opining, "Even one day in prison would be a cruel and unusual punishment for the 'crime' of having a common cold" (*Robinson v. California*, 1962). Finding it "unlikely that any State at this moment in history would attempt to make it a criminal offense for a person to be mentally ill, or a leper, or to be afflicted with a venereal disease," the Court stated that prisons could not be considered "curative" unless jail sentences were made "medicinal" and prisons provided treatment. Four decades later, treatment in prison is still sparse despite jails and prisons being filled with individuals on drug charges. In the late 1960s, civil commitment came about with passage of the Narcotic Addict Rehabilitation Act (1967) just as greater numbers of white, middle-class youth entered the ranks of heroin addicts. Law enforcement was lax in suburban settings, where heroin drug buys and use took place behind closed doors, unlike urban settings. New drugs including hallucinogens became available, and marijuana was deeply integrated into college life. The counterculture adopted these drugs and created new rituals centered on mind expansion.

During this time, racial-minority heroin users and returning Vietnam veterans came to attention on the streets. In a classic paper titled "Taking Care of Business," Edward Preble and John J. Casey observed that urban heroin use did not reflect apathy, lack of motivation, or laziness, but a different way to pursue a meaningful life that conflicted with ideas of the dominant social group. "Hustling" activities provided income and full-time, if informal, jobs where there were often no legitimate jobs in the formal economy. The lived experiences of drug users suggested that many people who got into bad relationships with drugs were simply self-medicating in ways designated by mainstream society as illegal. Members of this generation of heroin users suffered from the decline of social rituals and cultural solidarity that had once held drug-using subcultures together and enabled members of them to hold down legitimate jobs while maintaining heroin habits in the 1950s and early 1960s.

By the 1970s, heroin-using subcultures were more engaged in street crime than they had once been. The decline of solidarity became pronounced when crack cocaine came onto the scene in the mid-1980s at far lower cost than powder cocaine had been in the 1970s. Reading Preble and Casey's ethnographic work, which was done 30 years before the reemergence of heroin use among middle-class adolescents and the emergence of crack cocaine, we see how drug-using social networks met members' needs for a sense of belonging by forming social systems for gaining status and respect. In the 1970s, the Nixon administration focused the "war on drugs" on building a national treatment infrastructure of methadone clinics distributed throughout U.S. cities. Methadone maintenance has enabled many former heroin addicts to lead stable and productive lives. For a time, it appeared the "the opium problem" might be resolved through public health.

But there is always a "next" drug, and cocaine surfaced as the new problem in the 1980s. Powder cocaine had been more expensive than gold, so it was viewed as a "jet set" drug and used in combination with heroin. However, a cheaper form called crack cocaine became available in the poorest of neighborhoods during the 1980s. Mainstream media tend to amplify differences between drug users and nonusers, a phenomenon that was especially pronounced in the racialized representation of the crack cocaine crisis. Crack widened the racial inequalities of the War on Drugs at a time when social policy was cutting access to health care and service delivery and when urban African American communities were hit hard by economic and social crisis. The pregnant, crack cocaine–using woman became an icon of this moment. Women had long made up about one-third of illegal drug users (down from the majority status of white female morphine users in the early twentieth century), and little attention was paid to them. They were represented as a distinct public threat by the late 1980s and early 1990s, however. Despite so-called crack babies turning out not to have long-lasting neurobehavioral difficulties (especially in comparison with peers raised in similar socioeconomic circumstances), "crack baby" remains an epithet. Nor did so-called crack babies grow up to become crack users—like all drug "epidemics," the crack cocaine crisis waned soon into the 1990s.

Like fashion, fads, or earthquakes, drug cycles wax and wane, and policies swing back and forth between treatment and punishment. Policy is not typically responsible for declining numbers of addicts. Other factors, including wars, demographic shifts such as "aging out" or baby booms that yield large pools of adolescents, new drugs, and new routes of administration (techniques by which people get drugs into their bodies), change the shape of drug use. Social and personal experience with the negative social and economic effects of a particular drug are far better deterrents to problematic drug use than antidrug education and prevention programs; punitive drug policy; incarceration, which often leads to increased drug exposure; and even drug treatment. Although flawed in many ways, drug policy is nevertheless important because it shapes the experiences of drug sellers and users as they interact with each other.

Just as drugs have shaped the course of global and U.S. history, so have periodic "wars on drugs." The current U.S. drug policy regime is based on the Controlled Substances Act (1970), which classifies legal and illegal drugs onto five schedules that proceed from Schedule I (heavily restricted drugs classified as having "no medical use" such as heroin, LSD, psilocybin, mescaline, or peyote) to Schedule V (less restricted drugs that have a legitimate medical use and low potential for abuse despite containing small amounts of controlled substances). This U.S. law implements the United Nations' Single Convention on Narcotics Drugs (1961), which added cannabis to former international treaties covering opiates and coca. The Psychotropic Convention (1976) added LSD and legally manufactured amphetamines and barbiturates to the list. These treaties do not control alcohol, tobacco, or nicotine. They make evident the fact that drugs with industrial backing tend to be less restricted and more available than drugs without it, such as marijuana. Drugs that cannot be transported long distances such as West African kola nuts or East African qat also tend to remain regional drugs. Many governments rely heavily on tax revenue from alcohol and cigarettes and would be hard pressed to give them up. Courtwright argues that many of the world's governing elites were concerned with taxing the traffic, not suppressing it. Modernity brought with it factors that shifted elite priorities toward control and regulation as industrialization and mechanization made the social costs of intoxication harder to absorb.

Drug regulation takes many forms depending on its basis and goals. Hence there is disagreement among drug policy reformers about process and goals. Some seek to legalize marijuana and regulate currently illegal drugs more like currently legal drugs. Some see criminalization as the problem and advocate decriminalizing drugs. Others believe that public health measures should be aimed at preventing adverse health consequences and social harms, a position called harm reduction that gathered ground with the discovery that injection drug users were a main vector for transmitting HIV/AIDS in the United States. This alternative public health approach aims to reduce the risks associated with drug use.

Conflicts between those who advocate the status quo and those who seek to change drug policy have unfolded. Mainstream groups adhere to the idea that abstinence from drugs is the only acceptable goal. Critics contend that abstinence is an impossible dream that refuses to recognize the reality that many individuals experiment with drugs, but only a few become problematically involved with them. They offer evidence of controlled use and programs such as "reality-based" drug education, which is designed to teach people how to use drugs safely rather than simply avoiding them. Critics argue that the "just say no" and "drug-free" schools and workplaces have proven ineffective (see the entry on drug testing for a full account of how drug-free legislation was implemented). In arguing that the government should not prohibit consensual adult drug consumption, drug policy reformers have appealed to both liberal and conservative political ideals about drug use in democratic societies. Today's drug policy reform movement stretches across the political spectrum and has begun to gain ground among those who see evidence that the War on Drugs has failed to curb drug use.

See also Drugs and Direct-to-Consumer Advertising; Drug Testing; Medical Marijuana; Off-Label Drug Use; Tobacco.

Further Reading: Burnham, John. *Bad Habits: Drinking, Smoking, Taking Drugs, Gambling, Sexual Misbehavior, and Swearing in American History.* New York: New York University Press, 1994; Campbell, Nancy D. *Using Women: Gender, Drug Policy, and Social Justice.* New York: Routledge, 2000; Courtwright, David. *Forces of Habit: Drugs and the Making of the Modern World.* Cambridge, MA: Harvard University Press, 2001; DeGrandpre, Richard. *The Cult of Pharmacology.* Durham, NC: Duke University Press, 2006; DeGrandpre, Richard. *Ritalin Nation: Rapid-Fire Culture and the Transformation of Human Consciousness.* New York: Norton, 1999; Dingelstad, David, Richard Gosden, Brain

Martin, and Nickolas Vakas. "The Social Construction of Drug Debates." *Social Science and Medicine* 43, no. 12 (1996): 1829–38. http://www.uow.edu.au/arts/sts/bmartin/pubs/96ssm.html; Husak, Douglas. *Legalize This! The Case for Decriminalizing Drugs.* London: Verso, 2002; Inciardi, James, and Karen McElrath. *The American Drug Scene.* 4th edition. Roxbury, 2004; McTavish, Jan. *Pain and Profits: The History of the Headache and Its Remedies* New Brunswick, NJ: Rutgers, 2004; Musto, David. *The American Disease: Origins of Narcotics Control.* 3rd edition. New York: Oxford University Press, 1999; Preble, Edward, and John J. Casey. "Taking Care of Business: The Heroin Addict's Life on the Street." *International Journal of the Addiction* 4, no. 1 (1969): 1–24.

Nancy D. Campbell

DRUGS AND DIRECT-TO-CONSUMER ADVERTISING

In the 1990s, prescription drug manufacturers turned to the popular media—including television, radio, and magazines—to advertise their products. This phenomenon, known as direct-to-consumer advertising, helped to make blockbusters out of drugs such as Viagra and Allegra (which relieve impotence and allergies, respectively). As spending on direct-to-consumer advertising increased from 12 million dollars in 1989 to 4 billion dollars in 2004, such advertising became ingrained in popular culture, and spoof advertisements were common in comedy routines and on the Internet.

The success of pharmaceutical manufacturers in gaining visibility for their products (and increasing their profits) came at a cost, however. In 2004 a highly advertised painkiller, Vioxx, was removed from the market because of widespread reports of heart attacks and strokes. Critics alleged that its extensive marketing, including direct-to-consumer advertising, had led to overuse by extending prescribing to patients for whom the drug was inappropriate. The pharmaceutical industry came under fire for unethical behavior although it successfully staved off further regulation and scrutiny by introducing voluntary guidelines. The criticisms resonated with the long-held concerns of critics and public interest groups about the relationship between advertising and patient safety.

Drug safety regulations have long required drug companies to prove safety and efficacy, categorizing certain drugs as prescription-only. This need for a prescription led most major drug manufacturers to conclude as recently as 1984 that direct-to-consumer advertising was unwise, shortly after the antiarthritis drug Oraflex was recalled (like Vioxx) following widespread promotion and safety concerns.

Public interest groups—notably the Public Citizen's Health Research Group led by Sidney Wolfe—have drawn on examples such as Oraflex and Vioxx to caution against general advertising of prescription drugs. Advertising, in their view, has the goal of increasing sales and profits and simply cannot provide a balanced picture of drug risks and benefits. The Public Citizen's Health Research Group has been joined by the Women's Health Network, which advocates against unethical promotion of medicines, particularly contraceptives.

Patient safety is also a key concern of the government agency responsible for prescription drugs and their advertising, the Food and Drug Administration

(FDA). Until the mid-1990s, the FDA maintained a strict policy toward advertising to the public, asking that manufacturers submit their advertisements to the agency for preapproval. This changed, however, when the Washington Legal Foundation sued the FDA on First Amendment grounds in 1994.

In the 1970s and 1980s, a legal shift took place as the U.S. Supreme Court began to give First Amendment protection to advertising (known as "commercial speech"). Although courts had always recognized the importance of political speech, previously they had allowed blanket bans on advertising. In 1976 a pivotal case, *Virginia State Board of Pharmacy v. Virginia Citizens Consumer Council,* struck down a ban on the advertising of prescription drug prices. Over the following decade, bans on the advertising of alcohol and professional services were similarly deemed unconstitutional.

Although drug manufacturers contemplated advertising their products in this favorable legal environment, they were wary in the wake of the Oraflex controversy. They also wanted to comply with the FDA given its control over all aspects of drug regulation. The media and advertising industries, not the drug industry itself, were behind this policy change in favor of direct-to-consumer advertising.

Starting in the 1980s, the media corporation CBS sponsored meetings and research into consumers' health information needs. A powerful coalition was brought together, including advertising trade groups such as the American Advertising Federation, media corporations, and think tanks such as the American Enterprise Institute. This coalition had a powerful ally, the Washington Legal Foundation, a think tank that uses legal challenges to reduce government restrictions on speech. The Washington Legal Foundation successfully challenged FDA regulations on promotion of off-label uses (uses other than those for which the drug has been tested) in 1994 and thereby alerted the agency to First Amendment constraints. In 1997 the Food and Drug Administration announced a change in its enforcement of direct-to-consumer advertising regulations that enabled an explosion in advertising.

Unlike opponents of direct-to-consumer advertising, who question its impact on patient safety, free speech advocates emphasize the possibilities for *consumer empowerment*. They argue that advertising empowers consumers to make informed choices about prescription drugs. Although this perspective fails to acknowledge the differences between medicines and other consumer products, their opponents, in turn, have failed to emphasize that balanced, nonpromotional information could empower patients to make informed choices.

Backed by extensive resources and favorable legal doctrine, free speech advocates were successful in setting the terms of the policy debate in the 1990s. They have since produced data suggesting that direct-to-consumer advertising improves compliance with treatment regimens and have argued that direct-to-consumer advertising helpfully increases drug use. In contrast, advocates for increased regulation—who believe that drugs are inherently unsafe and overused—have fewer resources and a harder case to make. The link between advertising and inappropriate use is not obvious given that doctors control access to prescription drugs. Proponents of regulation argue, however, that doctors cannot always give consumers unbiased information themselves, considering that they also generally learn about new drugs from pharmaceutical sales representatives.

Problems with direct-to-consumer advertising are thus part of larger debates about the ways that pharmaceuticals are produced, developed, and marketed. Direct-to-consumer advertising creates incentives for the pharmaceutical industry to produce and market certain kinds of drugs and not others. Most new drugs are similar to drugs already available for a given condition, and pharmaceutical research prioritizes (often minor) medical conditions with large Western markets—a problem, critics argue, because no drug is without risks, and many seriously needy people go untreated.

See also Drug Testing; Drugs.

Further Reading: Angell, Marcia. *The Truth about the Drug Companies: How They Deceive Us and What to Do about It.* New York: Random House, 2004; Critser, Greg. *Generation Rx: How Prescription Drugs Are Altering American Lives, Minds, and Bodies.* Boston: Houghton Mifflin, 2005; Hilt, Philip J. *Protecting America's Health: The FDA, Business, and One Hundred Years of Regulation.* New York: Knopf, 2003; Hogshire, Jim. *Pills-a-go-go: A Fiendish Investigation into Pill Marketing, Art, History, and Consumption.* Los Angeles: Feral House, 1999.

Lorna M. Ronald

ECOLOGY

Ecology is the study of the interactions between organisms and their environments. Environment includes all biotic (living) and abiotic (nonliving) factors affecting the organism. The term *ecology* was first used by German biologist Ernst Haeckel in 1866 and was further developed by Eugenius Warming, a Danish botanist, in 1895. This is a relatively short time frame in comparison with other scientific disciplines. As a result, changes are still occurring in how this area of study is defined and understood, as well as in the manner in which it is applied.

Ecology is a broad-based discipline, with an extremely diverse area of study. The number of species or individuals and the number of their interactions involved in the study of ecology can be huge. As these numbers increase, so do the complexity of the interactions and therefore the complexity of the study. This broad base and the complex nature of the discipline are what make ecology both so interesting and so challenging.

Ecology draws upon other life or biological sciences but tends to work at the more complex level of biological organization. This means it is more inclusive of all biological entities because it works at the level of the organism and its ecosystem, rather than at a cellular or subcellular level.

Ecology is more holistic than atomistic. Traditional science tends to reduce the object of analysis to its elemental parts; these parts are then described, studied, and understood in order to understand the whole. The individual part is looked at in isolation from the whole, and its impact on the whole is derived from its removal. This means that the series of events that occurs when one part

DEEP VERSUS SHALLOW ECOLOGY

Deep ecology is a term first used by Norwegian philosopher Arne Naess early in the 1970s. The term was coined in an attempt to understand the differences in ecological thought and was used to illustrate the difference between "deep" and "shallow" or "reform" ecology. Deep ecology began as a philosophical idea that the entire biosphere, the living environment as a whole, has the same right to live and flourish as human beings do. The "deep" part of the name comes from the focus of this branch of ecology in asking "why" and "how" questions of environmental problems.

In order to understand the difference, we take a brief look at the pollution of a lake. Pollutant concentrations are greatest where one of the lake's tributaries empties into that lake. When scientists follow the river upstream, they find a large municipal waste disposal site. Studies are conducted and determine that this waste disposal site is the source of the contaminant. In shallow or reform ecology, the attempt to understand the problem would stop there, and a solution would be devised based on an administrative rule or regulation regarding the locating of waste disposal.

Deep ecology would continue to ask questions once the source of contamination had been discovered. What do the religious, political, economic, cultural, and other values and systems contribute to the problem? How do these systems contribute to the quantity of waste produced and the decision to dump it all in one place?

Essentially, deep ecology is an attempt to remove human beings as the central being and to place the integrity of all beings and Earth as a whole at the core. It is a questioning of our values as a culture and a species and provides an approach or attitude about the environment rather than a static set of principles and ideas.

Deep ecology places the whole greater than the sum of its parts and says that value is realized through the relationship of the organism to the whole, rather than placing value intrinsically on the individual. For many, deep ecology is a response to the modern human condition of isolation, separation, and lack of connection, in particular to the natural world.

Deep ecology is criticized for its notion of intrinsic value placed on the entire biosphere. Arguments have been made that because value is a human construct, the idea of intrinsic value in the environment is irrelevant. The environment derives its value from human interactions and perceptions, rather than value being an integral part of the environment.

As well, this movement has been criticized for the implications that the ideas put forth by its thinkers are "deeper" and therefore more meaningful than those of other branches of ecology and philosophy. Indeed, this difference between science and philosophy is another place of contention. Deep ecology does seem to be more of a philosophical and environmental approach than a branch of the science of ecology.

is removed is interpreted to be the opposite effect of the inclusion of that part in the system.

In comparison, the holistic approach exemplified in ecology focuses on the web of interactions of a system, in this case, for example, an ecosystem or an organism. The entire interactional system is important, rather than each or any individual part or component. The operation of the whole system is studied, and this is the basis for understanding the function of individual parts. Ecological research relies on making comparisons between similar systems or interactions in order to understand fundamental principles of those interactions.

The term *emergent properties* is a result of the holistic approach. Emergent properties are phenomena that are observable only in the intact, complete system. If one of the component parts is removed, that phenomenon would not occur. The nature of these properties cannot be predicted from an understanding of each of the parts individually; therefore, the whole system must be studied. For example, herd grazing is a phenomenon that arose from the interactions among plains bison, prairies plants, predators such as the plains grizzly bear and wolf, soil, weather, and a host of other organisms. In order to understand the typical size of bison herds, one must understand the interaction of all these (and likely other) factors.

The most important principle of ecology is that every living organism has a continuous relationship with its environment; those relationships are extremely important in order to understand the whole. One of the ways that ecology does this is through studying populations, communities, and ecosystems.

Populations are groups of individuals of the same species that together form a community. This could be a group of wolves that live within the same national park, a single species of tree that occurs in one forest, or the lichen on a rock face. These populations vary greatly from one to the next in terms of numbers of individuals, genetic diversity among individuals, and the spatial dimensions, geography, and topography of the occupied area.

There are always other living organisms that influence a particular population. Together, these populations within an area may be referred to as a biotic community. From one area to another, similar communities may be found, although they are rarely identical. At times, it may be difficult to clearly identify one community, given that several species may be part of more than one community. Within individual species in different communities, there may be some genetic variation that relates to the adaptations a certain population has made in order to function well within that community.

The biotic communities that interact and are connected by abiotic or non-living processes are termed ecosystems. These abiotic processes include such components of the external environment as weather, light, temperature, carbon dioxide and oxygen concentrations, soil, geology, and many others. In fact, any nonliving, external factor that may influence species distribution, survival, or ability to thrive may be considered part of the ecosystem.

Ecosystems interconnect through food chains or food webs. This is a hierarchical understanding of the flow of energy throughout the ecosystem. In a food web, each link feeds on the one below it and is fed upon by the one above. At the bottom of the food chain are photosynthetic organisms (green plants and algae) that can produce their own food (known as autotrophs), and at the top are large, carnivorous animals. Understanding the food web is important because it is a more easily managed method of working out the interaction between plants, animals, and the abiotic environment, and therefore the ecosystem as a whole.

ALDO LEOPOLD AND *A SAND COUNTY ALMANAC*

Aldo Leopold (1887–1948) was a U.S. environmentalist, with a career in ecology, forestry, and wildlife management. His influence has been felt over the development of modern environmental ethics and environmentalist movements, as well as in the scientific field of ecological inquiry.

Leopold spent his entire life in contact with the outdoors, which can be felt in all of his written work. He portrays natural environments with a directness showing his familiarity with and love for those environments. His professional life was spent in various natural environments from New Mexico and Arizona to Wisconsin, involved in forestry and wildlife management.

Leopold was an advocate for the preservation of wildlife and natural areas, a legacy still in effect today. He was not afraid to offer criticism of harm done to natural systems, particularly if that harm was a result of societal or cultural belief in a human superiority over those natural systems.

Leopold wrote *A Sand County Almanac,* beginning in 1937, out of a desire to take his message of environmental ethics to the broader public. This work has become pivotal to modern concepts of conservation and environmentalism, guiding policy and ethics. In his work he puts forth the view that the land is not a commodity that humans can possess. In order to not destroy the Earth, there must be a fundamental respect of it by all people: "That land is a community is the basic concept of ecology, but that land is to be loved and respected is an extension of ethics."

Aldo Leopold's book has been read by millions of people around the world, and his work has resonated in many areas of conservation and ecology. His encouragement for people to develop a land ethic and to allow that ethic to guide their actions and attitudes toward our natural spaces is an important message for individuals.

Ecology is closely linked to studies of adaptation and evolution. Evolution and adaptation involve genetic changes in a population over time as a result of external factors. Evolution is linked to ecology because it is the very nature of an organism's or population's ecology that will lead to adaptation and evolution. Evolutionary solutions are known as adaptations—genetic changes that lead to an organism being better suited its environment.

Along with the ideas of evolution and adaptation, the concept of succession is central to an understanding of population and community. During succession, all species successively appear and gradually alter the environment through their normal activities. As the environment changes, the species present will change and may replace the original species. Succession occurs following a disturbance, usually of a radical nature such as a fire or flood.

Succession can be seen as a cyclic series of events or as linearly progressive. No communities are completely stable, however, and at some point an event will occur in which the process begins again.

In general, scientists identify two stages of succession: primary and secondary. Primary succession involves the formation of the basis of most ecosystems: soil and its associated processes, before the introduction of plant and animal species. Geological weathering of parent material and the building of soil is a long-term process that may take hundreds or thousands of years. The first living organisms to be found are usually small types of plant material such as moss and lichens, adapted to long periods of drought and requiring small amounts of soil substrate. Over time, soil will collect around this small plant matter, as will dust and debris. This creates an environment that can support higher plant material as well as small animal species. As soil-building processes continue, larger and larger organisms will continually take over the environment, eventually reaching a more steady state. This may take thousands of years to complete. At this point, in order for succession to continue, a disturbance must occur.

When a radical disturbance occurs, and the biotic community returns to the level of those found earlier in the succession process, secondary succession will occur. This process is more rapid than primary succession because soils are already developed, and there are usually surviving species in the area. Succession of this type will reach a steady state much more quickly—perhaps in less than one hundred years. Fires, floods, and human disturbances are among the causes of secondary succession.

Ecology can also be applied as a tool of analysis. It is used to interpret and understand environmental problems and to point toward a solution. Environmental problems rarely hinge on one problem alone, so ecology is an ideal tool with which to discover the basis of those problems. Because ecology attempts to study an ecosystem or community as a whole, those parts of the ecosystem or community that are not functioning in concert with the whole can be discerned.

An interesting facet of ecology is that although it is a scientific discipline, it has also become a philosophy or mode of thinking. Environmental ethics and responsibility, made popular in the early part of the twentieth century, have become synonymous with ecology. It is at times difficult to separate the two, given that the science of ecology certainly leads toward a more holistic approach and attitude toward the environment. As a matter of fact, ecology as a science has provided the basis for many environmental movements, goals, and policies. The science of ecology, through its study of such things as biodiversity and species population dynamics, for example, has provided the impetus for the social focus on many environmental issues. It is through this type of research that changes in ecosystems are noticed along with whether those changes have a detrimental impact on some or all organisms living within them.

As crucial examples of how ecologists study whole systems, the primary focus or concern for ecologists in the twenty-first century is on the areas of global warming (the carbon cycle) and the loss of biodiversity felt worldwide. Each of these issues has the potential to be the cause of crises for human populations.

Life on Earth is based on carbon. Carbon molecules play an important role in the structure and chemistry of all living cells. The presence or absence of carbon helps define whether a substance is organic or inorganic. The carbon cycle is the movement of carbon between the atmosphere and the biosphere (including both terrestrial and aquatic parts of Earth).

Photosynthesis, the process by which green plants capture and store solar energy and carbon dioxide, is the fundamental energy source for life on Earth. Solar energy is stored as carbohydrate, a carbon-containing molecule. Some of this energy is used by plants for their own physiological processes. When energy is released from carbohydrate, the plant is respiring. One of the by-products of plant respiration is carbon dioxide released back to the atmosphere. This is essentially the reverse of photosynthesis. When energy is stored in plants rather than being used by them, it becomes available to other living organisms for food. In this manner, solar energy and carbon provide the basis of the food chain and energy for the entire world.

When animals (including humans) consume plants, the energy stored as carbohydrate is released in another form of respiration. Animals use energy for growth and development, and when they die and decompose, carbon is an end product, and carbon dioxide is again released to the atmosphere.

When photosynthesis exceeds respiration (in both plants and animals) on a global scale, fossil fuels accumulate. This occurs over a geological time frame of hundreds, thousands, or millions of years, rather than within a more human time scale. The last way carbon is returned to the atmosphere is through the burning of fossil fuels, and the gases released during this chemical reaction contain carbon.

It is important to remember that this process does not occur only on land but is a process of aquatic environments as well, in both fresh and salt water. Aquatic plants as well as algae photosynthesize and therefore fix carbon. Carbon dioxide from the atmosphere may mix with water and form carbonic acid, which in turn helps to control the pH of oceans. Oceans are major carbon sinks; once the carbon is fixed, it sinks to the ocean floor, where it combines with calcium and becomes compressed into limestone. This is one of the largest reservoirs of carbon within the carbon cycle.

The cycle is important to ecology because any disruption to the many interactions in the cycle will have an effect on all parts of the system. Concerns about global warming are directly related to the carbon cycle because increases in the use of fossil fuels, at a rate that is not replaceable, mean an increase in carbon dioxide in the atmosphere. Carbon sinks, those parts of the cycle that tie up more carbon than is released, are also disappearing at an alarming rate. The loss of forests, in particular temperate boreal forests, is a loss of a large carbon sink area. Reforestation is very important because young forests are particularly good at tying up carbon—better, in fact, than the older forests they are replacing.

Global warming is at times called the "greenhouse effect" because the buildup of carbon dioxide and other gases permits radiation from the sun to reach Earth's surface but does not allow the heat to escape the atmosphere. This warming may have far-reaching effects on such things as ocean levels, availability of fresh water, and global weather patterns. This can have potentially catastrophic impacts on plant and animal life, even threatening the survival of entire species.

Similarly, the loss of biodiversity can be felt worldwide. This includes the extinction as well as the extirpation of species. Extinction is the loss of a species on a global scale, whereas extirpation is the loss of a species from a specific limited geographic area. Botanists estimate that many plants are becoming extinct before their chemical properties are fully understood. The implications of this include the loss of species that might be used for human medicines or for food.

The loss or extinction of species is not a new phenomenon. For as long as there have been living organisms on this planet, various species have become extinct when their environments have changed as a result of changing climatic or other environmental factors. The current problem is that extinction rates have accelerated tremendously since the beginning of the twentieth century as habitats have been damaged or destroyed. The loss of biodiversity, although of concern in its own right, has far-reaching consequences for human life on Earth. All cultivated food, fiber, medicinal plants, and other useful plants have come from wild populations. If those sources are no longer available as valuable gene banks, human survival is at risk.

There are other contributing factors in the loss of biodiversity besides habitat damage or destruction. The introduction of exotic species to ecosystems and the planting of artificial monocultures have had a major impact on biodiversity levels. Exotic species are often more aggressive than local species, able to out-compete those species for food and space resources. The planting of vast monocultures—as in the millions of acres of agricultural land in North America or cities planted with one species of tree lining most streets—leaves these plants vulnerable to disease and pest infestation. Natural biotic communities are much more diverse in terms of numbers of species as well as genetic variation within a population.

When looking at biotic communities on a global scale, ecologists talk about uniform ecological formations or biomes. Each biome has typical flora, fauna, and climatic factors, as well as other homogeneous abiotic components. There are several biomes in North America. Biomes include different kinds of biotic species but are usually distinguished by the plant species they contain. These areas, which represent the biodiversity present worldwide, are often the focus of conservation efforts. The preservation of large areas of each biome would, in effect, preserve a sample of Earth's biodiversity.

Tundra is the northernmost biome, occupying land primarily north of the Arctic Circle. This biome is devoid of woody plant species, although there are miniature forms of some species. The presence of permafrost, or permanently frozen ground, at depths, in some places, as close as 10 centimeters (4 inches) to the surface, necessitates the growth of shallow-rooted plant species. Therefore, most floras are dwarves. There is little precipitation, and most species of plants and animals are at the limit of their survival potential. As such, this biome is very fragile, and small levels of human and larger animal disturbance can be seen for years after the event.

Taiga, or boreal forest, is located adjacent to and south of the tundra. Vegetation is dominated by coniferous tree species. The climate is characterized by long winters and warm summers, with reasonable rates of precipitation. A large majority of Canada is covered by this biome, and it represents a large carbon sink for the entire planet. A wide range of both plant and animal species is represented in this biome.

Temperate deciduous forests are characterized by broad-leaved tree species, and they occur in a large mass through the center of North America. There is a range in temperature, although not as wide a range as in the boreal forest. Rates of annual precipitation are higher, as are the depth and quality of soil. A unique and diverse range of flora may be found here, in turn supporting a large variety of animal life. A large amount of this forest has been disturbed or destroyed by European colonization; the large trees were felled for use as timber or cleared to make way for agriculture. The result is that large tree masses that shaded large areas of land are no longer present to provide their cooling effect.

Grasslands at one time dominated large portions of the continent as well. They were integrated with forest or desert biomes, depending on climatic and precipitation ranges. This biome was made up of three types of grassland: short grass, tall grass, and mixed grass prairies. Each prairie has distinct vegetation and animal life, and together they supported vast numbers of flora and fauna. The impressive herds of plains bison that once roamed the interior of North America were dependent on these grasslands and, because of extensive hunting and loss of habitat, are now confined to game preserves and zoos. Nearly all indigenous grasslands have been ploughed under for agricultural use, although there are some small remnants scattered throughout the continent.

There are desert biomes in North America, places where precipitation is consistently low and soils are too porous to retain water. There is usually a wide range in temperature fluctuation, although it rarely freezes. Plants and animals have developed unique adaptations to survive the extremes of heat and drought found in desert climates, and interestingly, many of these are similar to the adaptations found in the tundra. Many of the adaptations are to make species more nocturnal, when the extremes of heat are less problematic, and therefore more moisture is available.

Temperate rain forests, also known as coastal forests, exist along the west coast of North America. Trees in this biome tend to be huge, due in large part to the amount of precipitation that falls each season. The sheer size of these trees makes them desirable for logging, which has led to the loss of some of these beautiful forests. Again, a wide variety of flora and fauna can be found in temperate rain forests, and this biodiversity is valuable beyond the value of the timber.

This continent also supports a small amount of tropical rain forest, found in areas where annual rates of precipitation exceed 200 centimeters (80 inches) per year. Temperatures are warm, ranging usually from 25 degrees Celsius to 35 degrees Celsius (77 to 95 degrees Fahrenheit). These climatic conditions support the widest range of species found anywhere on the planet and represent a major source of genetic diversity. This is where the potential new medicines, foods, and fibers may be found that will continue to support human life on Earth.

How to manage these biomes, each of which comes with its own issues relating to human habitation and development, is not easily resolved. In part, this is because of the approaches to ecological issues reflected in two terms: *conservation* and *preservation*. Both lie behind, for example, the push to identify and protect areas of different biomes. The preservationist wants to keep areas in their "natural" state, to keep the local ecosystems from shifting and therefore destroying elements that should be preserved. Of course, in dealing with organic systems, change is inevitable; human effects, moreover, are global in nature, so that apart from placing a bubble over some areas, there will be human influences—such as PCBs showing up in the fat of polar bears in the Arctic or temperature changes associated with global warming—that are unavoidable. When the natural state involves the inevitable forest fire, are these fires permitted, if the area that is burning contains the last of one or more plant species? Further, population pressures make a preservationist position more and more difficult; ecotourism, for example, is not a good idea on any significant scale because the mere presence of people affects the behavior of the species found in the area.

The conservationist, on the other hand, recognizes the need to protect within the longer-term framework of conserving the resources of nature for future generations to enjoy. Whether conservation is in the interests of being able to use natural resources in the future—planting trees as well as cutting them down—or of preserving genetic or biodiversity against future needs (perhaps in medicine, as a biopreserve), the human element is inescapably present.

Conservation and preservation obviously both have merits in comparison with the wholesale destruction of habitats currently taking place. Yet the anthropocentric (human-centered) attitudes they reflect, according to deep ecologists, perpetuates the same problem that they supposedly address. We need to understand ecology within the wider perspective that includes humans as one species among many. Then we will see that the global system is designed (though not intentionally) not only for human use but also for the support of complex and diverse systems of flora, fauna, and their ecological niches. This makes clear why significant changes are needed in how human beings behave in relation to the global bioecological system.

See also Gaia Hypothesis; Global Warming; Pesticides; Sustainability; Waste Management; Water.

Further Reading: Merchant, Carolyn. *Radical Ecology: The Search for a Livable World.* New York and London: Routledge, 1992; Molles, Manuel C. *Ecology: Concepts and Applications.* 3rd ed. Toronto: McGraw Hill, 2005; Peacock, Kent A. *Living with the Earth: An Introduction to Environmental Philosophy.* New York and Toronto: Harcourt Brace, 1996; Pojman, Louis P. *Global Environmental Ethics.* Mountainview, CA: Mayfield, 2000; Ricklefs, R. E. *Economy of Nature.* 5th ed. New York: Freeman, 2001.

Jayne Geisel

Ecology: Editors' Comments

One of the implications of the study of ecology for religious groups has been cross-fertilization between ecology and spirituality or between ecology and theology. Particularly in the Christian tradition there has been an explosion of writing and discussion on the subject of "eco-theology," melding the required changes in attitudes toward nature and in environmental practice with new or revitalized interpretations of the Bible and of Christian theology. In other major religions, as well as in the religious and spiritual practices of indigenous peoples, there has been a growing awareness of the elements that emphasize a right relationship with the Earth. If our apparent inability to appreciate the consequences of poor environmental practice is the result of the dominance of mass consumer culture, then these new ways of perceiving nature and conferring meaning on the human relationship to the world around us may help both to articulate reasons for resistance and to motivate change.

Further Reading: Foltz, Richard C. *Worldviews, Religion, and the Environment.* Belmont, CA: Thomson/Wadsworth, 2003.

EDUCATION AND SCIENCE

Popular media, government officials, and advocacy groups are forwarding the claims that the United States is in an education crisis. Changes in school curriculums, underfunded educational institutions, scientific illiteracy, and the decline of the nation's global competitiveness in science and technology signal a potential cultural and economic disaster on the horizon. All this attention placed on preparing students for life outside of school should make one wonder what science education is and how it has changed over the centuries. Is science education the key for surviving in a technology-centered world? What debates and conflicts have shaped the nature of educating students and the broad citizenry over the centuries?

It has not always been this nation's position that science education should be a means of educating the population for a technologically advanced future. At times, science education in the classroom and in the public spheres has been utilized to impede social change or has encountered conflicts with the norms of society. In other historical moments in the United States, science education has been employed and manipulated to further political and cultural agendas ranging from the promotion of agriculture during the earliest moments of industrialization to the advancement of intelligent design (ID) as a valid scientific theory. Conflicts and debates over the construction and distribution of science education span historical time, geographical space, and diverse groups of invested actors.

The critical turning point for Western education occurred during the Enlightenment. Individualism, rationality, and the perfection of man became the guiding principles. Science education developed from these principles as a means of studying the laws of nature and through this process embedding logic and objectivity into students. Although Enlightenment thinking is still present in school curriculums, controversies have developed over religion and science, federal control of education, new conceptions of scientific practice, scholarly critiques of science education, and cultural tensions concerning the power of science education. These controversies occur against a backdrop of popular notions of science education and visions of technological progress.

Education in the United States has been an evolving entity since science was isolated as a particular field of education. Science education, and education generally, tended to be a secondary concern for most Americans. What little science education did occur in the first 100 years of the nation's history was based on memorization of facts, many of which were used in conjunction with religious teachings to forward a theistic education. Changes in the demographics of the nation—from rural farm to urban industry—created some of the initial reform movements in science education. A nature study movement attempted to keep nature, agriculture, and environmental health as primary concerns in science education. These reforms overlapped with previous Enlightenment thinking and started a move toward developing science as a separate field of study. Higher education had its own contentious reforms to manage in the early 1900s. Colleges were still bastions of classical studies and religious devotion. The practical matters of science were philosophically and theologically beneath the concerns of gentlemanly education. Industrialization, modernization, and two world wars finally shifted the alignment of science education from nature in early education and leisurely study in higher education to the advancement of the physical sciences as a contribution to the United States' economic and military success. The growing respect of scientists and technologists during and after World War II created a tense scenario in which the federal government and elite scientists attempted to influence education curriculums more than educators had experienced in the past. Arguments for the inclusion of science researchers in curriculum debates revolved around the notion that students who have acquired the techniques of the scientific method are better suited to participate in civil society as well as promote the security of the nation.

Continued external influences have converged on science education since this period, creating conflicts between the state and federal governments' interests and the interests of leaders of research institutions, grade school teachers, and a variety of interested community groups. One example of this confrontation is the relationship between government and higher education. For much of U.S. history, universities have strongly opposed the intervention of government in formulating curriculums for social goals. As a result of this conflict, the U.S. government proceeded to construct its own scientific bureaucracy to fill this gap in control over scientific research and regulation. Particularly after the launching of Sputnik and the Cold War arms race, funding programs were created for influencing the activities of research universities and institutions, including the creation of the National Institutes of Health and the Department of Defense's research wing. Federal manipulations of education have accelerated as nation states compete on a global level. Producing competitive scientists and engineers has become the nation's primary goal for science education rather than developing thoughtful and informed citizens.

Despite a new discourse about an impending crisis in science education, the public and many educators have simply not been swept up by these issues. Even though diverse groups have introduced new standards and tests that American students fail, and international comparisons on scientific literacy have placed the United States near the bottom of industrialized nations, interventions have not had popular support. Without revolutionary new curriculums that ground science education in the students' daily lives and concerns for an international conflict grounded in science and technology, such as the launching of Sputnik, it seems unlikely that the education crisis will be resolved.

Contemporary public debates have been less concerned with the supposed decline of science education than with the moral implications of science teaching. The classroom is argued to be a space for indoctrinating children to particular ways of thinking. In many regions of the United States, the call to return religion to science education has created controversies over giving equal time and credibility to the teaching of evolution and intelligent design (ID) in textbooks and classrooms. Although this is not a new debate (antievolution laws have been enacted since Charles Darwin's earliest publications), many scientists have helped make ID arguments more sophisticated and have lent credibility through their presence. Internally, since the eighteenth century, many conflicts have occurred in science education on teaching racism-tinged science. In the last several decades, science education has been repudiated for teaching eugenics and sociobiology and for introducing pseudoscience that links, for example, race and intelligence.

Further moral and cultural debates on the role of science education occurred during the radicalism of the 1960s, as it recreated the turn-of-the-century concerns over runaway technology, positioning science education as a main force in the subjugation of students to powerful elites. A recent example of such cultural indoctrination is found in the new America COMPETES Act (2007) signed into law as HR 2272. A portion of the stated goals of this act is to "promote better alignment of elementary and secondary education with the knowledge and skills needed for success in postsecondary education, the 21st century workforce, and the Armed Forces." In contrast to the Enlightenment tradition of creating better citizens through education, the new emphasis is on workforce creation and military preparedness.

This indoctrination of students in science education is more favorably presented in the work of Thomas Kuhn on worldview creation in science communities. His *The Structure of Scientific Revolutions* (1962) popularized a growing scholarly sentiment that science pedagogies are indoctrination tools that enable individuals to practice science but at the same time restrict the potential for alternative questions and techniques. If students do not assume the cultures of the sciences they pursue, it becomes impossible to be able to communicate and debate among one's peers. At the same time that the paradigms of learning and research become part of the students' personality, the students also become restricted in the types of questions that they can ask. Nonetheless, Kuhn was always a firm believer in scientific progress and the ability of scientists to overcome such biases, a contention that has been criticized by some science studies scholars.

In order to assimilate this new realization, a recent swing in science education circles argues that teachers should posit the activities of scientists as controversies. Realizing that the peer-review system in science creates a need for conflict between people and their ideas, education scholars argue that teaching the historical controversies between researchers and theories gives a more complete science education and also creates citizens with a perspective on how to become involved in future controversies. Science is becoming increasingly politicized, making it critical for citizens to be able to dissect the claims presented within government that are based on scientific intervention. Further, debates within the classroom should not only include the conflicts occurring among scientists but should also include evaluating the simplification of such myths as the scientific method and objectivity.

Less well-developed in education circles is how to manage the hurdles placed in front of women and minorities to join and fully participate in the sciences. The United States has recently experienced an increase in the number of women and minorities participating in higher education but has failed to determine the scope of this participation. The culture surrounding science and engineering, including the values and attitudes implicit within these branches of science education, limits accessibility and contributes to an invisible education crisis. Many women and minorities feel oppressed by the cultures surrounding the physical sciences, receiving direct and indirect comments from teachers, professors, media, and school officials that women and minorities are simply not as capable of succeeding in chemistry, math, and physics. Even when these hurdles are overcome in grade school and higher education, the perception (and perhaps the reality) remains that most elite positions within the science community and most of the research dollars go to white males.

Although most people dismiss the need to reinvent science education, a number of highly vocal groups have attempted to form a debate over what science education should be doing in the creation of a scientifically literate public. Two of the most visible movements arguing for reform are the Movement for Public Understanding of Science and Science for the People.

The Movement for Public Understanding of Science aligns most with the argument that if only people could think more like scientists, then society would proceed in a more logical and objective manner while contributing to the economic success of the nation. Claims of this sort are represented by the most powerful members of the scientific community and their advocacy institutions. Furthermore, these individuals argue for a scientifically informed society that utilizes the community of scientists in most realms of social life. Science education should then promote the notion of scientific thinking as the most appropriate option for all decision making.

Science for the People is a more radical movement attempting to break down the scientific community as an elite institution, advocating for more democratic science and one that is reflexive about the invisible assumptions, limitations, and dangers of scientific practice. Partly stemming from the Science for the People and other critical social movements of the 1960s and 1970s, the academic field of science and technology studies has developed entirely new paradigm shifts in science education. At the core of their concerns is the realization that classrooms teaching science are spaces that are culturally influenced and that tend not to teach the true complexity and controversies of science within society.

See also Creationism and Evolutionism; Math Wars; Religion and Science; Science Wars.

Further Reading: Bauer, H. H. *Scientific Literacy and the Myth of the Scientific Method.* Urbana: University of Illinois Press, 1992; Ceci, S. J., and W. M. Williams, eds. *Why Aren't More Women in Science? Top Researchers Debate the Evidence.* Washington, DC: American Psychological Association, 2007; Kuhn, T. *The Structure of Scientific Revolutions.* Chicago: University of Chicago Press, 1996; Majubdar, S. K., L. M. Rosenfeld, et al., eds. *Science Education in the United States: Issues, Crises and Priorities.* Easton: Pennsylvania Academy of Science, 1991; Montgomery, S. L. *Minds for the Making: The Role of Science in American Education, 1750–1990.* New York: Guilford Press, 1994; Traweek, S. *Beamtimes and Lifetimes: The World of High Energy Physicists.* Cambridge, MA: Harvard University Press, 1992; U.S. Congress. House. *America Creating Opportunities to Meaningfully Promote Excellence in Technology, Education and Science (COMPETES).* HR 2272 (2007); Yager, R. E., ed. *Science/Technology/Society as Reform in Science Education.* Albany: State University of New York Press, 1996.

Sean Ferguson

EPIDEMICS AND PANDEMICS

When an infectious disease appears in a location where it is not normally present and affects a large number of people, it is known as an epidemic. Epidemics can last weeks to years, but they are temporary and will eventually disappear. Epidemics are also localized, appearing in villages, towns, or cities. When an infectious disease with these characteristics spreads throughout a country, continent, or larger area, it is known as a pandemic. History has documented numerous epidemics and pandemics, many of which were fraught with controversy. For its long, varied, and at times dramatic history, smallpox, also known as variola, provides an excellent case study in epidemics and pandemics and the debates and issues that surround them.

In 430 B.C.E. the population of Athens was hit hard by an unknown plague. The plague, documented by Thucydides, claimed approximately one-third of the population. Some contemporary historians speculate that this unknown plague was actually smallpox. Similar plagues thought to be smallpox continued to appear throughout the Roman Empire from 165 to 180 B.C.E. and 251–266 B.C.E. What we now know as smallpox entered western Europe in 581 C.E., and eventually its presence became a routine aspect of life in the larger cities of Europe, such as London and Paris, where it killed 25 to 30 percent of those infected. By the eighteenth century smallpox was certainly endemic and responsible for an average of 400,000 deaths per year in Europe and the disfigurement of countless more.

In 1718 Lady Mary Wortley Montagu brought the practice of variolation to England from Turkey. The procedure was quite simple: a needle was used to scratch a healthy individual's skin, just breaking the surface; a single drop of the smallpox matter was added to the scratch and then loosely bandaged. If this was performed successfully, the individual would progress through an accelerated and mild case of smallpox, resulting in no scars and lifelong immunity.

The mortality rate for smallpox acquired in this manner was 1 to 2 percent, a considerable improvement over smallpox caught in the natural way, which had a mortality rate between 10 and 40 percent. When she returned to England, Lady Montagu variolated both of her children.

Most of London's well-to-do society recoiled in horror at the act of purposely giving an individual the pox. As a result, Lady Montagu was ostracized by all except her closest friends. Her actions sparked hot debates in the chambers of the London Royal Medical Society over the ethics of deliberately exposing an individual to smallpox, of the efficacy of the procedure, and of the methods of the procedure itself. Given the known mortality rate of smallpox and the success of Lady Montague's variation on her children, however, it was not long before others began requesting the procedure be performed on themselves and their children. After smallpox claimed the life of Queen Mary in 1692 and almost killed Princess Anne in 1721, members of the royal family became interested in the potential of variolation, influencing the opinions of the royal physicians.

Before members of the royal family could be subjected to the procedure, royal physicians demanded proof of the procedure's success through human experimentation. Several inmates scheduled to be hanged at Newgate Prison, London, who had not had smallpox, as well as one individual who had already had the pox, were chosen and subjected to the procedure. It is not known whether these subjects were chosen or if they volunteered, although it seems doubtful that they would have had a choice in the matter. The manner in which the experiment was performed would certainly be condemned by modern scientists as well as ethicists. The subjects were kept together in a separate cell and monitored daily by physicians. A constant stream of visitors, both medical and civilian, came to observe the infected prisoners in their cell. After all the subjects had made full recoveries, the procedure was considered successful, as well as morally acceptable. It is interesting to note that in England variolation required a specially trained physician, whereas in Turkey, where the practice originated, the procedure was generally performed by an elderly woman in the village.

Around the same time, medical controversy spread to America, specifically to Boston. The Reverend Cotton Mather is generally credited with bringing variolation to North America, having "discovered" the practice of variolation after a discussion with his slave who responded, "yes . . . and no" when asked if he had suffered the pox. This slave, Onesimus, provided Mather with the details of variolation as performed by his relatives in Africa. However, it was actually Dr. Zabdiel Boylston who performed the procedure. Whereas Mather might have publicly supported variolation, it was not until several months after it had been in practice that he allowed his children to be variolated, and then it was in secret. Boylston, on the other hand, was open with his actions and suffered from repeated threats of imprisonment from the government, as well as mob violence. The act of purposely giving an individual such a deadly infection was considered morally reprehensible by both citizens and public officials, regardless of its potential positive outcome. The uproar in Boston over variolation reached fevered levels, with some individuals supporting the practice and others supporting a ban. At various times the Selectmen of Boston banned individuals entering the city for the purpose of variolation and then banned the procedure itself. On at least one occasion Boylston's home was searched by authorities looking for individuals who had purposely been infected by smallpox through variolation, in an effort to find a legal reason to imprison Boylston. Eventually, fear of catching smallpox "naturally," combined with the apparent success of variolation and its popularity, forced the local government to legalize the practice. In fact, Boylston was even invited to England for an audience with the king, and he attended a number of variolation procedures during his visit.

Although variolation was a potent weapon against smallpox, it was an expensive procedure, equivalent to as much as $500 today and was initially available only to the wealthy. As a result, by the Revolutionary War, many Americans were still susceptible to the disease. This posed a problem for both America's soldiers and its civilians. Debates over variolation raged among the commanding generals of the American forces. Smallpox has a two-week incubation period during which the individual is asymptomatic but still contagious. The possibility of individuals who had undergone the procedure giving smallpox to their fellow soldiers during the infectious incubation period and triggering an epidemic among the American forces initially was considered too risky. In 1777, however, George Washington ordered the variolation of the entire Continental Army to prevent further outbreaks of the disease.

British forces were largely immune to smallpox, almost all having been exposed as children. Those who had not been exposed were quickly variolated. During the Revolutionary War the British crown promised freedom to any American slave who joined their forces. Being American, the majority of freed black slaves were not immune to smallpox. Many acquired it through variolation after joining British forces. During the contagious incubation period, black patients were allowed to wander the countryside, passing through American villages and towns, leaving smallpox in their wake. Some historians believe the British simply did not have the inclination or the resources to care for these individuals. Others, however, believe that this was the deliberate use of a biological weapon by the British to spread smallpox to American citizens and troops. In July of 1763 there was documented discussion among British forces during the French and Indian War of distributing smallpox-infected blankets to the local Native Americans. Whether the plan went into effect was never confirmed, but within six months of the exchange, a violent smallpox epidemic broke out among the local tribes.

The use of infectious diseases as weapons is not innovative. The oldest known use of a biological weapon occurred in the fourteenth century, when in an attempt to conquer the city of Kaffa, the khan of the Kipchak Tartar army ordered the bodies of plague (*Yersinia pestis*) victims catapulted over the city's walls. This event is cited as the catalyst of the Black Death plague, a pandemic that swept across Europe starting in the 1340s and lasting a century. The Black Death is believed to have killed as much as one-third of the European population. During World War II, the Japanese attempted to test the effectiveness of such illnesses as *Y. pestis,* smallpox, anthrax, and typhus as biological weapons through experimentation on an unsuspecting Chinese population. It is not out of the realm of possibility that smallpox, like other infectious diseases, could be weaponized and released, creating a pandemic. Smallpox vaccinations are effective for only 10 years; therefore almost all of the current world population has no immunity to smallpox and would be susceptible to such an attack.

The case of smallpox raises a number of issues concerning diseases that reach epidemic and pandemic levels. The introduction of a non-Western medical procedure by a non-professional, Lady Montagu, created a considerable amount of contention among physicians of the time. Although its long local history in Turkey, as well as its use by Lady Montagu's private physician, indicated that the procedure was successful, it was not until an "official" experiment, executed under the auspices of the London Royal Medical Society and royal physicians, that the procedure was considered both safe and effective. Individuals who sought to practice variolation put themselves at risk of bodily harm from citizens driven by fear and panic. This was the case until local authorities determined that the practice was safe.

In 1796, in an experiment that would never be permitted today, English doctor Edward Jenner purposely injected an eight-year-old boy with cowpox matter obtained from a pustule on a milkmaid's hand. Following this, he attempted to variolate the boy with smallpox. The results were astonishing. Cowpox, a relatively harmless infection passed from cows to humans, provided potent immunity from smallpox. From this experiment emerged *vaccinia virus,* the modern and more effective vaccine for smallpox. Although there were still skeptics, as illustrated by James Gillray's painting *The Cow Pock or the Wonderful Effects of the New Inoculation,* which depicted individuals who were half-human and half-bovine, individuals such as the British royal family who submitted to vaccination. By 1840 variolation was forbidden, and in 1853 vaccination against smallpox in Britain was mandated.

Even with these advancements in prevention, smallpox continued to rage into the twentieth century. According to the World Health Organization (WHO), a subcommittee of the United Nations, by the 1950s there were still 50 million cases of smallpox each year; in 1967 the WHO declared that 60 percent of the world's population was still in danger of being exposed to smallpox, with one in four victims dying.

Controversy continued to surround smallpox well into the twentieth century when an international group of scientists undertook the task of eradicating smallpox from the world permanently. In the 1950s the Pan American Sanitary Organization approved a program that allocated $75,000 annually toward the extermination of smallpox. In 1958 the WHO took over support of the program, but still no action was taken until 1967. At that time the WHO approved $2.4 million for a 10-year program aimed at total eradication of smallpox.

Although scientists involved had the support of several international organizations, contention surrounded their project. Some of the most vehement protests were based on religious grounds. Numerous religions, from Hinduism to Christianity, argued that smallpox was divine intervention and judgment and that humans had no right to interfere. During the WHO's quest to eradicate smallpox, individuals who feared that meddling would cause divine retaliation went so far as to hide those suffering from smallpox or who had not yet been vaccinated by Western doctors, making it extremely difficult to treat all cases as the program required. Others disliked the idea of mandatory vaccination, believing that freedom of choice should prevail. The program was ultimately successful, however, and the United Nations declared the world free of smallpox in 1979.

Even though there has not been a reported case of smallpox in almost 30 years, its well-guarded existence in two government facilities continues to generate attention. Governments and organizations argue over the destruction of the last known smallpox specimens. Those arguing for its elimination cite the potential for accidental release onto an unsuspecting public, as well as the need to create an environment where possession and use of smallpox are considered morally reprehensible. Those who argue for its preservation cite its potential in helping future scientists to understand viruses better and the potential for more effective and safer vaccines. Additionally, they question whether it is morally acceptable for humans to purposefully incite the extinction of another living organism. These questions have been debated for almost three decades, and the debate continues.

Disease and the possibility of epidemics and pandemics emerged at the same time that humans began to give up their hunter-gatherer way of life and settle into large communities and cities. Although the specific name of the disease might be in question, these events have been documented in some way since the beginning of written communication. Controversy over treatment has been widespread. Heated debates over the use of Eastern prevention and treatment methods in Western cultures resulted in new laws, fines, and in some cases arrests. At times religious opposition has helped to spread particular diseases when individuals have refused medical treatment. The use of infectious diseases as biological weapons is always a possibility; although this practice has been roundly condemned by the international community, it continues to create fear in the general public and affects decisions about how to manage a particular disease or virus. Once a lethal or contagious disease has been contained, ethical and moral questions inevitably arise as in how to manage the specimen.

See also Chemical and Biological Warfare.

Further Reading: Carrell, Jennifer Lee. *The Speckled Monster: A Historical Tale of Battling Smallpox.* New York: Plume, 2003; Fenn, Elizabeth Anne. *Pox Americana: The Great Smallpox Epidemic of 1775–82.* New York: Hill and Wang, 2001; Koplow, David. *Smallpox: The Fight to Eradicate a Global Scourge.* Berkeley: University of California Press, 2003; World Health Organization. "Smallpox." *Epidemic and Pandemic Alert and Response (EPR).* http://www.who.int/csr/disease/smallpox/en.

Jessica Lyons

ETHICS OF CLINICAL TRIALS

When new drugs and medical devices are developed, they need to be tested on humans to ensure their safety and effectiveness. Clinical trials—the tightly regulated and carefully controlled tests of pharmaceuticals in large groups of people—raise many ethical challenges. Some of these challenges revolve around the individuals participating in research: Are people being coerced? Are the clinical trials designed appropriately? Are researchers meeting their obligations and behaving ethically? Other challenges are more difficult to address because they are embedded in existing institutional practices and policies: Is it ethical to include or exclude certain groups as human subjects in clinical trials based on their nationality, income, or health insurance status? What are the responsibilities of researchers to human subjects and to communities after the clinical trials have concluded? Still further challenges arise as the location of clinical trials shift from university medical centers to profit-based research centers and as more studies are outsourced to developing countries. The history of abuses to human subjects in the United States has profoundly shaped the range of debates regarding ethical research practices and federal regulation of the research enterprise. Until the 1970s, deception and coercion of human subjects were common strategies used to enroll and retain individuals in medical research. A landmark case of deception and coercion was the U.S. Public Health Service's four decades of research on syphilis in rural African American men in Tuskegee, Alabama. In the Tuskegee study, the subjects were told that they were being treated for "bad blood"—the local term for syphilis—even through they were not actually receiving any treatment. Instead, the U.S. Public Health Service was interested in watching syphilis develop in these men until their deaths, to gain understanding about the natural course of the disease when left untreated. At the start of the study in the 1930s, there were no cures available for syphilis. During the course of the research, however, penicillin was identified as an effective treatment. Still, the men did not receive treatment, nor were they told that they could be cured of their disease.

In response to public outcry following an exposé on Tuskegee as well as other unethical uses of human subjects, the U.S. Congress passed the National Research Act of 1974. This act established the National Commission for the Protection of Human Subjects of Biomedical and Behavioral Research, a group charged with outlining ethical principles to guide research and recommending ways to regulate research. By the beginning of the 1980s, the U.S. government had enacted regulation to protect human subjects from potential research abuses. The regulation requires that all participants provide their informed consent before participating in any study, that the risks and benefits of each study are analyzed, and that all research protocols are reviewed and overseen by external reviewers. Today's institutional review boards (IRBs) are mandated by this regulation. IRBs are research review bodies at universities and hospitals and in the private sector that exist to ensure that researchers are following regulations, obtaining informed consent, and conducting ethical and scientifically rigorous research.

The requirement of informed consent is the primary means of protecting against the deception and coercion of human subjects. Researchers are required to provide detailed information about their studies, particularly about any potential risks and benefits, to all participants in the study. The participants, or their guardians, are required to sign the consent document confirming that they have read and understand the risks and benefits of the trial. Informed consent is meant to ensure that human subjects' participation in clinical research is voluntary. Unfortunately, informed consent has become largely procedural in many research contexts. Though the research trials are often long and complex, human subjects are often informed about the study and prompted for their consent only once, prior to the start of the trial. In response, many bioethicists are calling for a new configuration of informing participants and attaining their consent, a configuration that would treat informed consent as a process that is ongoing throughout the length of a clinical trial.

Although a revision of informed consent may certainly be necessary, it cannot address larger structural issues that must also be examined. Human subjects participate in research for myriad reasons. Some participate out of a belief that research can provide a cure for illness. Others participate because they have limited or no health insurance and can gain access to medicine while participating in the trial. Still others participate in the trials as a source of income through study stipends. These reasons often take precedence over the specific details contained within an informed consent form. In fact, there is currently much debate about the extent to which human subjects should be remunerated for their participation in clinical trials. Because cash incentives may be coercive, many bioethicists argue that the amount of money human subjects receive should only cover costs—such as transportation and parking, babysitters, time off from work—that they incur from their participation. In any case, the current regulatory environment is not structured to respond to the complex reasons that human subjects might have for enrolling in clinical trials.

The ethics of clinical trials extend beyond the voluntariness of human subjects' participation. The design of the clinical trials themselves is also subject to scrutiny for ethical concerns. Nowhere is this more obvious than in discussions about the use of the placebo—or an inert sugar pill with no inherent therapeutic properties—in clinical research. Placebos are valuable tools in clinical research because they provide a controlled comparison to the treatment or therapy being studied. In other words, clinical trials can compare how human subjects' conditions change based on whether they received the treatment under investigation or a placebo. This protocol design becomes problematic because there are instances when it might be considered unethical to give human subjects placebos. Some illnesses should not be left untreated regardless of the scientific merit of the study design. Other illnesses have multiple safe and effective products for treatment already on the market, and some argue that clinical trials should measure investigational products against these other treatments in order to provide the best possible care to human subjects.

In order to determine what is ethical, the medical establishment uses the principle of "clinical equipoise" to guide decisions about clinical trials. Within this framework, the design of clinical trials is considered ethical when the various arms of the study—investigational product, old treatment, placebo, and so on—are considered clinically equivalent. In other words, if researchers have no evidence that the new product is better than a placebo or an older treatment, then it is ethical to compare those groups. If, however, there is evidence that one product might be superior or inferior to another, then it is no longer considered ethical to give human subjects a product known to be inferior. Like many ethical principles, equipoise can be mobilized to guide the design of clinical trials. There are limitations, however, in its application. Importantly, the definition of what evidence counts to achieve equipoise is fairly loose, and the majority of clinical trials that are conducted are done using a placebo. Part of what shapes decisions regarding equipoise and even informed consent is the broader context of clinical trials, especially the funding sources for them. Since 1990 the pharmaceutical industry has shifted the location of clinical trials from university medical centers to private-sector settings, such as private practices and for-profit clinical research centers. While the bulk of most clinical research continues to take place in the United States, the pharmaceutical industry is outsourcing more and more studies to the developing world, including countries in Africa, Asia, eastern Europe, and Latin America. Both within the United States and abroad, the pharmaceutical industry relies on disenfranchised groups to become human subjects because of their limited access to medical care, their poverty, or their desperation for a cure for illnesses such as HIV/AIDS and other infectious diseases requiring treatment. As a result, the pharmaceutical industry's practices regarding human subjects can sometimes be highly exploitative. The ethical dilemma that is created concerns the distribution of risks and benefits. The populations most likely to enroll in clinical trials as human subjects are the least likely to benefit from the results of that research. Debates are currently ongoing about the need for researchers to provide care after the close of clinical trials in order make those relationships more reciprocal. Clinical trials create many ethical challenges. The challenges range from the ethical treatment of individual human subjects to the design and implementation of clinical studies and the distribution of risks and benefits of research within society. The design and conduct of clinical trials has been tightly regulated for several decades, but the changing profile of health care and developments in medical research give rise to new questions. Furthermore, as clinical research continues to grow as a profit-driven industry, ethical questions become increasingly challenging. Although there may not be straightforward or standardized answers to these questions, addressing them should be as important as the medical research that generates the need for clinical trials.

See also Drug Testing; Medical Ethics; Research Ethics.

Further Reading: Applebaum, P. S., and C. W. Lidz. "The Therapeutic Misconception." In *The Oxford Textbook of Clinical Research Ethics*, ed. E. J. Emanuel, R. A. Crouch, C. Grady, R. Lie, F. Miller, and D. Wendler. New York: Oxford University Press, 2006; Faden, R. R., and T. L. Beauchamp. *A History and Theory of Informed Consent.* New York: Oxford University Press, 1986; Halpern, S. A. *Lesser Harms: The Morality of Risk in Medical Research.* Chicago: University of Chicago Press, 2004; Jones, J. H. *Bad Blood: The Tuskegee Syphilis Experiment.* New York: Free Press, 1981; Shah, S. *The Body Hunters: How the Drug Industry Tests Its Products on the World's Poorest Patients.* New York: New Press, 2006.

Jill A. Fisher

EUGENICS

Eugenics was the popular science and associated political movement for state control of reproduction, controversial for its association with the Nazi Holocaust and forced sterilization and racist policies in the United States. In its day it was legitimate science but today it haunts any discussion of controlling fertility or heredity.

Broadly considered, eugenics represented not only the scientific study of human heredity and the potential controls of the heredity of the population, but also the policies that were created based on these scientific principles; because of this dual nature, eugenics remains hard to define. Eugenics was a dominant social, scientific, and political philosophy for thinking about differences in population and public health, and controversial as it was even at the time, it represented the state-of-the-art thinking in the 1920s through 1940s. Despite these difficulties in definition, one thing that eugenicists (scientists, philosophers, politicians, and even Christian clergy) had in common was a belief that reproduction should be controlled based on social considerations and that heredity was a matter of public concern. Although both the set of scientific theories and the associated social movement that aimed at the control of human heredity have since been discredited, they were considered acceptable and scientifically credible in their time and have had a lasting impact. Eugenicists were among those who pioneered in the mathematical evaluation of humans, and their influence in turning biology into the quantitative science it is today should not be underestimated.

The eugenics movement reached the zenith of its influence in the 1930s and 1940s, having influenced public health and population control policies in many countries. Its credibility only slowly faded away, even after being popularly associated with the doctrines of anti-Semitism and genocide of the National Socialist Party in Germany during World War II. Because of this connection to the atrocities of World War II, it is easy to forget the extent to which eugenics was accepted as an important science in the United States, which had enacted policies based on its precepts.

The word *eugenics* (from the Greek for "well bred") was coined by Francis Galton in 1883. It represented his participation in a broad cultural movement focused on breeding and heredity throughout the educated middle class of England and the United States. Galton was inspired to work on evolution and heredity by considering the writings of his cousin Charles Darwin and the economist Thomas Malthus, who both had been key contributors to the popular interest in population-level studies in biology during the nineteenth century. Darwin's theory of evolution stressed the importance of variation within populations, whereas Malthus's work focused on the dangers of overpopulation. From a synthesis of their works, Galton proposed a new science that would study variation and its effect in human populations. Though classification systems based on race and other factors existed, Galton's work advanced and popularized the idea of differing hereditable traits and their potential dangers.

JOSEF MENGELE

Josef Mengele (1911–79) is mainly remembered for his role as the "Angel of Death" in the Holocaust, supervising atrocities at the Auschwitz-Birkenau concentration camp during World War II, and then as a war criminal in hiding. What is less commonly known are his scientific motivations.

Prior to World War II, he had received his medical doctorate and researched racial classification and eugenic sciences in anthropology. Throughout the war, he provided "scientific samples" (largely blood and tissue samples from victims of the camp) to other scientists. Although he is singled out for his personal direction of the deaths of thousands, his participation in a community of scientists who are not considered war criminals remains controversial. Throughout the war, his position in the medical corps of the notorious military service of the SS kept him apart from colleagues at the prestigious Kaiser Wilhelm Institute, but many there, and in the scientific community in the United States, were in communication with him. His torture of prisoners was intended to expand German knowledge of such laudable topics as health and the immune system, congenital birth defects and the improvement of the species.

It is difficult today to balance the dedicated scientist and doctor with the monster capable of cruelties to those he regarded as less than human, but this contrast is often repeated in the history of eugenics and frequently appears in the media when contemporary scientists and doctors seem to cross the line between help and harm.

Although most well-known for his work in eugenics and genetics, Galton was a Renaissance man. He studied and did research in mathematics, meteorology, and geography; served with the Royal Geographical Society; traveled widely in Africa; and was a popular travel writer. His groundbreaking work on statistics is recognized as some of the earliest biometry (or mathematics of biological variation); his work was crucial in the early development of fingerprinting as a criminal science. Although these activities seem disconnected, Galton's commitment to the idea that mathematical analysis and description would provide deeper understanding has lived on in genetics and biology.

The goal of eugenics both as a scientific practice and as a social philosophy was to avoid what was considered to be the inverse of natural selection, the weakening of the species or "dysgenics," literally "bad birth." As humanity became better able to take care of the weaker, and as wars and revolutions were seen to take a greater toll on the elites and the intelligent, the population was believed to be diminishing in quality. The argument suggested that as the physically fit fought in the Great War and in World War II, the disabled remained at home receiving government support, and as the smartest struggled to learn, public schools and factory work allowed the least well-adapted to survive. Similarly, racial and economic differences were seen as promoting higher birth rates among these lower classes, whereas the "better born" were seen to be having too few children in comparison. Contemporary fears about birthrates in the developed world (i.e., Japan, France, and the United States) being lower than the

THE FIRST INTERNATIONAL EUGENICS CONGRESS

Even before the dominance of eugenics at its height in the interwar years, interest was widespread, and the First International Eugenics Congress exemplifies how broad participation was in conversation on eugenics. The congress opened July 24, 1912, a year after the death of Francis Galton, and was presided over by Major Leonard Darwin, the last living son of Charles Darwin. Although his father had carefully stayed away from discussion of eugenics, Leonard was an avid eugenicist, interestingly the only supporter among Charles's five sons, as well as the least accomplished scientist among them. There were more than a thousand registered participants, including luminaries such as Winston Churchill, Thomas Edison, and the Lord Mayor of London. The congress participants argued over new theories and data and the scientific nature and study of heredity as well as the appropriate actions it suggested. Though there was not general agreement on much, there was a shared assumption that some sort of intervention was needed in reproduction and heredity for fear that the weak and undesirable might outbreed the strong and fit.

birthrates in the less-developed world (i.e., India, China, and Latin America) suggest that these fears remain active.

For Galton and other eugenicists, the disparity between who was reproducing and who should be reproducing demanded intervention. Galton envisioned many ways to intervene, but drawing on the metaphor of domestication and breeding of animals that appeared in Darwin's work, Galton favored what would later be called positive, as opposed to negative, eugenics. The positive–negative model is based on the distinction between encouraging the increase of the reproduction of the favored and preventing the reproduction of the inferior. Galton proposed incentives and rewards to protect and encourage the best in society to increase their birthrate. In the end most national eugenics policies were based on the negative eugenic model, aiming to prevent some people from having children.

The control of reproduction by the state has a long history in practice and in theory, appearing in key political works since Plato's *Republic,* wherein the ruler decided which citizens would have how many children, and this history was often cited at the height of popular acceptance of eugenics. Public health, social welfare programs, and even state hospital systems were only beginning to be developed at the middle of the nineteenth century, and among the social and technological upheavals at the end of the nineteenth century, there was an increasingly strong movement to maintain public health through governmental controls, and there was widespread support in the United States for policies that were seen as progressive. In this context, an effort to promote the future health and quality of the population by encouraging the increase of good traits, while working to limit the replication of bad traits, seemed acceptable.

Broad movements throughout Europe and the United States gave rise to the first public welfare systems and stimulated continued popular concern over evolution. Widely held beliefs about the hereditary nature of poverty and other negative traits led to fear that these new social measures would throw off the natural selection of the competitive world. These debates about welfare and the effect on

MARGARET SANGER

Margaret Sanger (born Margaret Louise Higgins, 1879–1966) was a key figure in the birth and population control movement in the first half of the twentieth century. Revered as a central figure in moving the country toward legalizing access to birth control in the United States, she remains a contentious figure for her advocacy of eugenics. Sanger, a nurse, was horrified at seeing women's deaths from botched back-alley abortions. Her sympathy for the plight of women led her to found the American Birth Control League, which would later be known as Planned Parenthood, and open the Clinical Research Bureau, the first legal birth control clinic. A prolific speaker and author, her works include *Woman and the New Race* (1920), *Happiness in Marriage* (1926), *My Fight for Birth Control* (1931), and an autobiography (1938). Although her work on birth control would have been enough to make her contentious, her political support for eugenic policies such as sterilization has led to a fractured legacy, and these beliefs are frequently used as a reason to discredit her more progressive ones. She died only months after the federal decision in *Griswold v. Connecticut* officially protected the purchase and use of birth control in the context of marriage for the first time.

---

the population today still stimulate concern among citizens of the United States and elsewhere.

Because of popular acceptance and its utility in justifying a range of policies, eugenic science was agreed upon by a wide array of notables who might otherwise have been on different sides of issues. Among those who advocated some form of eugenic policy were President Franklin Delano Roosevelt, the Ku Klux Klan, and the League of Women Voters.

The complex relationship many public figures had with eugenics stems in part from the usefulness of using it as a justification because of its widespread support. Birth control advocate Margaret Sanger publicly supported a rational version of negative eugenics but may have done so only for the credibility she gained in doing so. She and other advocates for access to birth control were taken much more seriously by policy makers because they connected the issue with the more popular eugenics movement. In this light, Sanger's suggestion that the upper classes were able to get birth control despite the laws and that there was a need to change the laws to slow the breeding of the poor, who were unable to attain birth control, may be seen as a political as opposed to ideological choice.

Eugenics organizations and political movements were started in Germany in 1904, Britain in 1907, and the United States in 1910. At the height of the era of eugenics, there were more than 30 national movements in such countries as Japan, Brazil, and others throughout Europe. In some countries coercive measures were rejected, and in others policies were more limited, but in each country the adoption of national eugenics programs and popular movements represented an attempt to modernize and adopt scientific methods for advancing the health and well-being of the populace as a whole. Even the most notorious case of eugenics, the Nazi Germany eugenics program, was associated with discussion of the "greater good." It becomes easy to forget that the Nazi obsession

THE JUKES AND THE KALLIKAKS

Richard L. Dugdale's 1874 book *"The Jukes": A Study of Crime, Pauperism, Disease and Heredity* and Henry Herbert Goddard's 1912 account *The Kallikak Family: A Study in the Heredity of Feeble-Mindedness* are key examples of what were known as family studies, powerfully convincing stories of the danger of bad heredity that were widely circulated in the first half of the twentieth century. Both stories follow the troubles of the members of a family and the passage of harmful traits generation to generation. Dugdale was a progressive, and in the case of the Jukes family, he suggested the problem family was one that demanded rehabilitation, whereas Goddard was more closely associated with the eugenics movement, and he saw the problem as one of prevention. These stories were very important in the early days of genetics because they were influential in popularizing the heritability of traits regardless of environment. The comparison in the tale of the Kallikaks between the branch of the family who had been infected with the bad trait and their still pure and good relations resonated and spread the idea of a trait widely. In the end neither story has stood up to scrutiny, as historians have revealed manipulations and fabrications at their sources, but their influence is lasting nonetheless.

with a healthy nation led not only to genocide but also to national campaigns for healthy eating and the elimination of criminal behavior.

The German eugenics laws were capped by the three Nuremberg Laws in 1935 that signaled the beginning of the Nazi genocide, aimed at "cleansing" the German nation of bad blood through negative programs including sterilization and executions, while also promoting increased reproduction of those with good blood in positive eugenics programs. The Nazi German eugenics program sterilized nearly 400,000 people based on the recommendation of the Genetic Health and Hygiene Agency for what were considered hereditary illnesses, such as alcoholism and schizophrenia. Probably the most notorious manifestation of positive eugenics on record was the Nazi program that paired SS soldiers with unmarried women of good blood to increase the birthrate for the benefit of the nation.

The U.S. program was already underway when the German eugenics program was still beginning, and though the state governments in the United States eventually sterilized fewer people, they were used as a model by the German program. The center of the eugenics movement in the United States was the Eugenics Records Office (ERO) located at Cold Springs Harbor Research Center in New York. The ERO published the *Eugenical News,* which served as an important communications hub and was considered a legitimate scientific publication. By the late 1930s more than 30 states had passed compulsory sterilization laws, and more than 60 thousand people had been sterilized. In 1937 more than 60 percent of Americans were in favor of such programs, and of the remainder only 15 percent were strongly against them. In discussions of sterilization, a common consideration was the growing system of institutions and their populace. Sterilization was seen as a humane and cost-effective remedy for problems such as alcoholism when compared with lifelong incarceration, and these programs remained a key influence on the development of outpatient treatment for the mentally ill until well into the 1970s.

If there is any practice distinctly associated with the American eugenics movement, it is coerced and forced sterilization. Although Nazi German doctors performed these procedures in far greater numbers, in light of the Holocaust it loses its impact, but in the United States this same procedure remains shocking. Many of those sterilized were residents of mental hospitals and poorhouses who were forced to undergo the procedure. Others were voluntary or temporary patients at state hospitals. It is difficult to know how many sterilizations were performed and yet more difficult to confirm what percentage of those were coerced. Some patients intentionally sought sterilization as a form of birth control; others chose it as an avenue out of institutionalization; some percentage were tricked or forced. Today documents show that some institutions told patients who were to be sterilized that they were going to have their appendix removed, and in these and other institutions, we can see high rates of appendectomies. Forced or coerced surgery on a single individual today would seem shocking, but they were legally mandated in some states for more than 50 years, and because those most likely to have been sterilized were the mentally ill and the indigent, we are likely never to know the full story.

Numerous court decisions challenged the legality of state sterilization, and although several state laws were struck down in court, the Supreme Court decisions in two key cases upheld what was considered a legitimate state interest. In the 1927 case *Buck vs. Bell,* the Virginia statute requiring sterilization practices was upheld by the U.S. Supreme Court, and Chief Justice Oliver Wendell Holmes infamously wrote in the decision that the law was necessary because "three generations of imbeciles is enough." Carrie Buck, the plaintiff in the case, had been certified "feebleminded," as had her mother. When Carrie's daughter was "tested" at the age of one month and declared to be "feebleminded," Carrie did have the presence of mind to question the diagnosis and did not want to be sterilized. The court decision came down against Carrie. Although it was not publicized at the time, Carrie Buck's daughter received further intelligence testing when she was in her early teens and was determined to have above-average intelligence. Whereas many countries slowly rescinded eugenics laws over the course of the second half of the twentieth century, in others the laws remain on the books without implementation. The United States and most of the Scandinavian countries are among those nations that never officially eliminated their eugenics laws, and many others still have public health and hygiene laws from the eugenics period that have simply been modified.

From the 1890s until the late 1930s, a series of laws intending to limit the entry of immigrants into the United States was associated with eugenics, and the laws became increasingly harsh. Though these laws were widely popular among some groups, their explicit racism and isolationism became a growing source of concern for others. This legal link between eugenics and racist immigration policy was associated with the earliest anti-eugenics responses. Eugenics had initially been associated with the public good and reform, but this association too was tarnished by accusations of racism. Growing segments of the population recognized eugenics as biased against the poor, as non-eugenic reformers made social conditions of poverty public and advocated for institutional reform rather than hereditary control of poverty.

In the United States in the late 1930s, in light of the growing upset about the association between eugenics and racism, reformers tried to shift the eugenics movements to a more moderate stance, and many mainstream eugenics groups moved away from hard-line positions. By the late 1940s the increasing public awareness of Nazi atrocities pushed public opinion even more against eugenics, and the word started to lose its respectability. Eugenics laws were reframed by being called hygiene or public health laws. Many of the reform eugenicists joined other scientists working in the nascent field of genetics as it was forming, and some were founding members of the American Society of Human Genetics when it was formed in 1948. Although the growing anti-eugenics sentiment slowly turned eugenics from a dominant scientific field into a discredited memory, scientists who had worked on heredity as eugenicists embedded their study of hereditary diseases and mental and moral traits within Mendelian genetics.

Throughout the rise of eugenics, there was no clear understanding of the mechanism of inheritance within the intellectual community. Although today we have a scientific consensus on the workings of the cell and the importance of DNA, there was little known about the inner workings of reproduction and development at the turn of the century. Gregor Mendel (1822–1884) was a Czech monk and biologist whose experimental breeding of pea plants led to his developing a series of scientific laws regarding the segregation, parental mixing, and transfer of traits. The rediscovery and popularization of the work of Mendelian genetics offered an explanation based on finite internal properties of the cell, which appealed to some, but its laws did not appeal to Galton or many eugenicists who saw it as applying only to simple traits such as plant color. The emphasis in Galton's view was on formal Darwinism, the rate of reproduction, and the role of environment and external factors in sorting the fittest and removing the weak. Mendel's theory is no longer associated with eugenics, in part because one of its strongest supporters, geneticist Thomas Hunt Morgan, opposed eugenics, but many other key scientists involved in promoting the acceptance of Mendel's work were doing so because it so clearly defined heritability. It was a powerful argument for the lasting and finite specification of heritable traits, and it worked with the idea of eugenics, whereas other theories argued for more environmental impact and flexibility. Although today there is reason to believe that Mendel's laws oversimplify a more complicated phenomenon, the rediscovery and embrace of these ideas by eugenic science was instrumental in the founding of genetics.

In the early 1970s, around the time the last of the eugenics laws were enacted and only a few years after the latest forced sterilizations in the United States, references in popular press, media, and news sources that suggested a genetic cause for mental and moral defects were at an all time low. In the last 30 years, there has been a steady increase in the popular awareness of and interest in genetics and a dramatic resurgence of reference to genetic causes of traits. Between 1975 and 1985, there was a two hundred–times increase in public references that suggested a genetic cause for crime, mental capacity or intelligence, alcoholism, and other moral and mental traits that had been central concerns under eugenics. This increased by four times by the early 1990s and has not decreased. These issues are magnified today in areas where population growth adds to economic and social pressures. Where the use of technology for sex selection and choice of appropriate qualities of one's offspring becomes more active, it leads to controversy. In India and China, the perceived need to extend control to practices and technologies of heredity has garnered accusations of a new eugenics in media coverage.

Lasting interest and study of eugenics is due to its connection to two perennial questions. First, it asks how much of and what parts of who we are come from our heredity, often described as the debate between nature and nurture, and second, how a society should determine, react, and respond to undesirable traits of individuals. These two questions are interlinked in that a trait that is learned may be unlearned, but biological traits have been assumed to be innate and unchangeable, leading to different sorts of responses from society and law.

Today major news sources and media outlets eagerly publicize front-page stories on new scientific findings based on a widespread interest in genetics and biological traits such as "gay genes" causing homosexuality or "alcoholic genes" passed along from father to son, but few place the corrections and negative evaluations of these findings in view when they are discredited. Stories run about genes that cause diseases such as breast cancer, without discussing any connection to what can be done in response to these discoveries or their connection with the discredited science of eugenics. Little discussion takes place about why these genes are looked for or what good knowing about them does in a culture that emphasizes individual accomplishment as surpassing heredity in determining one's life story.

We do not often ask how a history of eugenics has contributed to the demand for genetic explanations and medical testing today, but the idea of heredity, of unchangeable inherited traits, continues to hold particular power despite, or because of, its importance at the founding of genetics. One explanation is to be found in the American ethos and legends of the self-made individual. The idea that all people start from a clean slate is ingrained into American society, and the American dream of the ability of anyone to work hard and get ahead is challenged by the failure of so many hard workers to get ahead. The persuasiveness of inherited cause for success or failure shifts the discussion away from systemic environmental constraints on success, such as racism, sexism, and class, allowing the focus to remain on the individual. Another concept frequently connected to eugenics and to contemporary genetics is the idea of the easy solution, as exemplified in the lasting presence of the 1950s "better living through chemistry" mentality of the single-drug cure. How much easier to imagine fixing one gene, one trait, than to think through the myriad of causes that might otherwise contribute to something we want to change.

With the successes and promises for the future of molecular biology and genetic engineering, we are offered new avenues and a new reason to rekindle interest in heredity. The eugenicists believed that heredity was important as a predictive and evaluative tool but did not have the means to alter the traits they attempted to study, whereas contemporary innovations promise to offer the potential to act upon those traits determined to be harmful.

Today approximately 1 in every 16 babies in the United States is born with some birth defect, and although the impacts range in severity, the common conception is that any abnormality or defect creates a victim and represents part of a public health problem. Thinking about the victims of genetic disease, it is very tempting to consider a return to state control or even a voluntary eugenics where parents make the choice presented by their doctor. It is this eugenics of choice that has emerged today. As prenatal tests have been improved and are more widely practiced, they are sometimes compared to eugenics. Amniocentesis, in which genetic testing of unborn babies is performed, has been frequently connected to this history because for most anomalies found there is no treatment, leaving parents only with the choice to abort or not to abort. Abortion has been connected to eugenics since Margaret Sanger and others championed birth control legalization at the turn of the century. Medical methods of abortion have gotten more sophisticated, but fertility control methods have been a presence in most human societies in one form or another and always involve the question of what sort of person the child will be and what sort of life the child will have. Explicit mentions of eugenics in contemporary discussions of abortion appear on both sides: pro-choice advocates are concerned about excessive government control of fertility, and antiabortion activists attempt to use eugenic associations with abortion to compare it to the Holocaust. The language of eugenics is used on both sides to discuss the differential access and use of abortion between the wealthy and poor, between black and white, as questions of what sort of people are having abortions, discouraged from having or encouraged to have children.

The hygiene laws of the first half of the century have faded, and today public health regulations in many states require blood tests before marriage so that couples may be better prepared to choose in having children when they carry some traits. But who decides what traits are to be tested for? If the core of eugenics was a belief that society or the state has an interest in heredity, do we still practice eugenics?

Contemporary premarital blood-test regulations parallel some of the aims and content of the eugenic hygiene laws, though frequently the underlying motivation may be different. In the early part of the twentieth century, these rules were enacted based on eugenic arguments against urbanization and growing populations of immigrants and poor and on notions of social purity that we no longer articulate. In recent years, fear of HIV/AIDS and conceptions of personal risk may have taken their place. More than 30 states have evaluated legislation requiring premarital HIV screening, and states including Illinois, Louisiana, Missouri, and Texas made them the law. Although later concerns over privacy and the damage done by false positives led all these states to eliminate the laws, some of the state laws had gone so far as to ban marriage for those who had

AIDS, and while the fear at the heart of this social crisis passed, we cannot say what is yet to come. Neither were these HIV/AIDS laws unusual; many states still require blood tests for other diseases to receive a marriage license, and in an echo of eugenics, some regulations exempt those who are sterile or prevent marriage until treatment for sexually transmitted diseases has been received.

How will recent court decisions that have legally limited parental rights during pregnancy, for instance criminalizing drug use as child abuse, be expanded as society maintains its claim on control of fertility and heredity, and through them the definition of acceptable people in society?

See also Cloning; Genetic Engineering; Reproductive Technology; Research Ethics.

Further Reading: Curry, Lynne. *The Human Body on Trial: A Handbook with Cases, Laws, and Documents.* Santa Barbara, CA: ABC-CLIO, 2002; Duster, Troy. *Backdoor to Eugenics.* New York: Routledge, 1990; Engs, Ruth C. *The Eugenics Movement: An Encyclopedia.* Westport, CT: Greenwood Publishing Group, 2005; Forrest, Derek Williams. *Francis Galton: The Life and Work of a Victorian Genius.* New York: Taplinger, 1974; Gould, Stephen Jay. *The Mismeasure of Man.* New York: Norton, 1981; Kerr, Anne, and Tom Shakespeare. *Genetic Politics: From Eugenics to Genome.* Cheltenham, UK: New Clarion Press, 2002; Kevles, Daniel J. *In the Name of Eugenics: Genetics and the Uses of Human Heredity.* New York: Knopf, 1985; Knowles, Lori P., and Gregory E. Kaebnick. *Reprogenetics: Law, Policy, Ethical Issues.* Baltimore, MD: Johns Hopkins University Press, 2007; Paul, Diane B. *Controlling Human Heredity: 1865 to Present.* Amherst, NY: Humanity Books, 1995.

Gareth Edel

Eugenics: Editors' Comments

It is interesting to note that eugenicists were prominent in the development of the concept of "profession" in sociology. Sir Alexander Carr-Saunders (1886–1966) was one of these eugenicist-sociologists. The model of the profession has many analogues to eugenics, bringing into play certain social practices and policies for distinguishing experts from nonexperts, acceptable authorities from unacceptable ones. It is as if upon recognizing that it was impossible scientifically and politically to implement eugenics programs, some eugenicists tried to develop a social eugenics that would accomplish the same sorts of classifications and discriminations they could not pursue biologically or genetically.

F

FATS

Butter and vegetable oils are some of the most important products of American agriculture. At one time such fats were uncontroversial: they were simply part of every meal, in the form of butter, lard, and various vegetable oils and in meats and some vegetables. But during the 1970s and 1980s, the public began digesting the conclusions of nutrition scientists: saturated fats from animal sources, in foods such as lard, red meats, and butter, appeared to greatly increase the chances of developing heart disease and strokes if consumed in large quantities. This caused a major consumer backlash against animal fats and prompted the food industry to develop alternatives. Soon partially hydrogenated vegetable oils, produced through an industrial chemical process that adds extra hydrogen atoms to the fat molecule, were being produced in huge amounts to fill the role formerly played by saturated fats such as lard and beef tallow. Science and progress seemed to have saved the day, replacing dangerous natural fats with a safer industrially altered fat. This caused many people to sigh with relief, stop buying blocks of butter, and switch to margarine and also caused fast food chains to abandon tallow and other animal fats and switch to partially hydrogenated vegetable oils for deep frying.

It became clear in the late 1980s and early 1990s, however, that the partially hydrogenated oils, which like some naturally occurring animal fats are referred to by food scientists as "trans fats," can be equally damaging to human health, causing heart and artery problems just like the supposedly bad fats they replaced. Now science seemed to have developed, and the industrial food production system seemed to have promoted, fats that killed people.

With the addition of decades of preaching by some nutritionists against fats in general, fats themselves often came to be seen as health dangers, as something to be avoided at all costs. While a general anti-fat backlash was developing, medical and scientific knowledge kept evolving, offering a more complex understanding of the role of fats in the diet and helping develop new products that could apparently avoid the dangers of both saturated fats and partially hydrogenated fats.

Nutritionists began to understand that a certain amount of fat is necessary for the human body to properly develop and function. Restricting all fat from the diet began to appear to be more dangerous than consuming limited amounts of "good fats."

Some ancient vegetable oils, such as flax oil, sunflower oil, and olive oil, began to be seen as "good fats" because they contain low saturated fat levels and contain other beneficial compounds. These oils, previously available mostly in gourmet restaurants and foods and in health food stores, exploded in popularity in the 1990s and became commonly available. A relatively new oil—canola—came to be seen as among the healthiest because of its extremely low saturated fat level. Canola appeared to be a triumph of science because it was created by university researchers in Canada who managed to create a healthy oil out of the seeds of the rapeseed plant, which had formerly produced oil fit only for industrial uses because of high amounts of inedible oil content. (Rapeseed oil is still commonly consumed by people in China.) Traditional crop breeding techniques were used to develop edible rapeseed oil, and the plants producing it were renamed "canola." After the anti–trans fat campaign began gaining momentum, most notably with the New York City ban on restaurant use of trans fats implemented in 2007, canola oil once more gained prominence—and scientists could again claim a triumph—because specialized varieties of the crop were developed that produced oils that could simply replace partially hydrogenated oils in restaurant deep fryers with little effect on taste. Soybean breeders also began developing "high stability" types of soybean oil.

In recent years proponents of the much-maligned red meat oils—butterfat, beef fat, and so on—have begun making a comeback. Nutritional research began revealing that certain elements in butter, beef fat, and other animal-based oils contained compounds beneficial to human health. While accepting the dangers of too-high levels of saturated fats, dairy and red meat producers could plausibly claim that the fat in their products could be seen as healthy foods.

Although there have been wild gyrations in the views of nutritionists about which fats are healthy and which are not and wild swings of opinion about how much fat should be included in the human diet, the evolving science of nutrition appears to have found important places in the diet for both scientifically developed and longtime naturally occurring fats. As with much progress, there is sometimes a "two steps forward and one step back" phenomenon. After the anti–trans fat campaign became widespread and food processors and fast food restaurants moved quickly to abandon trans fats, while some processors and chains adopted oils such as sunflower and canola, others moved to cheaper palm oil. Palm oil does not contain trans fats—but is extremely high in saturated fat.

For part of the food industry, the anti–trans fat campaign meant a return to higher saturated fat levels in their products.

See also Obesity; Cancer.

Further Reading: Mouritsen, Ole G. *Life–As a Matter of Fat; The Emerging Science of Lipidomes.* New York: Springer-Verlag, 2005.

Edward White

FOSSIL FUELS

In the understanding of the material world provided by physics, energy is defined as the ability to do work. Work in this sense means the displacement of an object ("mass") in the direction of a force applied to the object. In everyday life, we use the word *energy* more generally to indicate vitality, vigor, or power. Here, we focus on another definition of energy: a source of usable power. Our homes, our industries, and our commercial establishments from the supermarket to the stock exchange can work because they are provided with sources of usable power. The same is true for hospitals and medical services, schools, fire and police services, and recreational centers. Without usable energy, TVs, the Internet, computers, radios, automobiles and trucks, construction equipment, and schools would not work. To put it transparently and simply, energy is important because doing anything requires energy.

There are many sources of usable power on our planet, including natural gas, oil, coal, nuclear energy, manure, biomass, solar power, wind energy, tidal energy, and hydropower. These energy sources are classed as renewable or nonrenewable. For renewable energy (for example, wind, biomass, manure, and solar power), it is possible to refresh energy supplies within a time interval that is useful to our species. If well planned and carefully managed on an ongoing basis, the stock of energy supplied is continuously renewed and is never exhausted. The fossil fuels are forms of nonrenewable energy. For each, the planet has a current supply stock that we draw down. These nonrenewable supplies of fossil fuels are not replaceable within either individual or collective human time horizons. It is not that the planet cannot renew fossil fuels in geologic time—that is, over millions of years—but for all practical purposes, given human lifespan and our limited abilities, renewal is far beyond our technical and scientific capabilities.

For almost all of human history, our species used very little energy, almost all of it renewable, usually in the form of wood fires for heating and cooking. If we had continued in that fashion, the human population might easily be in the low millions and would be approximately in a steady state in relation to the environment. As the first animal on this planet to learn how to use fire, we were able to establish successful nomadic tribes, small farm settlements, early cities, and the kinds of kingdoms that were typical of medieval and preindustrial civilizations. The associated rise in human numbers and our spread across the planet was also associated with the early use of coal, for example to make weapons. But, it is only with the rise of industrial civilization, and specifically and increasingly in the last five hundred years, that we learned how to use concentrated forms of energy in industrial processes and to exploit various nonrenewable forms of energy massively for uses such as central heating, industrial processes, and the generation of electricity.

The rise of our current global civilization was dependent on abundant and inexpensive concentrated energy from fossil fuels. All of this energy ultimately derives from sunlight, but as we learned how to use coal (replacing wood as fuel) and then oil and natural gas, there was little concern for energy conservation or for developing rules for limiting energy use. Up until the middle of the last century, and somewhat beyond, the typical discussion of energy would have linked usable energy with progress, as is the case today. The spirit of the presentation would have been celebratory, however, celebrating the daring risks taken and the hard work of miners and oil and gas field workers in dominating nature to extract resources. In addition, it would have celebrated the competence of the "captains of industry" whose business skills and aggressive actions supplied energy to manufacturing industry, and would have implied that this pattern of resource exploitation could go on forever without taking limits into account. Today that era of celebration belongs to the somewhat quaint past, and we are now much more aware of the cumulative damage to the environment from aggressive exploitation of limited fossil fuel resources. We now know that we face an immediate future of global warming, shortages of usable energy, and rising prices. From a material perspective, the planet is a closed system, and the dwindling stocks of nonrenewable but usable energy are critically important. For each fossil fuel, what is left is all we have.

There is currently no social convention to limit the use of nonrenewable energy to essential production or essential services. Under the rules of the neoliberal market system, resources are provided to those who have the ability to pay for them. This is the kind of human behavior that an unregulated or weakly regulated market system rewards. Because the stocks of fossil fuels took millions of years to create, the ability to extract them is inherently short-run when there is no strong social planning to provide for a human future on other than a very short-range basis. We commit the same error with fossil fuels that we commit with fish stocks—as ocean fish dwindle in numbers, and species after species significantly declines, the main response has been to develop more and more efficient methods and machines to kill and extract the remaining fish. The same is true with fossil fuels. As abundance disappears, and the cost of extraction continues to increase, the primary response has been to find more efficient methods of extraction and to open up previously protected areas for extraction. As a general pattern, Georgesçu-Roegen and others have pointed out that resources are exploited sequentially, in order of concentration, the easy sources first. After the easy sources of fossil fuels are exhausted, moderately difficult sources are exploited. Then more difficult sources are exploited. Each more difficult source requires the input of more energy (input) in order to extract the sought-after energy resource (output).

In the material world, the process of the energy extraction from fossil fuels requires more and more input energy. And as extraction proceeds to more difficult sources, it is also associated with more and more impurities mixed in with the energy resources. These impurities are often toxic to our species (and other species). Examples include the acidic sludge generated from coal mines and the problem of sour gas in oil and gas drilling (sour gas contains hydrogen sulfide and carbon dioxide). As more and more input energy is required per unit of output energy, we also need to do more work with more and more impurities and toxic waste. Remember now that from our species standpoint the planet is a closed system with respect to nonrenewable forms of usable energy. In physics, the change in internal energy of a closed system is equal to the heat added to the system minus the work done by the system. In this case, *more energy has to be added* to develop a unit of energy output, and more and more work has to be done. For example, as coal, gas, and oil become harder to reach, increasing gross amounts of waste materials are generated.

Beyond this, all of our processes for extracting energy from fossil fuels are inefficient in that energy is lost in the process of doing work. In physics, the measure of the amount of energy that is unavailable to do work is called entropy. (Entropy is also sometimes referred to as a measure of the disorder of a system.) Georgesçu-Roegen and others have developed a subfield of economics based on the priority of material reality over conventional economic beliefs. The fundamental insight grounding this subfield of economics is that the earth is an open system with very small (residual) usable energy input. So, like a closed system, it cannot perform work at a constant rate forever (because stocks of energy sources run down).

So if we look at the extraction of energy from finite stocks (of coal, oil, or natural gas), the extraction process must become more and more difficult per unit of energy extracted, become more and more costly per unit of energy extracted, and generate more and more waste per unit of energy extracted.

This understanding, which follows from physics and the nature of the material reality of the planet, does not fit with the conventional capitalist economic theory that currently governs world trade, including the extraction of energy resources. Market economics, sometimes called the "business system," typically advises arranging life so as not to interfere with the operations of markets. This advice comes from a perspective that regularly disregards the transfer of "externalities," costs that must be suffered by others, including pollution, health problems, damage to woodlands, wildlife, waterways, and so on.

Conventional economic thinking employs economic models that assume undiminished resources. That is why it seems reasonable to advise more efficient means of extraction of resources (e.g., with fish and coal) as stocks of resources diminish. Another feature of conventional economic thinking is that it (literally) discounts the future. Depending on the cost of capital, any monetary value more than about 20 years in the future is discounted to equal approximately nothing. These features of conventional economics mean that the tools of economic calculation operate to coach economic agents, including those who own or manage extractive industries, to act for immediate profit as if the future were limited to the very short term.

This is in contrast to a material or engineering viewpoint, the perspective of community-oriented social science, and the humane spirit of the liberal arts. All of these are concerned not simply with the present but with the future of the human community and with the quality of human life and of human civilization in the future as well as today. Outside of the limited focus of conventional economics, most disciplines place a high value on the quality of the human community and sustaining it into the distant future. Practical reasoning in everyday life often puts a higher value on the future—most of us would like things to get better and better. One way to understand this difference is to contrast the interest of today's "captains of industry" with the perspective of a student finishing secondary school or beginning college, just now. For the "captains," everything done today has a certain prospect for short-term profit, and the future is radically discounted (progressively, year by year) so that 20 years out, its value is essentially zero. For the student, the point in time 20 years out has a very high value because the quality of life, the job prospects, the environment (including global warming), the prospects for having a family, and the opportunities for children 20 years out will be of direct personal relevance. The student might argue that the future is more important than today (and should be taken into account without discounting), as would most families that would like a better future for their children. Today's student has a strong interest in having usable energy resources available and the disasters of global warming avoided or lessened. Conventional market economics does not do this; it takes strong regulation, strong central planning, and an engineer's approach to nonrenewable resources to best use and stretch out resources for the future, rather than a conventional economist's approach.

Currently the growth curve of the planetary economy continues to increase. India and China are undergoing rapid economic growth, and the Western economies continue to follow traditional consumption patterns. Capitalist strategies abound in these economies; companies make money by engineering built-in obsolescence into their products. Not only does this require regularly replacing products with new or upgraded versions; it also leaves openings for replacing obsolete products with entirely new lines of products. The computer industry offers numerous examples of this obsolescence imperative. The demand for products of all kinds is soaring in comparison with past decades or centuries. At the same time the human population has increased dramatically over past centuries. All of this requires more and more energy.

Current industry projections for fossil energy suggest that there may be about 250 more years of coal, 67 years of natural gas, and 40 years of oil. These kinds of industry projections change from year to year and are much more generous than projections made by independent university scientists and conservation groups. Several scientists believe we have passed the time of peak oil. The point here, however, is not the specific numbers (it is easy to find more on the Internet) but that these numbers provide a rough indication of remaining stocks. Also, note that the optimistic industry projections are not for millions or thousands of years into the future. From your own perspective, if you knew there were perhaps 250 years of coal left or 40 years of oil, would you want fossil energy carefully rationed for specific uses that cannot be easily met by renewable energy (so that it might last one or two thousand years)? This is an alternative to the current system of neoliberal market rules that destroy or weaken the institutions of social planning in many small states. Coal, oil, and natural gas are forced onto world markets (by military force, if market pressures and diplomacy do not suffice) with ever more intense extraction for use by those who can afford it (to use as quickly as they like). Which policy is best for you, your family, your community, your country, and the world?

What makes the number of years of remaining stock estimates tricky is that sometimes new resources are found (though this does not happen much anymore), new technical improvements can sometimes increase extraction, and the more optimistic projections tend to use bad math. That is, sometimes the math and statistics fail to take into account factors such as dwindling supply with more and more difficult access, increased percentage of impurities mixed into remaining stocks, increased waste streams, and the entropy factor. When we interpret these estimates, we need to keep in mind that it is not simply that we will "run out" of coal and oil but that remaining stocks will become more and more expensive to extract.

Energy is important because doing anything requires energy. Any area of human civilization largely cut off from fossil fuels (oil, natural gas, or coal in current quantities) will fail to sustain human carrying capacity. Jobs will be lost, businesses will have to close down, and home energy supplies for heating, cooling, and cooking will become sporadic as energy costs spiral beyond people's means. As a secondary effect, the same thing happens to food supplies that are gradually made too costly for increasing numbers of people.

We are currently watching income in the lower and middle to upper-middle sections of society decrease or not increase. By contrast, income in the upper 1 and 5 percent of households is growing rapidly. We are witnessing, in other words, a resurgence of a class division similar to that of the Middle Ages, with a relative handful of privileged households at the apex (enjoying access to usable energy and food supplies) and a vast surplus population and marginalized population of different degrees below them. We have a choice in planning for a long and well-balanced future for the human community in our use of fossil fuel stocks or continuing with neoliberal economics and conventional market rules (supported by military force), which will allow small elites to live well for a while and surplus most of the rest of us.

As important as they are, conservation and renewable energy are insufficient to countervail this future unless we make significant changes in lifestyle and gently reduce the number of humans to a level close to that sustainable by renewable technologies. This will take more mature thinking than is typical of the business system or of conventional market economics. In particular, we need an economics in which beliefs are subordinated to the realities of the physics of the material world.

See also Biodiesel; Coal; Geothermal Energy; Nuclear Energy; Sustainability; Wind Energy.

Further Reading: Beard, T. Randolph, and Gabriel A. Lozada. *Economics, Entropy and the Environment; The Extraordinary Economics of Nicholas Georgesçu-Roegen.* Cheltenham, UK, and Northampton, MA: Edward Elgar, 1999; Jensen, Derrick, and Stephanie McMillan. *As the World Burns; 50 Simple Things You Can Do to Stay in Denial, A Graphic Novel.* New York: Seven Stories Press, 2007; McQuaig, Linda. *It's the Crude, Dude: War, Big Oil, and the Fight for the Planet.* Rev. ed. Toronto: Anchor Canada, 2005; Odum, Howard T., and Elisabeth C. Odum. *A Prosperous Way Down; Principles and Policies.* Boulder: University Press of Colorado, 2001. For statements and analysis by leading scientists and analysts relating to peak oil and the general problems of current patterns of use of the limited stocks of fossil fuels, see http://dieoff.org.

Hugh Peach

GAIA HYPOTHESIS

The Gaia hypothesis proposes that earthly life has evolved in coexistence with the environment, to form a complex geophysiological system, or "superorganism," able to reestablish homeostasis when unbalanced, much the way bees cool a hive. Life produces and maintains the environmental conditions that promote more life. First proposed by English atmospheric chemist James Lovelock in 1968, the hypothesis met with vociferous opposition for being unscientific and teleological and for presuming that planetary biota (terrestrial life) had anthropomorphic-like foresight and planning abilities to create their own environment. Today, the premise that the earth's planetary biosphere and atmosphere belong to a single complex system is generally accepted, but the efficacy of feedback mechanisms and the stability of long-term planetary temperature and carbon dioxide levels are debated.

Gaia is based on the premise that the metabolic processes associated with life-like respiration, nutrient ingestion, and waste production facilitate the circulation of materials and chemistry in the environment on a global scale. By-products of life's metabolism effectively determine the composition and concentration of elements in the atmosphere, soil, and water. Moreover, the composition of the atmosphere, oceans, and inert terrestrial surfaces with which living organisms exchange metabolic gases and by-products are modulated by feedback mechanisms that ensure the continuation of favorable conditions for life. Empirical evidence gathered since Lovelock's proposal has lent the theory credibility and has contributed useful data for understanding the increase in atmospheric carbon dioxide and its relationship to climate change.

For Gaian theory, life is not separate from the geophysical elements of Earth, a concept that resonates in interdisciplinary sciences such as geophysiology, and Earth systems science. The latter, according to Lovelock's longtime collaborator, microbiologist Lynn Margulis, is identical to Gaian research with the "unscientific" language removed. Named after the ancient Greek Earth goddess, the Gaia hypothesis alludes to animistic beliefs in a living earth, a humanistic-cultural association popularly embraced as a techno-ethical guide. A Gaian framework demonstrates how scientific specializations obscure complicated systems of feedback loops and homeostatic processes. Accumulating geo-evolutionary and biological evidence of organismic strategies such as symbiosis and mutualism within colonies of protoctists and microbial cells challenges the neo-Darwinian emphasis on competing gene pools and puts life on a more social foundation. Linked to the geological sciences, biology moves into a position of scientific centrality once occupied by physics. Gaian science introduces new debates from the scale of natural selection to the role of human activity in global climate change.

Evolutionary geologist James W. Kirchner claims that Gaia, which is impossible to verify, is not even really a hypothesis. Science progresses through the proposition and testing of theories, but because Gaia-influenced research emphasizes the importance of complex cybernetic systems of which life is only a part, it turns away from studying organisms in isolation, troubling science with the limits encountered in a system too large to allow full comprehension. Kirchner points out that we cannot turn lights on to illuminate this "stage," in order to be scientific, that is, to attain the needed distance for objective observation.

James Lovelock came to the Gaia hypothesis while looking for life on Mars for NASA. He compared Earth's atmosphere to that of Mars. Lovelock was struck by the differences; the Martian atmosphere, composed largely of carbon dioxide, was chemically inert and stable and could be entirely understood through chemistry and physics. On a so-called dead planet, all potential gaseous reactions have already been exhausted. The Earth's atmosphere contained reactive gases such as oxygen (21%) and nitrogen (79%) and smaller amounts of argon and methane and was liable to volatile reactions and disequilibrium. Based on the anomalies Lovelock found when analyzing Earth, he accurately predicted there would be no life on Mars.

Seeking to learn why, despite this gaseous volatility, the Earth's atmosphere maintains a dynamic yet constant profile, Lovelock began to consider how biological inputs contribute to this apparent equilibrium. He studied the 300 billion-year stability of Earth's surface temperature at a mean of 13 degrees Celsius, despite a 40 percent increase of solar luminance. Further questions addressed homeostatic regulation of atmospheric gases. Complex life survives only in a narrow range of atmospheric oxygen; increasing oxygen only 4 percent would cause massive conflagrations. How does oxygen remain stable at 21 percent? Consulting geological records, Lovelock conjectured that the mix of atmospheric gases on Earth can be traced to chemical by-products of planetary life, as if the atmosphere were the circulatory system of the biosphere. The process of evapotranspiration, moving water from soil to trees to water vapor and across great distances, suggests the scale of atmospheric exchanges proposed by Gaia.

Lovelock asserts that Earth's atmosphere functions as it does because of the abundance of local life. Combined biotic activity, from a forest's respiration to bio-geochemical cycles such as rock weathering, sedimentation, and oxidation, integrated with the metabolic activity of millions of species of microscopic organisms, all generatively contribute to a global atmospheric mix. Air, water, and soil are not only substances in themselves but also conveyors for supplying materials to different layers of the atmosphere, producing a fluctuation of gases with a capacity for self-correction.

Addressing accusations of teleology, Lovelock developed, with Andrew Watson, the Daisyworld model to demonstrate how planetary homeostasis is achieved when organisms operate according to individual interests. In Daisyworld, clusters of light and dark daisies adapted to specific temperature ranges to compete for solar exposure and impact the planetary surface temperature. Dark daisies tolerate lower temperatures and will proliferate until they warm the planetary surface through their absorptive capability. Less tolerant of cool temperatures, light, heat-loving daisies take over until they cover so much surface that their white surfaces, through the albedo effect, reflect back the solar energy. Back to a chilling environment, black daisies begin to increase again. The model demonstrates how positive and negative feedback mechanisms change environmental conditions.

Ongoing research includes investigations into how self-regulatory and mutually dependent mechanisms arise from a coevolutionary standpoint. Work in geophysiology, ecology, climatology, biochemistry, microbiology, and numerous hybridized disciplines continues to produce data on whether or not the fluctuating properties of the atmosphere are controlled by the sum total of the biosphere.

See also Ecology; Sustainability.

Further Reading: Lovelock, James. *Gaia: A New Look at Life on Earth.* Oxford: Oxford University Press, 2000; Lovelock, James. *The Ages of Gaia; A Biography of Our Living Earth.* New York: Norton, 1988; Lovelock, James. *The Revenge of Gaia: Why the Earth Is Fighting Back—and How We Can Still Save Humanity.* New York: Penguin, 2006; Lovelock, J. E., and L. Margulis. "Biological Modulation of the Earth's Atmosphere." *Icarus* 21 (1974): 471–89; Margulis, Lynn, and Dorion Sagan. *Slanted Truths.* New York: Springer-Verlag, 1997; Molnar, Sebastian. "Gaia Theory." http://www.geocities.com/we_evolve/Evolution/gaia.html; Schneider, Stephen H., Jones R. Miller, Eileen Crist, and Pedro Ruiz Torres, eds. *Scientists Debate Gaia.* Boston: MIT Press, 1991.

Sarah Lewison

GENE PATENTING

The general public seems comfortable with the notion of copyright. Almost no one would think it was acceptable to go to the bookstore, buy the new best-selling novel by Danielle Steele, retype all the words into a computer, and print it out and sell it to other people. People generally understand that someone—perhaps the author herself or the publishing company—owns the right to publish that work, and other people cannot simply reproduce it and make money from it. This form of "intellectual property" protection is known as copyright. Similarly, few people would think that it would be allowable to make their own oat-based cereal and sell it to others under the name Cheerios. That name is owned as a trademark by a company. Most people also understand that many new inventions cannot simply be copied and sold, given that many innovative technologies are covered by patents that give the inventors the right to control the use of the invention for a limited period of time.

What if someone managed to find a way to identify genes or gene sequences in plants, animals, or even human beings that performed useful functions; discovered a way to manipulate, transfer, or remove these genes; and then applied for a patent that would give that person control of the deliberate use of those gene structures? Could someone be given a patent that gave him or her control of the use of genes that exist in nature, in plants, animals, or human beings? Not very long ago these might have seemed like strange questions or crazy ideas, but for decades now inventors have been applying for—and receiving—patent protection for the manipulation and use of specific genes that exist in living creatures.

Some say this is just a natural and logical extension of the notion of intellectual property protections that have long been covered by concepts and legal structures such as copyright, trademarks, and patents. Others say it is a dangerously radical extension of concepts designed for one sort of scientific development that are inappropriate in the field of genetic engineering, which is a science that manipulates the building blocks of life.

Whether the use of naturally occurring or existing genes should be covered by patent laws was thrust upon first the American patent and legal system and then that of the rest of the world, by the genetic engineering revolution. Inventors began filing patent claims with patent offices to cover genetic engineering innovations they had made, and the system was forced to react, to decide whether to grant patents to methods of accessing, moving, and manipulating elements of life forms.

Ananda Chakrabarty, a U.S.-based scientist (Indian by birth) pioneered in this area by discovering ways to manipulate and develop new strains of naturally occurring bacteria to break down oil from oil spills so that the material left afterward was essentially harmless. He and the company he worked for, General Electric, filed a patent claim in 1971. Years of legal wrangling followed, but in the 1980 U.S. Supreme Court case, Chief Justice Warren Burger delivered a very clear ruling that stated that not only was the patenting of genetic innovations allowable; it was not even a fundamentally new application of the patenting system. Burger wrote in the decision that the "relevant distinction" was not about whether something was alive, but whether an inventor had somehow changed something in a way that made it a human-made invention and therefore eligible for patent protection. Because patents could be issued for "anything under the sun made by man," and Chakrabarty's bacteria had clearly been changed from the naturally occurring forms, they could be patented. This ruling was not unanimous, with five judges supporting it and four dissenting. Since 1980 the ruling has been upheld; it has become the bedrock of laws regarding genetic patenting in the United States and an oft-quoted piece of legal reasoning by legal authorities around the world.

The biotechnological revolution has meant a huge increase in the number of patents applied for and received every year. In 1990 fewer than 2,000 biotechnology patents were granted in the United States, but in 2002 close to 8,000 patents were granted.

The rationale for the patent system is that when patent rights to inventors are allowed, inventors are given an economic incentive to invest time and money in innovation. This, the theory argues, creates much more research and development than would otherwise exist. Other inventors are also benefited because they get access to the information that will allow them to use the patented process for a fee or other consideration during the period of protection and then free access once the protection has expired. Some researchers have argued that the actual effect is not so clearly positive. In fact, some argue that the quest for patent rights to potentially lucrative innovations can delay or block advancements.

In the United States there has been much discussion of the effect of the 1980 Bayh-Dole Act, which laid out the legislation governing the use of government-funded research by nonpublic partners and other parties. Some critics have argued that American taxpayers are not being fairly reimbursed for the inventions for which public researchers are at least partially responsible. Although the act was passed at the birth of the biotechnology revolution and was not focused on biotech when it was written, biotechnological research and development, including pharmaceutical development, is the biggest area affected by Bayh-Dole.

Most legal systems add their own wrinkles to the situation, making the worldwide acceptability of patent rights for genetic manipulations something that needs to be studied on a country-by-country basis. For example, the famous "Harvard Mouse" case has led to fairly similar but differing rulings in the United States, Canada, and the European Union. The Harvard Mouse is a genetically engineered mouse that develops cancer extremely easily because of a gene inserted by genetic engineering. It is known by scientists as the "oncomouse" and is useful in experiments in which researchers want to discover what substances cause various forms of cancer and then what can be done to control or change the progression of the disease. The inventors, one of whom worked at Harvard University, applied for a U.S. patent to cover their technique of creating "transgenic" animals but eventually received a patent covering only rodents created by the technique. This occurred in 1988, and patents soon followed in the European Union, Canada, and Japan. An extensive legal struggle has resulted, however, in the oncomouse case, and its biotechnology has different—and generally more restricted—patent protections in these other places. Since then legal systems around the world have had to deal with an escalating number of genetic patent claims for animals, plants, and humans, with quite different approaches being applied in various countries. For instance, the genetic manipulation of human genes has created one set of legal rules and restrictions in the United States, another for the European Union, and specifically tailored rules in individual countries such as France and Germany. In Germany special restrictions are placed not just on the use of human genes, but also on genes from primates.

Some may wonder why governments and regulatory systems have not simply imposed a moratorium on the issuing of patents for genetic modifications given that their scope, impact, and potential costs and benefits are still not fully realized. Any deliberate banning of patents for elements of life forms, however, would have to grapple with the legal reasoning expressed by Burger that they do not require or justify special treatment; most governments have left the overall issue to the courts to decide rather than dealing with it in broad legislation. In addition, stopping the issuing of patents for altered life forms would potentially put the brakes on much scientific research and development, something few countries appear to be willing to contemplate. Countries may have so far imposed limited restrictions and regulations on the patenting of genetically altered life forms, but blanket bans are uncommon. Although caution in dealing with new scientific innovations appears prudent to many people, many also want access to the improved medicines, treatments, foods, and other products that genetic engineers say could be produced by altered life forms. If a gene-based lifesaving drug was never invented because the potential inventors could not get a patent and profit from their work, would humanity be better off?

Patents are the bedrock of the innovation world. Without them, many believe that few companies or individual inventors would bother to invest the huge amounts of time and money required to develop new technologies, leaving government virtually alone to invent new technologies. Thomas Edison, for example, was determined to profit from his invention of electrical devices. Most governments have decided it is in the best interest of their citizens to grant patents to the creators of new technologies. The U.S. Constitution provides the framework for American patents, in Article 1, Section 8, Clause 8, which states that Congress has the power "to promote the progress of science and useful arts, by securing for limited times to authors and inventors the exclusive right to their respective writings and discoveries."

Although patents give their owners the exclusive right to control the use of their innovation for a limited period—generally 20 years in the United States—the patent also gives the public access to the inventor's knowledge. Part of the process of receiving a patent is to explain in writing all the details of how the invention is made. This information is made public when the patent is granted. The inventor may have a limited period when he or she can control access to the innovation, charging others for the right to do it themselves, for example, but the public and other researchers and inventors are able to see what the inventor has invented and to learn how to do it themselves. When the patent protection expires, competitors can launch their own versions of the invention, or inventions that use the formerly patented invention as part of their process. Patents are granted because they both encourage innovation by the inventor and give other inventors knowledge they need to invent further technologies and techniques. Entire industries are based on patents, such as the pharmaceutical and biomedical companies. It is estimated that almost 200,000 people are directly employed in the United States by the biomedical research industry. Some economists have also claimed that technological progress is responsible for up to half of all U.S. economic growth and is the main factor responsible for the improvement of Americans' lives over the decades. In recent years an explosion in patent-protected gene-based inventions has created large industries that produce everything from gene therapies for humans to altered crops for farmers to pharmaceuticals for people. Without the ability to patent these inventions, few believe these industries would be anywhere near as large as they are today.

With so many apparently positive economic consequences flowing from the patenting of genetic innovations, why is the issue of gene patenting contentious? The answer has to do with the "apparently" part of the last sentence. The benefits from patent rights that are obvious to inventors and believers are not obvious to others. Critics of the extension of patent rights to gene-based inventions see a number of problems arising from them. Some of their concerns are moral or spiritual. Other concerns emerge from what the desire for patents does to research in general and to government- and university-funded research in particular. Further concerns are expressed about the dominance patents may give individuals, companies, or governments in the field of genetic engineering, a field that has so many potential social and environmental (as well as economic) effects.

To Warren Burger of the U.S. Supreme Court, there may have been nothing particularly startling about the idea of allowing patent rights to the genetic engineers who managed to find ways to alter life forms. To many others, a gut feeling leads them to consider the patenting of elements of life forms to be a stunning overextension of property rights to things that were never considered by most people to be a form of property. How can someone obtain the legal right to control the use of genes and gene sequences that have arisen naturally, in the world's many forms of life, even if used in a different way? Even if the manipulation, changing, or triggering of these genes can only be done by an "inventive step" developed by a scientist, does anyone have the moral right to prevent someone else from using the same methods to affect the genes of a living organism, such as a human? Do humans have the right to control and do what they like with their own genes? Are not plants growing and reproducing in the field, and animals living, breathing, and reproducing in pastures and barns, whether they are genetically modified or not, a completely different matter from, say, a specialized gizmo used in an automobile engine?

Supporters of patent rights are quick to point out that patent rights do not generally give their owner the right to control the existence of genes or gene sequences in creatures such as humans. No one can own a person's genes. In general, what patent rights do is protect a particular method of creating or using genes and gene sequences. In British law, for example, a naturally occurring human gene or protein in people cannot be patented, but human genetic material removed from an individual and then refined or reproduced in the laboratory can be patented. It is the non-naturally occurring deliberate use of the genetic material that can be controlled by patents. Also, specific tests or ways of manipulating genes within a living person can be patented, but this is not a patent on the genes themselves, but rather on techniques for manipulating and using them.

For crops in general, a patent on a variety of a genetically-altered plant does not give automatic ownership of any plant existing anywhere that contains the altered genes to the patent holder, but rather stops anyone from knowingly using the patented genetically-altered plants without permission. It is the deliberate use of the genetically-altered material that is controlled by the patent. In a case that occurred in western Canada, but that has been watched around the world, multinational corporation Monsanto sued a Saskatchewan farmer, Percy Schmeiser, for infringing its patent to a genetically engineered form of canola by deliberately growing the crop in his fields without having paid a license fee for doing so. After being found to have many fields containing the patent-protected canola, Schmeiser claimed that the genetically engineered crop had ended up in his field by accident—by the wind blowing pollen, by seeds blowing off trucks, and by other paths—and by a convoluted path had ended up covering his fields of canola that summer. Monsanto said there was no legitimate way that his canola fields could have innocently become almost entirely dominated by its patent-protected crop. The Federal Court of Canada judge found that it did not matter whether Schmeiser had deliberately planted the genetically engineered canola and did not bother to determine whether Schmeiser was telling the truth about how his fields became seeded by the legally-protected crop. The judge determined that Schmeiser was legally guilty of infringing on Monsanto's patent because he knew the crop he grew was mostly of the type covered by the Monsanto patent. It was the knowing use of the crop that made him liable for damages.

The qualifications of the rights held by patent holders may be sufficient to allay the concerns of many patent attorneys, legal authorities, and scientists, but to many people there still seems to be something different about using patents to control living genes, and the treatment of genetically-modified organisms as simply another invention raises rather than allays concerns.

One concern that some critics have raised is the possibility that publicly funded researchers, at places such as universities, are delaying the publication of their research until they can obtain patent rights for what they have invented. Instead of public money helping to get innovative research into public hands for the benefit of all, they worry that it is being used by researchers to obtain patents for themselves or their institutions and that the researchers are actually holding back from public knowledge their discoveries for longer than if they were not seeking patents. Some researchers have admitted to holding back research information from colleagues while seeking patents and to delaying the publication of their research results until they had patents or other commercial arrangements in place.

Many university and government research institutions have recognized these potential problems and have attempted to develop rules and regulations to make sure the public interest is protected. Many universities have offices for "technology transfer," which is the process of getting information and inventions out of the laboratory and into the public realm. For example, the National Institutes of Health forces researchers using its money to publish their results.

Another concern about the possibility of patents slowing down scientific advancement deals with the extent of patents granted for inventions. A "narrow" patent that covers a limited area can help other researchers by revealing information and knowledge that they need to make further advancements in the area. But a "broad" patent covering too much of an area of research can have, the critics worry, the effect of blocking innovations by others. If other researchers are working in the same area and going along the same path, a patent obtained by one researcher can be used as a method of stopping or discouraging the other researchers from continuing to work in the area. That is not the goal of the patent system.

Supporters of the patent system argue that there is little evidence of this "blocking" effect, however, and if there is some obstruction, it often encourages researchers to find other ways and means of inventing the same thing, by "inventing around" the area patented by another. This then provides the public with a number of methods of achieving the same result, which is a benefit.

These factors can all be seen in the debate around the oncomouse. The inventors and the company holding the patent rights say their invention is a wonderful advancement for science, allowing researchers to better study, understand, and develop therapies for cancer. Critics contend that the patent holders charge high rates for use of the patented mice, discouraging research by many scientists, and that the patent prevents researchers from developing similar genetically altered research animals without paying exorbitant fees.

Some critics are concerned that the wealth offered by successful patents might lead some public researchers and universities to focus on inventions that will make them money, rather than create the most benefits for the public. If something is valuable and can be patented, is it more likely to be researched than something that is valuable but cannot be patented? With many universities and government research programs now requiring that researchers find nonpublic partners such as companies to partially fund their research, is general-interest scientific development suffering? As many governments reduce their share of research spending, leaving more to industry, will researchers stop thinking about the overall benefit to society of their research and focus instead on what can bring them or their employer the most economic gain? When it comes to gene-based research, is the commercial focus going to overwhelm the social concern?

There is much debate about this, with no clear conclusion. There has been a long-standing debate over the difference between "pure" science and science focused on pragmatic, industrial, or commercial results. Some believe that scientists need to feel free to pursue whatever path of research they find most interesting and rewarding and that society will benefit from the social, commercial, and industrial impacts of their discoveries, however unintentionally they are made. Others wonder why so much public money is being invested in university and government research that does not have a direct and demonstrable benefit to the public.

Given that legal systems have allowed patents to be granted for the inventive use of genetic elements of life forms, and some of these uses and modifications can be very lucrative, are publicly funded researchers going to ignore ethical, moral, and social concerns in a desire to make money for themselves or their employers? Science is often seen as amoral because it is focused on simply working out the facts and mechanics of elements of nature free of moral judgments, but can scientific development become immoral if the lure of money causes scientists to ignore concerns that they would otherwise consider more seriously?

A major concern expressed by many critics of genetic patenting is that patents covering elements of life forms may give individuals, companies, or governments greatly expanded powers over human individuals and human societies. For instance, if a farmer wants access to the best crops and in fact needs the best crops simply to compete with his neighbors and remain a viable farmer, and those crops are all covered by patents owned by companies, has he become economically weaker than during the days when crops were covered by far fewer legal restrictions? Some critics suggest the situation of many relatively poor and weak people such as farmers in a technologically advanced and patent-dominated area is similar to that of serfs in the medieval world. They rely utterly on the lordly authorities to be allowed to subsist on their humble plot. Rather than giving them more useful tools, the patent controlled innovations strip away almost all of their ability to be independent. Patents, to these critics, can drive people such as farmers to dependency on companies that will allow them just enough profit within the system to survive, but not enough to flourish.

Similarly in medicine, if a person has a dangerous disease and could benefit from a gene therapy protected by patents, does he become utterly dependent on the patent holder, from whom he needs access to a lifesaving treatment? Some might say it does not matter. Without the therapy the person would be much worse off. Others would say the relative weakening of the individual and the relative strengthening of the patent holder is something to be concerned about because power shifts within societies can cause grave political and social stresses.

This concern is not restricted to individuals within a society. It also applies from society to society. If companies in wealthy places such as the United States, the European Union, and Japan, or the governments themselves obtain patents for important and essential gene-based human medicines and therapies, or for crops and livestock, that gives these wealthy nations even more power than ever before over the poor nations of the developing world. The developing nations seldom have well-funded universities or government research institutions, and few major companies are based in developing nations: therefore, does the advent of gene patenting and the genetic engineering revolution put poor nations at an even greater disadvantage than they were at previously? If farmers in these nations want access to the best crops and livestock in the future, they may feel compelled to use the patent-protected crops of the developed nations and their companies. If citizens in these countries want access to the best medications and therapies in the future, they may need to pay for access to patent-protected genetically-altered medicine. With money from these farmers and citizens in the developing world flowing to companies and shareholders in the developed world, are these already disadvantaged people falling even further behind?

Some say the poorer nations, though not getting direct, financial gain from most of the patent rights held by developed nations and their companies, gain by getting access to the innovations created by the patent system. If they get better medicines, better crops, and better methods because of the patent-focused innovations of the developed nations, they are better off than they were before. When the patents expire in the developed nations, the poorer nations will get all the benefit of the inventions without having had to incur the enormous costs borne by developed nations' citizens and companies. In this light, the money transferred from poorer nations to wealthier ones, or the lack of access, during the period of patent protection is paid back many times by free access to the developments after the period of patent protection.

Still, critics say the utilitarian approach fails to recognize the relative weakening of the state of developing nations caused by developed nations having control of innovations that quickly become necessary in a competitive world. The innovations may not bring a relative advancement for the poor and weak but instead create a greater situation of desperation, in which the developing world appears to be falling further behind, rather than catching up to the developed nations.

See also Biotechnology; Genetic Engineering; Genetically Modified Organisms; Human Genome Project; Intellectual Property.

Further Reading: The Chartered Institute of Patent Attorneys. http://www.cipa.org.uk; Council for Responsible Genetics. http://www.gene-watch.org; The European Commission's official legal journal. http://eur-lex.europa.eu; Greenpeace. http://www.greenpeace.org; Schacht, Wendy H. *The Bayh-Dole Act: Selected Issues in Patent Policy and the Commercialization of Technology.* Congressional Research Service of the Library of Congress, 2006; Suzuki, David T., and Peter Knudtson. *Genethics: The Ethics of Engineering Life.* Toronto: Stoddart, 1988.

Edward White

GENETIC ENGINEERING

Genetic engineering has plunged the world into a stunning technological revolution, one that brings great promise, spurs grave fears, and has unquestionably changed humanity's relationship with the very blueprint of life and physical existence. The problem with being in the midst of a revolution is that one can have little idea where one will end up when the revolution is complete.

So far, genetic engineering and gene-based knowledge have lifted biological science from a relatively crude state of inexactitude, have allowed humans to crack the genetic code, and have given researchers the tools to alter human, animal, and plant life to serve human goals.

Already the products of genetic engineering and genetic science are common throughout the developed world: gene therapies to treat human disease, genetically modified foods for people and animals, and pharmaceuticals for humans produced through genetically engineered bacteria.

The wave of potential products is stunning: organs from pigs transplanted into sick humans, drugs for humans produced in cow's milk, plastics produced by plants rather than with fossil fuels, gene therapies that could extend human life.

Many people worry about the implications of this revolution, however. Not only is it a radically new science with little proof that its many innovations will be entirely safe, but in addition, no one is in control of it. Like all revolutions of knowledge, once the scientific breakthroughs have been achieved and the information widely disseminated, human individuals and societies, with all their virtues and vices, will be free to use the knowledge as they see fit. There is presently nobody to say yea or nay to genetic engineering developments on behalf of the human species. Human history does not suggest that all human beings are either entirely altruistic or completely competent when embracing the possibilities of radical new technology.

What exactly is genetic engineering? In essence, it involves the manipulation of genes using recombinant DNA techniques to modify what the gene does, either by itself or in combination with other genes. "Recombinant" means combining genes from different sources in a different manner than occurs naturally. Genes are the units formed by combinations of the nucleotides G (guanine), A (adenine), T (thymine), and C (cytosine), which lie in two equally long and twisting strings (the famous "double helix") that are attached to each other throughout their length. G, A, T, and C nucleotides combine in pairs, across the space between the two strings. About three billion pairs form the human genome—the string of genes that make up each individual human's genetic structure. (Other biological life forms have different numbers of genes.) A gene is a stretch of A-T and C-G pairs that, by their complex arrangement, lay out the instructions for a cell to produce a particular protein. Proteins are the basic agents, formed from amino acids, that determine the chemical reactions in the cell. This incredibly long and complex genome is also incredibly small—it is contained in every cell in the body as a microscopic molecule. Although all of the genetic code is included in each cell in the body, each cell performs only a relatively tiny number of highly specialized functions, with only a comparatively few genes being activated in the functioning of a cell's production and use of proteins. Each cell may produce thousands of proteins, each the product of a different gene, but most of the genome's genes will never be employed by each cell. The genome can perhaps be understood as an instruction manual both for the construction of a life form and for its functioning once it has formed. It is like a computer operating system that also contains the information that a tiny piece of silicon could use to build itself into the computer that will use the operating system.

Because genes determine what cells do within an organism, scientists realized that by altering, adding, or deleting genes, they could change the functioning of the larger life form of which they are a part. To do so they need to use genetic engineering to alter and switch genes.

What scientists have been able to do with genetic engineering is (1) make it possible to "see" the genes in the DNA sequence, (2) understand the functions of some of those genes, and (3) cut into the DNA and remove or add genes and then reform it all as a single strand. Often the genes that are added come not from members of the same animal, plant, or bacterial species, but from entirely different species.

How is genetic engineering done? Again, there are very simple and exceedingly complex answers to this question, depending on how much detail one wants about the underlying processes.

The recombinant DNA revolution began in the 1970s, led by three scientists from the United States: Paul Berg, Stan Cohen, and Herb Boyer. They knew that certain bacteria seemed to be able to take up pieces of DNA and add it to their own genome. They discovered that even recombinant DNA created in the lab could be taken up by these bacteria. By 1976 scientists had successfully created the production of a human protein in a bacterium and later managed to produce human insulin in bacteria. Bacterially produced human insulin, produced using this bacteria-based process, is now the main form of insulin supplied to human diabetics.

Genetic engineers have discovered ways to isolate a gene in one species that they think could have a useful function in another, insert that gene (with others that make it "stick" to the rest of the DNA strand) into a cell's nucleus, and then make that cell develop into an entire life form. It is comparatively easy for scientists to introduce genes and comparatively much harder to get the altered cell to develop into a larger life form.

Genetic engineering can be seen as radically new, but to some it is merely a continuation of humanity's age-old path of scientific development. Some see it as an unprecedented break with age-old methods of human science and industry and fundamentally different; others see it as the next logical step in development and therefore not fundamentally radical at all. One's general outlook on scientific development can also color one's view as to whether these developments seem generally positive or negative. Do you see scientific progress as opening new opportunities and possibilities for humans to improve their situation and the world, or do you see it as opening doors to dangers against which we need to be protected? To some degree these different perspectives determine whether one is alarmed and cautious about this new science or excited and enthusiastic about it.

The overall contemporary positives-versus-negatives situation of genetic engineering and gene-based science can be summed up in a paraphrase of a former U.S. Secretary of Defense (talking about a completely different situation):

> There are "known knowns." Those are the present products and methods of genetic engineering, with their so far discovered benefits and dangers. For example, crops designed to kill the corn borer pest can also kill insects that people appreciate, such as butterflies.
>
> There are "known unknowns." Those are the elements and implications of the technology and science that we know we don't fully understand yet, but that we realize we need to discover. If a genetic modification of an animal or a plant makes it much stronger and more competitive compared with unaltered relatives in the environment, will those unaltered relatives be wiped out? Will genetically altered life forms become like the kudzu that covers so much of the U.S. South?
>
> Then there are the "unknown unknowns." Those are the elements and implications of this radical new science that we haven't even thought of yet, but which might have a big positive or negative effect in the future. This includes . . . Well, that's the point. Unknown unknowns cannot be anticipated.

As humanity lives through this stunning revolution, the number of known knowns will increase, but few believe we are anywhere near the peak of the wave of innovations and developments that will occur because of the ability of scientists and industries to use genetic engineering to alter life. Indeed, most scientists consider this to be a scientific revolution that is only just beginning.

Humanity began its social evolution when it began manipulating its environment. Hunter-gatherer peoples often burned bush to encourage new plant growth that would attract prey animals. At a certain point in most cultures, early hunters learned how to catch and domesticate wild animals so that they would not have to chase them or lure them by crude methods such as this. The ex-hunters would select the best of their captured and minimally domesticated animals and breed them together and eliminate the ones that were not as good. Eventually the animals became very different from those that had not been domesticated. The earliest crop farmers found plants that provided nutritious seeds and, by saving and planting some of those seeds, created the first intentional crops. By selecting the seeds from the plants that produced the biggest, greatest number or nutritionally most valuable seeds, those early farmers began manipulating those plant species to produce seeds quite different from the uncontrolled population.

The plants and animals created by selective breeding were the result of a very primitive form of genetic engineering, by people who did not know exactly what they were doing (or even what a gene was): the attractive animals and plants with heritable characteristics were genetically different from the ones that did not have those characteristics, so when they were bred together, the genes responsible for the attractive qualities were concentrated and encouraged to become dominant, and the animals and plants without the genes responsible for the attractive characteristics were removed from the breeding population and their unattractive genes discouraged.

Over centuries and thousands of years, this practice has produced some stunningly different species from their natural forebears, as deliberate selection and fortuitous genetic mutations have been embraced in the pursuit of human goals. For example, it is hard to imagine today's domestic cattle at one time being a smart, tough, and self-reliant wild animal species capable of outrunning wolves and saber-tooth tigers, but before early humans captured and transformed them, that is exactly what they were. (Consider the differences between cattle and North American elk and bison. Even "domesticated" elk and bison on farms need to be kept behind tall wire mesh fences because they will leap over the petty barbed wire fences that easily restrict docile cattle. But in 100 years, after "difficult" animals are cut out of the farmed bison and elk herds, will these animals still need to be specially fenced?)

Wheat, one of the world's most common crops, was just a form of grass until humans began selective breeding. The fat-seeded crop of today looks little like the thin-seeded plants of 7,000 years ago. Under the microscope it looks different too: although the overall wheat genome is quite similar to wild grass relatives, the selective breeding over thousands of years has concentrated genetic mutations that have allowed the farmers' wheat to be a plant that produces hundreds of times more nutritional value than the wild varieties. Did the farmers know that they were manipulating genes? Certainly not. Is that what they in fact did? Of course. Although they did not understand *how* they were manipulating the grass genome, they certainly understood *that* they were manipulating the nature of the grass called wheat.

In the past few centuries, selective and complex forms of breeding have become much more complex and more exact sciences. (Look at the stunning yield-increasing results of the commercialization of hybrid corn varieties beginning in the 1930s.) But it was still a scattershot approach, with success in the field occurring because of gigantic numbers of failed attempts in the laboratory and greenhouse. Scientists were able to create the grounds for genetic good fortune to occur but could not dictate it. They relied on genetic mutations happening naturally and randomly and then embraced the chance results.

This began to change after the existence and nature of DNA (deoxyribonucleic acid) was revealed by scientists in the 1950s. Once scientists realized that almost all life forms were formed and operated by orders arising in DNA, the implications began to come clear: if elements of DNA could be manipulated, changed, or switched, the form and functions of life forms could be changed for a specific purpose.

It took decades to perfect the technology and understanding that allows genes and their functions to be identified, altered, and switched, but by the 1990s products were rolling out of the laboratory and into the marketplaces and homes of the public. In animal agriculture the first big product was BST (bovine somatotropin), a substance that occurs naturally in cattle but that is now produced in factories. When it is given to milk-producing cows, the cows produce more milk. Farmers got their first big taste of genetic engineering in crops when various Roundup Ready crops were made available in the mid-1990s. Dolly, a cloned sheep, was revealed to the world in 1997. (Generally, cloning is not considered genetic engineering because a clone by definition contains the entire, unaltered gene structure of an already existing or formerly existing animal or cell. The genes can be taken from a fully developed animal or plant or from immature forms of life. Genetic engineering is generally considered to require a change in or alteration of a genome, rather than simply switching the entire genetic code of one individual with another. Although not fitting the classic definition of "genetic engineering," cloning is a form of genetic biotechnology, which is a broader category.)

With all the promise and potential, a wave of beneficial products appears set to wash over the human species and make human existence better.

Since the beginning of the genetic engineering revolution, however, some people have been profoundly concerned about the implications and possible dangers of the scientific innovations now occurring in rapid succession.

From its beginning, genetic engineering has prompted concerns from researchers, ethicists, and the public. For example, Paul Berg, the genetic engineering pioneer, called for a moratorium on molecular genetic research almost simultaneously with his team's early discoveries, so that people could consider the consequences of the new methods they had developed. Since then, scientists have debated the positives and negatives of their new scientific abilities, while also overwhelmingly embracing and employing those abilities. Many—but not all—of the scientific worries have been alleviated as scientists have improved their knowledge, but the worries of the public and nonscientists conversely have greatly increased.

Some of the concerns of critics about genetic engineering are practical. Is it safe to move genes around from one individual to another? Is it safe to move genes from one species to another? For example, if organs from a pig were genetically altered so that humans could accept them as transplants, would that make that person susceptible to a pig disease? And if that pig disease struck a human containing a pig organ, could that disease then adapt itself to humans in general and thereby become a dangerous new human disease? The actual nuts and bolts of genetic engineering often include many more strands of genetic material than just the attractive characteristic that scientists want to transfer. Different genetic materials are used to combine and reveal changes in genetic structure. What if these elements bring unexpected harm, or if somehow the combination of disparate elements does something somehow dangerous?

Some fear that ill-intended people, such as terrorists or nasty governments, might use genetic engineering to create diseases or other biological agents to kill or injure humans, plants, or animals. For instance, during the years of apartheid, a South African germ warfare program attempted to find diseases that could kill only black people and attempted to develop a vaccine to sterilize black people. During the Cold War, both NATO and Warsaw Pact nations experimented with biological warfare. The program of the Soviet Union was large and experimented with many diseases, including anthrax and smallpox. In one frightening case, an explosion at a Soviet germ warfare factory caused an outbreak of anthrax in one of its cities, causing many deaths. If scientists become able to go beyond merely experimenting with existing diseases to creating new ones or radically transformed ones, the threat to human safety could be grave. Australian scientists alarmed many people when they developed a form of a disease that was deadly to mice. If that disease, which is part of a family that can infect humans, somehow became infectious to humans, science would have created an accidental plague. What if scientists deliberately decided to create new diseases?

This fear about safety is not limited just to humans intentionally creating dangerous biological agents. What if scientists accidentally, while conducting otherwise laudable work, create something that has unexpectedly dangerous characteristics? What if humans simply are not able to perceive all the physical risks contained in the scientific innovations they are creating?

This concern has already gone from the theoretical to the real in genetic engineering. For instance, British scientists got in trouble while trying to develop a vaccine for hepatitis C after they spliced in elements of the dengue fever genome. Regulators disciplined the scientists for breaching various safe-science regulations after some became concerned that a frightening hybrid virus could arise as a result. The scientists had not intended any harm, and no problem appears to have arisen, but potential harm could have occurred, and any victims might have cared little about whether the damage to them was caused deliberately or by accident. Once a disease is out of the laboratory and floating in an ocean of humanity, it might be too late to undo the damage.

Responding to this concern, some argue for an approach they refer to as the "precautionary principle." This suggests that innovations and developments not be allowed out of the laboratory—or even created in the laboratory—until their safety or potential safety has been exhaustively demonstrated. Critics of genetic engineering often claim that the absence of long-term tests of genetic engineering innovations means that they should not be introduced until these sorts of tests can be conducted. This sounds like a good and prudent approach, but if actually applied across the spectrum, this approach would have prevented many innovations for which many humans now are profoundly grateful. If organ transplantation had been delayed for decades while exhaustive studies were conducted, how many thousands of Americans would not be alive today because they could not receive transplants? If scientists were prevented from producing insulin in a lab and forced to obtain it from human sources, how many diabetics would be short of lifesaving insulin? If scientists develop ways to produce internal organs in pigs that could save the many thousands of people who die each year because they cannot obtain human transplant organs in time, how long will the public wish to prevent that development from being embraced? The "precautionary principle" may appear to be an obvious and handy way to avoid the dangers of innovations, but it is difficult to balance that caution against the prevention of all the good that those innovations can bring.

Some of the concerns have been political and economic. Regardless of the possible positive uses of genetic engineering innovations, do they confer wealth and power on those who invent, own, or control them?

Many genetic engineering innovations are immediately patented by their inventors, allowing them to control the use of their inventions and charge fees for access to them. If an innovation makes a biological product such as a crop more competitive than non-engineered varieties, will farmers be essentially forced to use the patented variety in order to stay competitive themselves? Will the control of life forms changed by genetic engineering fall almost entirely into the hands of wealthy countries and big companies, leaving poor countries and individuals dependent on them? If a researcher makes an innovation in an area that other researchers are working in and then gets legal control of the innovation, can he prevent other researchers from developing the science further? The latter is a question American university researchers have often debated.

Humanity has had grave concerns about new science for centuries. These concerns can be seen in folk tales, in religious concepts, and in literature. Perhaps the most famous example in literature is the tale of Dr. Victor Frankenstein and the creature he creates. Dr. Frankenstein, driven by a compulsion to discover and use the secrets to the creation of life, manages to create a humanoid out of pieces of dead people but then rejects his living creation in horror. Instead of destroying it, however, he flees from its presence, and it wanders out into the world. The creature comes to haunt and eventually destroy Dr. Frankenstein and those close to him. The story of Dr. Frankenstein and his creature can be seen as an example of science irresponsibly employed, leading to devastating consequences.

Another tale is that of the sorcerer's apprentice. In order to make his life easier, the apprentice of a great magician who has temporarily gone away improperly uses magic to create a servant out of a broomstick. Unfortunately for the apprentice, he does not have the skill to control the servant once it has been created, and a disaster almost occurs as a result of his rash employment of powerful magic.

Both of these tales—popular for centuries—reveal the long-held uneasiness of those hesitant to embrace new technology.

On a practical and utilitarian level, many people's concerns focus on a balance of the positives versus the negatives of innovations. They are really a compilation of pluses and minuses, with the complication of the known unknowns and unknown unknowns not allowing anyone to know completely what all the eventual pluses and minuses will be.

Balancing complex matters is not an easy task. Innovations in life forms created by genetic engineering can have a combination of positive and negative outcomes depending on what actually occurs but also depending on who is assessing the results. For instance, if genetically altered salmon grow faster and provide cheaper and more abundant supplies of the fish than unaltered salmon, is that worth the risk that the faster-growing genetically engineered salmon will overwhelm and replace the unaltered fish?

A helpful and amusing attempt at balancing the pluses and minuses of genetic engineering's achievements was detailed in John C. Avise's 2004 book *The Hope, Hype and Reality of Genetic Engineering.* In it he introduces the "Boonmeter," on which he attempts to place genetic innovations along a scale. On the negative extreme is the "boondoggle," which is an innovation that is either bad or has not worked. Closer to the neutral center but still on the negative side is the "hyperbole" label, which marks innovations that have inspired much talk and potential, but little success so far. On the slightly positive side is the "hope" label, which tags innovations that truly seem to have positive future value. On the extreme positive pole is the "boon" label for innovations that have had apparently great positive effects without many or any negative effects. Throughout his book Avise rates the genetic engineering claims and innovations achieved by the time of his book's publication date using this meter, admitting that the judgments are his own, that science is evolving and the ratings will change with time, and that it is a crude way of balancing the positives and negatives. It is, however, a humorous and illuminating simplification of the complex process in which many people in society engage when grappling with the issues raised by genetic engineering.

Ethical concerns are very difficult to place along something as simplistic as the "boonmeter." How does one judge the ethics of a notion such as the creation of headless human clones that could be used to harvest organs for transplanting into sick humans? Is that headless clone a human being? Does it have rights? Would doctors need the permission of a headless clone to harvest its organs to give to other people? How would a headless clone consent to anything? This sounds like a ridiculous example, but at least one scientist has raised the possibility of creating headless human clones, so it may not be as far-off an issue as some may think. Simpler debates about stem cells from embryos are already getting a lot of attention.

As scientific genetic engineering innovations create more and more crossovers of science, industry, and human life, the debates are likely to intensify in passion and increase in complexity. Some biological ethical issues do appear to deflate over time, however. For example, in the 1980s and 1990s, human reproductive technology was an area of great debate and controversy as new methods were discovered, developed, and perfected. Notions such as artificial insemination and a wide array of fertility treatments—and even surrogate motherhood—were violently divisive less than a generation ago but have found broad acceptance now across much of the world. Although there is still discussion and debate about these topics, much of the passion has evaporated, and many young people of today would not understand the horror with which the first "test tube baby" was greeted by some Americans.

Some of these concerns, such as in vitro fertilization, appear to have evaporated as people have gotten used to novel ideas that are not fundamentally offensive to them. Other debates, such as those surrounding sperm and egg banks, remain unresolved, but the heat has gone out of the debates. Other concerns (like those regarding surrogate motherhood) have been alleviated by regulations or legislation to control or ban certain practices. Whether this will happen in the realm of genetic engineering remains to be seen. Sometimes scientific innovations create a continuing and escalating series of concerns and crises. Other crises and concerns tend to moderate and mellow over time.

Even if genetic science is used only to survey life forms to understand them better—without altering the genetic code at all—does that allow humans to make decisions about life that it is not right for humans to make? Some are concerned about prenatal tests of a fetus's genes that can reveal various likely or possible future diseases or possible physical and mental problems. If the knowledge is used to prevent the birth of individuals with, for example, autism, has society walked into a region of great ethical significance without giving the ethical debate time to reach a conclusion or resolution? A set of ethical issues entirely different from those already debated at length in the abortion debate is raised by purposeful judging of fetuses on the grounds of their genes. A simple, non–genetic engineering example of this type of issue can be seen in India. Legislators have been concerned about and tried to prevent the use of ultrasounds on fetuses to reveal whether they are male or female. This is because some families will abort a female fetus because women have less cultural and economic value in some segments of Indian society. Similar concerns have been expressed in North America. Humans have been concerned about eugenics for a century, with the profound differences of opinion over the rights and wrongs of purposely using some measure of "soundness" to decide when to allow a birth and when to abort it yet to be resolved. Genetic engineering is likely to keep these issues alive indefinitely.

One school of concerns is not worried about the utilitarian, practical, concrete, and measurable results or about straight ethical concerns. These are the spiritual and religious concerns, which can be summed up as the "playing God" question: by altering the basic building blocks of life—genes—and moving genes from one species to another in a way that would likely never happen in nature, are humans taking on a role that humans have no right to take? Even if some genetic engineering innovations turn out to have no concrete and measurable negative consequences at all, some of a religious frame of mind might consider the very act of altering DNA to produce a human good to be immoral, obscene, or blasphemous. These concerns are often raised in a religious context, with discussants referring to religious scriptures as the basis for moral discussion. For example, the Christian and Jewish book of Genesis has a story of God creating humans in God's image and God creating the other animals and the plants for humanity's use. Does this imply that God's role is to be the creator, and humans should leave creation in God's hands and not attempt to fundamentally alter life forms? If so, what about the selective breeding humans have carried out for thousands of years? On the other hand, if humans are created in God's image, and God is a creator of life, then is not one of the fundamental essences of humanity its ability to make or modify life? Because God rested after six days of creation, however, perhaps the creation story suggests there is also a time to stop creating.

The advent of the age of genetic engineering has stirred up a hornet's nest of concerns about the new technology. Some of these concerns are practical and utilitarian. Some are ethical, and some are religious in nature. Regardless of whether one approves of genetic engineering, it is doubtless here to stay. The knowledge has been so widely disseminated that it is unlikely any government, group of governments, or international organizations could eliminate it or prevent it from being used by someone, somewhere. The genie is out of the bottle, and it is impossible to force him back in, it appears. Humans will need to ensure that they are developing their ethical considerations about genetic engineering as quickly and profoundly as scientists are making discoveries and developing their methods if they wish to find acceptable approaches before changes are thrust upon them, rather than be forced to deal with ethical crises after they have arisen.

See also Chemical and Biological Warfare; Cloning; Eugenics; Genetically Modified Organisms; Human Genome Project.

Further Reading: Avise, John C. *The Hope, Hype and Reality of Genetic Engineering.* New York: Oxford University Press, 2004; LeVine, Harry. *Genetic Engineering: A Reference Handbook,* 2nd ed. Santa Barbara, CA: ABC-CLIO, 2006; McHughen, Alan. *Pandora's Picnic Basket—The Potential and Hazards of Genetically Modified Foods.* New York: Oxford University Press, 2000; Sherwin, Byron. *Golems among Us—How a Jewish Legend Can Help Us Navigate the Biotech Century.* Chicago: Ivan R. Dee, 2004; Steinberg, Mark L., and Sharon D. Cosloy. *The Facts on File Dictionary of Biotechnology and Genetic Engineering.* New York: Checkmark Books, 2001; Vogt, Donna U. *Food Biotechnology in the United States: Science, Regulation and Issues.* Washington, DC: Congressional Research Service of the Library of Congress, 2001.

Edward White

GENETICALLY MODIFIED ORGANISMS

Genetically modified plants, microbes, and animals have been a source of controversy since the development of genetic engineering techniques in the

1970s, intensifying with the growth of the life sciences industry in the 1990s. A wide range of critics, from scientists to religious leaders to antiglobalization activists, have challenged the development of genetically modified organisms (GMOs). Controversies over GMOs have revolved around their environmental impacts, effects on human health, ethical implications, and links to patterns of corporate globalization.

A GMO is a plant, microbe, or animal whose genetic material has been intentionally altered through genetic engineering. Other terms often used in place of "genetically modified" are *transgenic* or *genetically engineered* (GE). Genetic engineering refers to a highly sophisticated set of techniques for directly manipulating an organism's DNA, the genetic information within every cell that allows living things to function, grow, and reproduce. Segments of DNA that are known to produce a certain trait or function are commonly called genes. Genetic engineering techniques enable scientists to move genes from one species to another. This creates genetic combinations that would never have occurred in nature, giving the recipient organism characteristics associated with the newly introduced gene. For example, by moving a gene from a firefly to a tobacco plant, scientists created plants that glow in the dark.

Humans have been intentionally changing the genetic properties of animals and plants for centuries, through standard breeding techniques (selection, crossbreeding, hybridization) and the more recent use of radiation or chemicals to create random mutations, some of which turn out to be useful. In this broad sense, many of the most useful plants, animals, and microbes are "genetically modified."

The techniques used to produce GMOs are novel, however. To produce a GMO, scientists first find and isolate the section of DNA in an organism that includes the gene for the desired trait and cut it out of the DNA molecule. Then they move the gene into the DNA of the organism (in the cell's nucleus) that they wish to modify. Today, the most common ways that this is done include the following: using biological vectors such as plasmids (parts of bacteria) and viruses to carry foreign genes into cells; injecting genetic material containing the new gene into the recipient cell with a fine-tipped glass needle; using chemicals or electric current to create pores or holes in the cell membrane to allow entry of the new genes; and the so-called gene gun, which shoots microscopic metal particles, coated with genes, into a cell.

After the gene is inserted, the cell is grown into an adult organism. Because none of the techniques can control exactly where or how many copies of the inserted gene are incorporated into the organism's DNA, it takes a great deal of experimentation to ensure that the new gene produces the desired trait without disrupting other cellular processes.

Genetic engineering has been used to produce a wide variety of GMOs. Following are some examples:

- Animals: Genetically modified (GM) animals, especially mice, are used in medical research, particularly for testing new treatments for human disease. Mosquitoes have been genetically engineered in hopes of slowing

the spread of malaria. Farm animals, such as goats and chickens, have been engineered to produce useful substances for making medicines. Salmon DNA has been modified to make the fish grow faster. Pet zebra fish have been modified to have a fluorescent glow.
- Microbes: GM microbes (single-celled organisms) are in use in the production of therapeutic medicines and novel GM vaccines. Research is underway to engineer microbes to clean up toxic pollution. GM microbes are being tested for use in the prevention of plant diseases.
- Plants: Scientists have experimented with a wide variety of GM food plants, but only soybeans, corn, and canola are grown in significant quantities. These and a small number of other crops (e.g., papaya, rice, squash) are engineered to prevent plant disease, resist pests, or enable weed control. Some food crops have been engineered to produce pharmaceutical and industrial compounds, often called "molecular farming" or "pharming." Other, nonfood plants have also been genetically engineered, such as trees, cotton, grass, and alfalfa.

The research and development of GMOs and other forms of biotechnology have occurred in both universities and corporations. The earliest technologies and techniques were developed by professors in university laboratories. In 1973 Stanley Cohen (Stanford University) and Herbert Boyer (University of California, San Francisco) developed recombinant DNA (rDNA) technology, which made genetic engineering possible.

Although the line between "basic" and "applied" research has always been fuzzy, GMO research has all but eliminated such distinctions. The first release of a GMO into the environment resulted directly from a discovery by Stephen Lindow, a plant pathologist at the University of California–Berkeley. His "ice-minus bacteria," a GM microorganism that could be sprayed on strawberry fields to resist frost damage, was tested by Advanced Genetic Sciences (a private company)

UNIVERSITY–INDUSTRY PARTNERSHIPS

Biotechnology firms have begun to invest heavily in university research programs. Such university–industry partnerships have been quite controversial. In one example, the Novartis Agricultural Discovery Institute (a private corporation) and the Department of Plant and Microbial Biology at University of California–Berkeley formed a research partnership in 1998. Supporters of the agreement praised the ability of a public university to leverage private assets for the public good during a time of decreasing governmental support of research and celebrated the opportunity for university researchers to access proprietary genetic databases. Meanwhile, critics warned of conflicts of interest, loss of autonomy of a public institution, and research trajectories biased in the direction of profit-making. An independent scholarly evaluation of the agreement by Lawrence Busch and colleagues at Michigan State University found that neither the greatest hopes nor the greatest fears were realized but recommended against holding up such partnerships as models for other universities to mimic.

in 1986 amid great controversy. In many cases, university professors have spun off their own companies to market and develop practical uses for their biotechnology inventions. Herbert Boyer, for example, cofounded Genentech (NYSE ticker symbol: DNA) in 1976, a biotechnology company that produced the first approved rDNA drug, human insulin, in 1982. Such entrepreneurial behavior by academics has become common, if not expected, but has also attracted criticism from those who mourn what some have called the "commercialization of the university."

The early 1990s witnessed a growth of "life science" companies—transnational conglomerations of corporations that produced and sold agricultural chemicals, seeds (GM and conventional), drugs, and other genetic technologies related to medicine. Many of these companies began as pharmaceutical companies or as producers of agricultural chemicals, especially pesticides (e.g., Monsanto, Syngenta). Companies combined and consolidated in the hope of taking advantage of economic and technological efficiencies, and they attempted to integrate research, development, and marketing practices. By the late 1990s, however, many life science companies had begun to spin off their agricultural divisions because of concerns about profit margins and the turbulent market for GM crops and food. Today there are a mixture of large transnational firms and smaller boutique firms, the latter often founded by former or current university researchers.

The biotechnology industry is represented by lobby groups including the Biotechnology Industry Organization (BIO) and CropLife International. There are also a variety of organizations that advocate for continued research and deployment of GMOs, such as the AgBioWorld Foundation and the International Service for the Acquisition of Agri-Biotech Applications (ISAAA).

GMOs SLIPPING THROUGH THE REGULATORY CRACKS?

In the United States, three agencies are primarily responsible for regulating GMOs: the Department of Agriculture (USDA), the Environmental Protection Agency (EPA), and the Food and Drug Administration (FDA). The USDA evaluates the safety of growing GM plants—for instance, to see if GM crops will become weedy pests. The EPA deals with GMOs when they involve herbicides or pesticides that may have an impact on the environment and also reviews the risks of GM microorganisms. The FDA is responsible for the safety of animals, foods, and drugs created using genetic engineering. Some believe that the U.S. system of regulation of GMOs is not stringent enough. Food safety advocates often criticize the FDA because most GM foods are exempted from the FDA approval process. In certain cases, the U.S. government provides no regulatory oversight for GMOs. For example, the "GloFish," a GM zebra fish, has not been evaluated by any U.S. government agencies yet is now commercially available at pet stores across the United States. The USDA, EPA, and Fish and Wildlife Service all said that the GloFish was outside of their jurisdiction. The FDA considered the GloFish but ruled that it was not subject to regulation because it was not meant to be consumed.

Opposition to GMOs has emerged from many different sectors of society and has focused on various aspects and consequences of biotechnologies. The following list captures the breadth and some of the diversity of critique, although there are too many advocacy organizations to list here.

- Consumers (Consumers Union, Organic Consumers Association): Both as individuals and as organized groups, some consumers have opposed GM food by boycotting products and by participating in campaigns against politicians, biotechnology companies, and food distributors. Reasons include the lack of labeling of GM foods and ingredients (a consumer choice or right-to-know issue), health concerns (allergies, nutritional changes, unknown toxic effects), and distrust of the regulatory approval process (especially in the European Union).
- Organic farmers (Organic Trade Association, California Certified Organic Farmers): Organic agricultural products demand a premium that stems from special restrictions on how they are grown and processed. Under most organic certification programs (e.g., USDA organic), the presence of transgenic material above certain very low thresholds disqualifies the organic label. Organic farmers have therefore sustained economic losses because of transgenic contamination of their crops. Routes of contamination include pollen drift (from neighboring fields), contaminated seeds, and post-harvest mixing during transport, storage, or processing. Some conventional farmers have also opposed GM crops (especially rice) because significant agricultural markets in Asia and the European Union (EU) have refused to purchase grains (organic or conventional) contaminated with transgenic DNA.
- Antiglobalization groups (International Forum on Globalization, Global Exchange, Peoples' Global Action): Efforts to counter corporate globalization have frequently targeted transnational biotechnology companies—GM food became a kind of rallying cry at the infamous World Trade Organization protests in Seattle in 1999. Critics oppose the consolidation of seed companies, the loss of regional and national variety in food production and regulation, and the exploitation of human and natural resources for profit.
- Scientists (Union of Concerned Scientists, Ecological Society of America): Scientists critical of GMOs (more commonly ecologists than molecular biologists) tend to emphasize the uncertainties inherent in developing and deploying biotechnologies. They criticize the government's ability to properly regulate GMOs, highlight research that suggests unwanted health or environmental effects, and caution against unchecked university–industry relations.
- Environmental organizations (Greenpeace, Friends of the Earth): Controversy exists over the realized and potential benefits of GM crops. Critics emphasize the negative impacts, dispute the touted benefits, disparage the regulatory process as too lax and too cozy with industry, and point out that yesterday's pesticide companies are today's ag-biotech companies.

- Religious groups (United Church of Canada, Christian Ecology Link, Eco Kosher Network, Directors of the Realm Buddhist Association): Faith-based criticism of GMOs may stem from beliefs against tinkering with life at the genetic level ("playing God"), concerns about inserting genes from "taboo" foods into other foods, or social justice and environmental principles.
- Sustainable agriculture/food/development organizations (ETC Group, Food First/Institute for Food and Development Policy): These nongovernmental organizations (NGOs) bring together ethical, technological, cultural, political, environmental, and economic critiques of GMOs, often serving as clearinghouses of information and coordinating transnational campaigns.
- Indigenous peoples: Because many indigenous groups have remained stewards of eco-regions with exceptional biodiversity, scientists and biotechnology companies have sought their knowledge and their genetic resources ("bioprospecting"). At times, this has led to charges of exploitation and "biopiracy." In some cases, indigenous peoples have been vocal critics of GMOs that are perceived as "contaminating" sacred or traditional foods, as in a recent controversy over GM maize in Mexico.

Ever since researchers first began to develop GMOs, governments around the world have had to decide whether and how to regulate them. Controversies around GMOs often refer to arguments about the definition, assessment, and management of risk. Promoters of GMOs tend to favor science-based risk assessments ("sound science"), whereas critics tend to advocate the precautionary principle.

Calls for science-based risk assessments often come from stakeholders who oppose increased regulation and want to see GM technologies developed and marketed. Specifically, they argue that before a technology should be regulated for possible risks, those risks must be demonstrated as scientifically real and quantifiable. Although the definition of "sound science" is itself controversial, proponents state that regulatory agencies such as the EPA and FDA have been too quick to regulate technologies without good evidence—arguing that such government interference not only creates financial disincentives for technological innovation but actually causes social harm by delaying or preventing important technologies from becoming available. Such a perspective views government regulation as a risk in itself.

By contrast, advocates of the precautionary principle stress the existence of scientific uncertainties associated with many modern environmental and health issues. They have proposed a framework for decision making that errs on the side of precaution ("better safe than sorry"). Major components include the following: (1) anticipate harm and prevent it; (2) place the burden of proof on polluters to provide evidence of safety, not on society to prove harm; (3) always examine alternative solutions; and (4) include affected parties in democratic governance of technologies. Critics argue that the precautionary principle is little more than a scientific disguise for antitechnology politics.

In line with a precautionary approach to regulation, some governments (England, for example) have focused on genetic engineering as a process that may pose novel environmental or health risks. Other governments (for example, the United States and Canada) focus instead on the product, the GMO itself. Such countries generally do not single out GMOs for special regulation, beyond what is typical for other products. In addition, some governments have restricted the use of GMOs because of concerns about their social, economic, and ethical implications. Austria, for example, requires GMOs used in agriculture to be "socially sustainable."

International law also reflects controversy over regulating GMOs. The agreements of the World Trade Organization, the international body that develops and monitors ground rules for international trade, initially set out an approach similar to that of the United States. In 2000, however, more than 130 countries adopted an international agreement called the Cartagena Protocol on Biosafety, which promotes a precautionary approach to GMOs. This conflict has been a matter of much speculation and will likely feature in trade disputes over GM foods in the future.

Labeling of GM foods represents another contentious regulatory issue. Some governments take the position that if GMOs are found to be "substantially equivalent" to existing foods, they do not need to be labeled. In the United States, for example, food manufacturers may voluntarily label foods as "GMO-free," but there is no requirement to note when foods contain GMOs. The European Union and China, on the other hand, require foods made with GMOs to be labeled as such. In countries where labeling is required, there are typically fierce debates about tolerance levels for trace amounts of GMOs in foods meant to be GMO-free.

One dimension of the public debate about GMOs that is difficult to resolve is the question of whether it is morally, ethically, and culturally appropriate to manipulate the genetic makeup of living things. Some people respond with revulsion to the idea that scientists can move genes across species boundaries, putting fish genes into a strawberry, for instance. For some, this feeling stems from a philosophical belief that plants and animals have intrinsic value that should not be subordinated to human needs and desires. Unease with gene transfer may also be based on religious belief, such as the conviction that engineering living things is a form of playing God. But where is the line between divine responsibilities and human stewardship of the earth? Some religious leaders, such as the Vatican, have taken the position that if GMOs can be used to end world hunger and suffering, it is ethical to create them.

Evolutionary biologists point out that boundaries between species are not as rigid, distinct, and unchanging as critics of genetic engineering imply. All living things have some genes in common because of shared common ancestors. Furthermore, the movement of genes across species boundaries without sexual reproduction happens in a process called horizontal gene transfer, which requires no human intervention. Horizontal gene transfer has been found to be common among different species of bacteria and to occur between bacteria and some other organisms.

Regardless of the scientific assessment of the "naturalness" of genetic engineering, it is highly unlikely that all people will come to agreement on whether it is right to create GMOs, and not only for religious reasons. Those with philosophical beliefs informed by deep ecology or commitment to animal rights are unlikely to be persuaded that genetic engineering is ethical. Furthermore, many indigenous peoples around the world understand nature in ways that do not correspond with Western scientific ideas.

Given the diversity and incompatibility of philosophical perspectives, should we bring ethics, morality, and cultural diversity into policy decisions, scientific research, and the regulation of GMOs? If so, how? Some have proposed that labeling GMOs would enable people with religious, cultural or other ethical objections to avoid GMOs. Others see widespread acceptance of GMOs as inevitable and judge philosophical opposition as little more than fear of technology. These issues often become sidelined in risk-centered debates about GMOs but remain at the heart of the controversy about this technology.

As the world's population continues to grow, many regions may face food shortages with increasing frequency and severity. A variety of groups, including the Food and Agriculture Organization of the United Nations, anticipate that genetic engineering will aid in reducing world hunger and malnutrition, for instance, by increasing the nutritional content of staple foods and increasing crop yields. Such claims have encountered scientific and political opposition. Critics point out that conventional plant-breeding programs have vastly improved crop yields without resorting to genetic engineering and that GMOs may create novel threats to food security, such as new environmental problems.

Whether or not GMOs will increase agricultural productivity, it is widely recognized that greater yields alone will not end world hunger. Food policy advocacy groups such as Food First point out that poverty and unequal distribution of food, not food shortage, are the root causes of most hunger around the world today. In the United States, where food is abundant and often goes to waste, 38 million people are "food insecure," meaning they find it financially difficult to put food on the table. Similarly, India is one of the world's largest rice exporters, despite the fact that over one-fifth of its own population chronically goes hungry.

GOLDEN RICE

For over 15 years, Swiss researchers have been developing "Golden Rice," a type of GM rice that contains increased levels of beta-carotene, which is converted by the human body into vitamin A. The aim of the research is to combat vitamin A deficiency (a significant cause of blindness among children in developing countries), yet the project has drawn criticism. Some critics see Golden Rice as a ploy to gain wider enthusiasm for GMOs rather than a genuine solution to widespread malnutrition. Advocates of sustainable agriculture argue that vitamin A deficiency could be ended if rice monocultures were replaced with diverse farming systems that included production of leafy green vegetables, sweet potatoes, and other sources of beta-carotene. Scientists also continue to investigate whether Golden Rice would provide sufficient levels of beta-carotene and whether Asian farmers and consumers would be willing to produce and eat the bright orange rice.

Distribution of GM crops as emergency food aid is also fraught with controversy. Facing famine in 2003, Zambia's government refused shipments of corn that contained GMOs, citing health worries and concerns that the grains, if planted, would contaminate local crop varieties. U.S. government officials blamed anti-GMO activists for scaring Zambian leaders into blocking much-needed food aid to starving people. A worldwide debate erupted about the right of poor nations to request non-GMO food aid and the possibility that pro-GMO nations such as the United States might use food aid as a political tool.

Patents are government guarantees that provide an inventor with exclusive rights to use, sell, manufacture, or otherwise profit from an invention for a designated time period, usually around 20 years. In the United States, GMOs and gene sequences are treated as inventions under the patent law. Laws on patenting GMOs vary around the world, however. Many legal issues are hotly debated, both in national courts and in international institutions such as the World Trade Organization and the United Nations Food and Agriculture Organization. Should one be able to patent a living thing, as though it were any other invention? Unlike other technologies, GMOs are alive and are usually able to reproduce. This raises novel questions. For instance, do patents extend to the offspring of a patented GMO?

Agricultural biotechnology companies stress that they need patents as a tool for collecting returns on investments in research and development. Patents ensure that farmers do not use GM seeds (collected from their own harvests) without paying for them. Monsanto Company, for instance, has claimed that its gene patents extend to multiple generations of plants that carry the gene. The biotechnology industry argues that the right to patent and profit from genes and GMOs stimulates innovation in the agricultural and medical fields. Without patents, they say, companies would have little incentive to invest millions of dollars in developing new products.

Complicating the issue, however, is evidence that biotechnology patents increasingly hinder scientific research. University and corporate scientists sometimes find their work hampered by a "patent thicket," when the genes and processes they wish to use have already been patented by multiple other entities. It can be costly and time-consuming to negotiate permissions to use the patented materials, slowing down research or causing it to be abandoned.

Advocacy groups, such as the Council for Responsible Genetics, argue that patents on genes and GMOs make important products more expensive and less accessible. These critics worry that large corporations are gaining too much control over the world's living organisms, especially those that provide food. Some disagree with the idea that societies should depend on private companies to produce needed agricultural and medical innovations. Such research, they say, could be funded exclusively by public monies, be conducted at public institutions, and produce knowledge and technology freely available to anyone.

Furthermore, a wide variety of stakeholders, from religious groups to environmentalists, have reached the conclusion that "patenting life" is ethically and morally unacceptable. Patenting organisms and their DNA treats living beings and their parts as commodities to be exploited for profit. Some say this creates a slippery slope toward ownership and marketing of human bodies and body parts.

Many controversies over GMOs center on their perceived or predicted environmental impacts. Although both benefits and negative impacts have been realized, much of the debate also involves speculation about what might be possible or likely with further research and development.

With respect to GM crops, there are a variety of potential benefits. Crops that have been genetically engineered to produce their own pesticides (plant-incorporated protectants, or PIPs) eliminate human exposures to pesticides through hand or aerial spray treatments and may reduce the use of more environmentally harmful pesticides. Crops that have been genetically engineered with tolerance to a certain herbicide allow farmers to reduce soil tillage, a major cause of topsoil loss, because they can control weeds more easily throughout the crop's life cycle. If GMOs increase agricultural yields per unit of land area, less forested land will need to be converted to feed a growing population. Finally, some believe that GMOs represent a new source of biodiversity (albeit human-made).

The potential environmental harms of GM crops are also varied. PIPs may actually increase overall pesticide usage as target insect populations develop resistance. PIPs and herbicide-tolerant crops may create non-target effects (harm to other plants, insects, animals, and microorganisms in the agricultural environment). GM crops may crossbreed with weedy natural relatives, conferring their genetic superiority to a new population of "superweeds." GMOs may reproduce prolifically and crowd out other organisms—causing ecological damage or reducing biodiversity. Finally, because GMOs have tended to be developed for and marketed to users that follow industrial approaches to agriculture, the negative environmental impacts of monocultures and factory farming are reproduced.

With regard to GM microorganisms, proponents point to the potential for GMOs to safely metabolize toxic pollution. Critics emphasize the possibility of creating "living pollution," microorganisms that reproduce uncontrollably in the environment and wreak ecological havoc.

MONARCH BUTTERFLIES

Protesters dressed in butterfly costumes have become a regular sight at anti-GMO demonstrations. What is the story behind this ever-present symbol of anti-GMO activism? In 1999 John Losey and colleagues from Cornell University published a study that suggested that pollen from GM corn could be lethal to monarch butterflies. The corn in question had been genetically modified to express an insecticidal protein throughout the plant's tissues, including the pollen grains. The genetic material for this modification came from bacteria that are otherwise used to create a "natural" insecticide approved for use on organic farms. The GM corn thus represented both an attempt to extend a so-called organic method of crop protection to conventional agriculture (an environmental benefit) and a potential new threat to a beloved insect already threatened by human activities. Controversy erupted over the significance and validity of the Losey study, and the monarch butterfly remains symbolic of the controversy over the environmental pros and cons of GMOs.

GM animals also offer a mix of potential environmental harms and benefits. For example, GM salmon, which grow faster, could ease the pressure on wild salmon populations. On the other hand, if GM salmon escape captivity and breed in the wild, they could crowd out the diversity of salmon species that now exist.

No long-term scientific studies have been conducted to measure the health impacts of ingesting GMOs. As a result, there is an absence of evidence, which some proponents use as proof of GMOs' safety. Critics counter that "absence of evidence" cannot serve as "evidence of absence" and accuse biotechnology corporations and governments of conducting an uncontrolled experiment by allowing GMOs into the human diet. Several specific themes dominate the discussion:

- Substantial equivalence. If GMOs are "substantially equivalent" to their natural relatives, GMOs are no more or less safe to eat than conventional foods. Measuring substantial equivalence is itself controversial: Is measuring key nutrients sufficient? Do animal-feeding studies count? Must every transgenic "event" be tested, or just types of GMOs?
- Allergies. Because most human allergies are in response to proteins, and GMOs introduce novel proteins to the human diet (new sequences of DNA and new gene products in the form of proteins), GMOs may cause novel human allergies. On the other hand, some research has sought to genetically modify foods in order to remove proteins that cause widespread allergies (e.g., the Brazil nut).
- Horizontal gene transfer. Because microorganisms and bacteria often swap genetic material, the potential exists for bacteria in the human gut to acquire transgenic elements—DNA sequences that they would otherwise never encounter because of their non-food origin. Debate centers on the significance of such events and whether genetic material remains sufficiently intact in the digestive tract to cause problems.
- Antibiotic resistance. Antibiotic-resistant genes are often included in the genetic material that is added to a target organism. These DNA sequences serve as "markers," aiding in the selection of organisms that have actually taken up the novel genetic material (when an antibiotic is applied, only those cells that have been successfully genetically modified will survive). Some fear that the widespread production of organisms with antibiotic resistance and the potential for transfer of such traits to gut bacteria will foster resistance to antibiotics that are important to human or veterinary medicine.
- Unpredictable results. Because the insertion of genetic material is not precise, genetic engineering may alter the target DNA in unanticipated ways. Existing genes may be amplified or silenced, or novel functioning genes could be created. A controversial study by Stanley Ewen and Arpad Pusztai in 1999 suggested alarming and inexplicable health effects on rats fed GM potatoes, despite the fact that the transgenic trait was chosen for its non-toxic properties. Unfortunately, most data on the health safety of GMOs remains proprietary (privately owned by corporations) and unavailable to the public for review.

- Second-order effects. Even when GMOs are not ingested, they may have health consequences when used to produce food. For example, recombinant bovine growth hormone (rBGH) was approved for use in increasing the milk production of dairy cows. No transgenic material passes into the milk, but rBGH fosters udder inflammation and mastitis in cows. As a result, milk from cows treated with rBGH includes higher-than-average levels of pus and traces of antibiotics, both of which may have human health impacts.

Given that most GMOs retain their biological ability to reproduce with their conventional counterparts, there exist a number of reasons to segregate GMOs (to prevent mixing or interbreeding). First, some consumers prefer to eat food or buy products that are made without GMOs. Second, some farmers wish to avoid patented GM crops, for instance, in order to retain the right to save their own seeds. Third, there may be a need for non-GMO plants and animals in the future—for instance, if GM foods are found to cause long-term health problems and must be phased out. Fourth, it is essential that unauthorized GMOs or agricultural GMOs that produce inedible or medicinal compounds do not mix with or breed with organisms in the food supply.

For all of these reasons, the coexistence of GMOs and non-GMOs is a topic of heated debate around the world. There are a variety of possibilities for ensuring that GMOs and conventional organisms remain segregated. One possibility for the food industry is to use "identity preserved" (IP) production practices, which require farmers, buyers, and processors to take special precautions to keep GM plants segregated from other crops, such as using physical barriers between fields and using segregated transportation systems. Thus far, such efforts have proven unreliable, permitting, in some instances, unapproved transgenic varieties to enter the food supply. The biotechnology industry has advocated for standards that define acceptable levels of "adventitious presence"—the unintentional comingling of trace amounts of one type of seed, grain, or food product with another. Such standards would acknowledge the need to segregate GMOs from other crops but accept some mixing as unavoidable.

Critics of biotechnology, on the other hand, tend to see the mixing of GMOs with non-GMOs at any level as a kind of contamination or "biopollution," for which the manufacturers should be held legally liable. Because cross-pollination between crops and accidental mixture of seeds are difficult to eliminate entirely, critics sometimes argue that GMOs should simply be prohibited. For this reason, some communities, regions, and countries have declared themselves "GMO-free zones" in which no GMOs are released into the environment.

One possible technical solution to unwanted breeding between GMOs and their conventional relatives is to devise biological forms of containment. The biotechnology industry has suggested that Genetic Use Restriction Technologies (GURTs), known colloquially as "Terminator Technologies," may aid in controlling the reproduction of GM plants by halting GMO "volunteers" (plants that grow accidentally). GURTs make plants produce seeds that will not grow. Critics have mounted a largely successful worldwide campaign against Terminator Technology, calling attention to its original and central purpose: to

194 | Genetically Modified Organisms

GMOs ON THE LOOSE

In August 2006, the U.S. Department of Agriculture (USDA) announced that an unapproved variety of GM rice (Liberty Link Rice 601), manufactured and tested by Bayer CropScience a number of years earlier but never approved for cultivation, had been discovered to be growing throughout the U.S. long-grain rice crop. Despite the USDA's attempts to reassure the public of the safety of the unapproved variety of rice, when it was found in food supplies around the world, major importers stopped buying rice from the United States, causing prices for American exports to plummet. Hundreds of U.S. farmers filed a class action lawsuit against Bayer. Although it remains unclear how, exactly, the GM rice got into the seed supply, one possible explanation that the company offered is that it became mixed with "foundation" seeds, used to develop seeds that are sold to farmers, at a Louisiana State University rice breeding station. Rice breeders there had collaborated on the field trials for the experimental rice. From there, it seems, the rice was reproduced, spreading throughout the food system.

force farmers to purchase fresh seeds every year. Other research efforts aim at controlling pollen flow, not seed growth. For instance, a number of EU research programs (Co-Extra, Transcontainer, and SIGMEA) are currently investigating ways to prevent GM canola flowers from opening; to use male-sterile plants to produce GM corn, sunflowers, and tomatoes; and to create transplastomic plants (GM plants whose pollen cannot transmit the transgenic modification).

Should there be more GM crops? Advocates of GMOs argue that currently marketed technologies (primarily herbicide-tolerant and pest-resistant corn, rice, and soy) represent mere prototypes for an expanding array of GMOs in agriculture. Three directions exist, with some progress in each area. First, genetic engineers could focus on incorporating traits that have a more direct benefit to consumers, such as increased nutrition, lower fat content, improved taste or smell, or reduced allergens. Second, existing technologies could be applied to more economically marginal crops, such as horticultural varieties and food crops important in the Global South. Third, traits could be developed that would drastically reduce existing constraints on agriculture, such as crops with increased salt and drought tolerance or non-legume crops that fix their own nitrogen. It remains to be seen how resources will be dedicated to these diverse research paths and who will benefit from the results.

Should there be GM animals? With animal cloning technology possible in more and more species, and some signs of acceptance of cloned animals for the production of meat in the United States, conventional breeding of livestock could veer toward genetic engineering. Scientists around the world are experimenting with genetic modification of animals raised for meat, and edible GM salmon are close to commercialization. GM pets may also be in the future, with one GM aquarium fish already commercially available.

Should there be GM "pharming"? Some companies are pursuing the development of GM crops that manufacture substances traditionally produced by industrial processes. Two directions exist. First, if vaccines or medications can be genetically engineered into food crops, the cost and ease of delivery of such pharmaceuticals could decrease dramatically, especially in the global South (the developing world). Second, crops might be modified to produce industrial products, such as oils and plastics, making them less costly and less dependent on petroleum inputs. A California-based company, Ventria Biosciences, already has pharmaceutical rice in production in the United States. Animals are also being genetically engineered to produce drugs and vaccines in their milk or eggs, raising questions about the ethics of using animals as "drug factories."

Should there be GM humans? Genetic technologies have entered the mainstream in prenatal screening tests for genetic diseases, but the genetic modification of humans remains hypothetical and highly controversial. "Gene therapy" experiments have attempted to genetically modify the DNA of humans in order to correct a genetic deficiency. These experiments have remained inconclusive and have caused unpredicted results, including the death of an otherwise-healthy 18-year-old (Jesse Gelsinger). Even more controversial are calls for "designer babies," the genetic modification of sex cells (sperm and eggs) or embryos. Some advocate for such procedures only to correct genetic deficiencies, whereas others see attractive possibilities for increasing intelligence, improving physical performance, lengthening the life span, and choosing aesthetic attributes of one's offspring. Several outspoken scientists even predict (with optimism) that GM humans will become a culturally and reproductively separate species from our current "natural" condition. Critics not only doubt the biological possibility of such developments but also question the social and ethical impacts of embarking on a path toward such a "brave new world."

See also Agriculture; Ecology; Genetic Engineering; Gene Patenting; Organic Food; Pesticides; Precautionary Principle.

Further Reading: Charles, Daniel. *Lords of the Harvest: Biotech, Big Money, and the Future of Food.* Cambridge, MA: Perseus, 2001; Cook, Guy. *Genetically Modified Language: The Discourse of Arguments for GM Crops and Food.* London: Routledge, 2004; Kloppenburg, Jack Ralph, Jr. *First the Seed: The Political Economy of Plant Biotechnology 1492–2000.* 2nd ed. Madison: The University of Wisconsin Press, 2005; Miller, Henry I., and Gregory P. Conko. *The Frankenfood Myth: How Protest and Politics Threaten the Biotech Revolution.* Westport, CT: Praeger, 2004; Nestle, Marion. *Safe Food: Bacteria, Biotechnology, and Bioterrorism.* Berkeley: University of California Press, 2003; Schacter, Bernice. *Issues and Dilemmas of Biotechnology: A Reference Guide.* Westport, CT: Greenwood Press, 1999; Schurman, Rachel, and Dennis D. Kelso. *Engineering Trouble: Biotechnology and Its Discontents.* Berkeley: University of California Press, 2003.

Jason A. Delborne and Abby J. Kinchy

GEOTHERMAL ENERGY

Geothermal energy is energy derived from beneath the surface of the earth. It takes two main forms, either the transfer of heat into some form of power generation (like a steam turbine) or a heat exchange between the surface and some point below it (using a combined heat pump/heat sink).

The question "Energy Forever?" in the title of the book *Energy Forever?: Geothermal and Bio-Energy* by Ian Graham makes us wonder about natural energy and how to use it most effectively. At first glance, geothermal energy seems to have all the answers. It relies on a renewable energy source, and unlike the energy converted from burning fossil fuels, geothermal units also produce little in the way of harmful gases or dangerous waste materials. Countries around the world produce electricity from energy stored deep underground; once installed, geothermal energy is cost-efficient and can run for years without extensive repair. On a global scale, geothermal power plants are kinder to the environment. Geothermal energy has the benefit of being local; unlike oil or coal, which has to be removed from the ground, transported to a refining facility, and then shipped around the world to its point of use, geothermal power is generated on site, where it is intended to be used. Power from geothermal energy can be used for heating large spaces such as greenhouses or roads, as well as for heating (and cooling) individual homes.

Is this energy really "forever"? Or are there hidden conflicts and concerns that make geothermal energy merely one of a series of more environmentally friendly energy sources, not a significant answer to energy supply problems in the future?

Although both forms of retrieving energy from the ground are tagged "geothermal," the different technologies involved raise very different issues.

In the first instance, geothermal energy intended to power steam turbines relies on direct sources of heat stored beneath the surface. Naturally heated water can be drawn to the surface (as in hot springs), or originally aboveground water can be pumped beneath the ground, where it is heated to a high-enough temperature that when it returns to the surface, it can also power a turbine. The hot water from geothermal activity was believed to be good for one's health; the Romans built hot spring baths from North Africa to northern England and from Spain to Turkey. Capitalizing on this form of heat energy, the world's first geothermal plant was built in 1903 in Larderello, Italy, on the site of the healing waters of such ancient baths. Electricity was produced, and the system still generates enough to power a small village. Although in most places in the world (with the exception of Iceland), there are only occasional hot springs that can be tapped, geothermal fields can be created by pumping water underground and heating it to usable temperatures.

Although the construction costs of smaller units relying on hot springs are reasonable, the amount of power generated is unlikely to be enough to justify the costs of transmission to a significant distance from the plant. Larger units capable of generating enough electricity to make transmission economically viable would require an increase in construction costs that would make the facility more expensive than a fossil fuel–burning generating plant and probably equivalent to a hydroelectric generating plant.

On a small scale, such geothermal projects are well within the capacity of the earth's mantle to redistribute the heat required to make these geothermal units function. What remains to be seen, however, is what happens if the earth's thermal energy is tapped to a significant extent from a few places and what this might mean for a shifting of the earth's crust. Although cheap and clean renewable energy is environmentally desirable, the benefits are quickly erased if this causes an increase in volcanic activity or earthquakes as the subsurface shifts to accommodate a significant heat loss.

For homeowners, the second form of geothermal energy comes in many shapes and sizes. From drilling a shaft that is filled with water, acting as a heat pump in the winter and a heat sink in the summer, to laying a circuit on a bed of a body of water, to open loop systems, there is something for everyone in almost any environment. The science is quite simple and straightforward, and very little highly technical engineering is required to build, install, and maintain a system that provides heating and cooling. There is additional cost up-front, however, because such a system is much more expensive to install than a conventional forced-air heating system.

Although for nations, there seem to be many benefits and few disadvantages, for the average North American homeowner, the issue is about replacing home heating and cooling systems with geothermal energy. The average homeowner does not benefit directly from the power generated from the overall power grid system in the form of electricity. While the benefits for countries and nations with high levels of geothermal activity, such as from volcanoes or geysers, are quite clear, it is not quite clear what direct advantage a geothermal system might have for individual homeowners, who would sustain a significantly higher cost for installation against lower energy costs in the longer term. (In similar fashion, while it is possible to use solar energy to create a completely self-sustainable home "off the grid," the costs to do so are enormous.)

Although increasing energy costs related to fossil fuels will obviously change the proportion, in the absence of government incentives, a homeowner with a geothermal heating and cooling system in the province of Manitoba in Canada (for example) would expect to pay 35 percent of the heating costs of an electric furnace and 45 percent of the cost of a high-efficiency natural gas furnace (every regional power company will have its own chart for comparison). Factoring in the additional costs of geothermal installation—particularly the costs of conversion in an older home as opposed to a new home under construction—however, the homeowner is unlikely to reach the break-even point for 15–20 years. (Although the longevity of these systems varies with the type, maintenance schedule, and climatic conditions, individual home geothermal units could reasonably be expected to last as long as it would take the homeowner to reach this point.) Of course, the additional use of the unit for cooling (if it could be substituted for air conditioning) would reduce the payback time for the system somewhat, but for homeowners who do not intend to live in a particular location for a long time, the economic costs outweigh the economic benefit.

Thus, while geothermal energy systems involve some fascinating technology, they are unlikely to provide a major power source for electricity or for home heating and cooling in the near future and require more thought and effort for their potential to be maximized.

See also Fossil Fuel; Global Warming.

Further Reading: Graham, Ian. *Energy Forever? Geothermal and Bio-energy.* Austin, TX: Steck-Vaughn Company, 1999; Manitoba Hydro Web site (as an example, check out similar numbers with a power company in your local area). http://www.hydro.mb.ca/.

Gordon D. Vigfusson

GLOBAL WARMING

Since the 1980s, global warming has been a hotly debated topic in the popular media and among the general public, scientists, and politicians. The debate is about whether global warming has been occurring, whether it is an issue with which the global community needs to be concerned, and whether the current global warming is part of natural cycles of warming and cooling. Currently, the nature of the debate has begun to focus on whether there is anything we can do about global warming. For some, the problem is so insurmountable, and there seems to be so little we can do, that it is easier to entirely forget there is a problem.

In order to understand the changes that need to be made to have any meaningful and lasting impact on the level of global warming, the science behind the greenhouse effect must be understood.

The average temperature on Earth is approximately 15 degrees Celsius. The surface of Earth stays at such a consistent temperature because its atmosphere is composed of gases that allow for the retention of some of the radiant energy from the sun, as well as the escape of some of that energy. The majority of this energy, in the form of heat, is allowed to leave the atmosphere, essentially because the concentrations of gases that trap it are relatively low. When solar radiation escapes the atmosphere, it is largely due to the reflection of that energy from clouds, snow, ice, and water on the surface of Earth. The gases that trap heat are carbon dioxide, methane, nitrous oxides, and chlorofluorocarbons. These gases are commonly known as greenhouse gases.

In the last 60 years, the percentage of greenhouse gases (in particular, carbon dioxide) has begun to climb. Although the global increase in these gases has been noticed since the beginning of the Industrial Revolution approximately 200 years ago, the increase since the 1950s has been much more dramatic. Carbon dioxide comes from such sources as plant and animal respiration and decomposition, natural fires, and volcanoes. These natural sources of carbon dioxide replace atmospheric carbon dioxide at the same rate it is removed by photosynthesis. Human activities, however, such as the burning of fossil fuels, pollution, and deforestation, add excess amounts of this gas and therefore disrupt the natural cycle of carbon dioxide.

Scientists have discovered this increase in carbon dioxide and other greenhouse gases by drilling into ice caps at both the north and south poles and in glaciers and by taking ice-core samples that can then be tested. Ice cores have rings, similar to the rings found in trees, which allow for accurate dating. When snow and water accumulate each season to form the ice in these locations, air bubbles are trapped that are now tested for the presence of greenhouse gases. These studies have shown drastic changes in the levels of carbon dioxide.

Global warming is significantly impacted by the burning of fossil fuels and the massive loss of vegetation. First, the loss of vegetation removes photosynthetic plants that consume carbon dioxide as part of their life cycle, and second, the burning of fossil fuels releases carbon dioxide that has been stored for thousands of years in decayed plant and animal material into the atmosphere. These two processes have increased significantly globally in the last 100 years.

Although the rate of warming seems small and gradual, it takes only minor temperature fluctuations to have a significant effect on the global scale. During the last ice age, temperatures were less than 5 degrees Celsius cooler than they are today. This small change in temperature is so significant because of the properties of water. Water has a high specific heat, meaning it takes a large amount of heat energy to warm water. The result of this is that it takes a long time to warm or cool large bodies of water. This effect can be noticed in the temperate climate experienced in coastal areas. Once the oceans begin to warm, they will stay warm for an extended period of time. This is critical for life that has adapted to the temperatures currently experienced in the oceans.

The other important and alarming factor related to global warming and the warming of the oceans is the loss of the ice caps at both poles. This melting of ice has the potential to raise the level of the oceans worldwide, which will have potentially disastrous effects for human populations. The largest urban centers worldwide are located in coastal areas, which have the potential to flood. This will displace millions, and possibly billions, of people.

These changes are only gradual when considered within a human time frame. In terms of geological time, the change is extremely fast. This precipitous change will have far-reaching affects on both flora and fauna because most species will not have time to adapt to changes in climate and weather patterns. The result of this will be extinctions of species on a scale that is difficult to predict. It is certain that changes that have already taken place have had an impact on polar species, such as polar bears, because that habitat is where the changes are most strongly felt right now.

One of the largest issues in the debate on global warming is the difference in the ability to deal with mitigation and the large disparity in the consequences felt between developing and developed nations. The reality faced by many developing nations of poverty and subsistence living means that those populations do not have the ability to withstand some of the changes with which the world is faced. The most vulnerable people living in developed countries will not be able to adapt as easily.

These people, who generally do not contribute as much to the problems associated with an increase in greenhouse gases, will suffer the consequences most severely. Their contributions to global warming are less because many in this segment of the global population do not own cars, do not have electricity or refrigerators with chlorofluorocarbons, do not use air conditioning, and so on. Their lives are generally more closely tied with climate than those more fortunate, however. Their work may involve physical labor outside, they usually are involved in agriculture, or they may not be able to access health care for the inevitable increase in climate-related diseases such as malaria. The large and growing populations of many developing nations live mainly in coastal areas; less privileged people will not have the resources needed to move away from rising water levels. This means there will be a large refugee population that the international community will not easily be able to help.

Rapidly developing nations such as China and India, playing catch-up with the West, are becoming, if they are not already, major contributors to global warming. Older technologies, outdated equipment, and the nature of developing an industrial sector are largely to blame. In the development stage of industry, high carbon dioxide–emitting sectors such as shipping and manufacturing are predominant. Worldwide, work is needed to assist nations in developing their economies without sacrificing the environment to do so.

Global warming is not merely an issue of science and environmental protection; it is also a humanitarian and ethical concern. The methods of mitigation are being debated, and there is no clear answer to the questions concerning the appropriate measures to take. There are generally two appropriate responses. The first is to take any and all steps to immediately reduce the amount of pollution and greenhouse gas emission worldwide, or there will be no life on Earth. The second approach is based on the thought that nothing we do will have a lasting effect on the amount of pollution, so we must better equip the people of the world to deal with the consequences of this crisis. This means breaking the poverty cycle, addressing such issues as disease and access to good food and water, and providing appropriate education on a global scale.

KYOTO PROTOCOL

The Kyoto Protocol, sometimes known as the Kyoto Accord, is an international agreement requiring the international community to reduce the rate of emission of greenhouse gases causing global warming. It was signed in Kyoto, Japan, in 1997, to come into effect in 2005.

The Kyoto Protocol was initiated by the United Nations Framework Convention on Climate Change (UNFCCC) in order to extract a commitment from developed nations to reduce greenhouse gas emissions. The hope was that with developed countries leading the way, businesses, communities, and individuals would begin to take action on climate change.

The Kyoto Protocol commits those countries that have ratified it to reduce emissions by certain amounts at certain times. These targets must be met within the five years from 2008 to 2012. This firm commitment was a major first step in acknowledging human responsibility for this problem, as well as taking a step toward rectifying the crisis. Not all developed countries have ratified the protocol, however, with the United States and Australia among those that have not. This is of great concern, given that the United States is the worst offender when it comes to greenhouse gas emissions.

Criticisms of the protocol are that it puts a large burden for reduction of greenhouse pollution on developed nations, when developing nations are set to far surpass current levels of emissions. As well, the protocol does not specify what atmospheric levels of carbon dioxide are acceptable, so reduction is not a concrete enough goal to have any real, lasting effect. Finally, the Kyoto Protocol is seen as a bureaucratic nightmare, too expensive a solution for this problem, when compared to the amount of gain that results.

Another debate surrounding mitigation of global warming is whether individual effort will have an effect on rates of carbon dioxide and other greenhouse gases. Will one person choosing to ride his or her bike or take public transit reduce the level of emissions across the globe? If one person uses electricity generated by wind instead of coal, is that enough? Critics say that public apathy is so high, and there is such a strong sense of entitlement to resources, that there will never be enough people making the so-called green choice to make any kind of a difference at all. Others feel that all change must happen at a grassroots level and that every step counts and is important. If every single person in North America cut the number of hours they spend driving in half, of course there would be a significant decrease in pollution.

See also Coal; Ecology; Fossil Fuels; Gaia Hypothesis; Sustainability.

Further Reading: *An Inconvenient Truth.* Documentary. Directed by David Guggenheim, 2006; Dow, Kirstin, and Thomas E. Downing. *The Atlas of Climate Change: Mapping the World's Greatest Challenge.* Berkeley: University of California Press, 2007; Flannery, Tim. *The Weather Makers: How We Are Changing the Climate and What It Means for Life on Earth.* Toronto: HarperCollins, 2006; Monbiot, George. *Heat: How to Stop the Planet from Burning.* Toronto: Random House, 2006.

Jayne Geisel

GLOBALIZATION

One big planet, a global community, the vision of everyone and everything together that reflects those pictures of the Earth from space first sent back by Apollo 8—globalization can be romantically portrayed as any of these. From the dark side, it can also be seen as something that shatters local communities, takes away individual autonomy, destroys local cultures, and renders everyone helpless in the face of overwhelming power from somewhere else.

That globalization can be seen as both the happy inevitability of a bright future and the dismal gray of a grinding disaster reflects the reality of a significant conflict between opposing perspectives. Globalization can be represented in economic; cultural; sociopolitical; and environmental terms, each of which has its own means of measuring the difference between heaven and hell.

In a history of globalization, looking to identify the means by which people or cultures have sought to spread around the planet and why, the primary means has been military, conquering the world through the use of force. For historical examples, we can look to Alexander the Great, the emperors of Rome, Genghis Khan, and so on. In such instances, the means becomes the object; there is no particular value to be gained by conquest, yet the conquest continues because the military machine, so unleashed, has no particular boundary or end to its use. Like a forest fire, globalization by such means continues until it reaches some natural boundary—like a river or an ocean—or it runs out of "fuel" to sustain it.

On the heels of military globalization, the means by which the gains of conquest are maintained and the benefits accrue to the state or group that initiated the conquest are primarily political. One of the reasons for the failure of Alexander's empire was the fact he absorbed the local political structures, virtually unchanged, into his own; when he died, of course, that was the end of the empire. The Roman Empire, by contrast, brought with it Roman forms of government and social organization, structures that tended to be imposed on the local populations that were controlled and directed by Roman law and institutions. Caesars and other leaders came and went, but the Empire continued until the center fell apart, and the institutions—though not the roads—also fell apart. Political organization may be combined with religious organization, however, and although certain Roman institutions lost their sway in the outlying areas, the religion that was propagated through the military and political structures continued and spread.

With military and political impulses to globalization come economic considerations. In the first instance, to the victor the spoils, for the fruits of conquest are inevitably monetary—someone, after all, has to pay the costs of the operation and make it possible for further conquest. In the second instance, the establishment of political institutions makes an economic return on conquest more than the immediate spoils of war; a steady flow of money back to the state at the center of the empire enables the maintenance of a structure from whose stability everyone benefits, at least to some extent. Trade flourishes in the context of political stability, and military power protects such trade from the natural depredations of those who want to profit through force and not commerce.

Naturally, to maintain this kind of structure in the longer term requires both common currency and common language; in the wake of military and political conquest inevitably comes the standardization of currency (the coin of the empire) and some common language for the exercise of political and economic power. Latin—and particularly Latin script—became the language of the Roman Empire to its farthest reaches, providing a linguistic uniformity and continuity that outlasted the Empire itself by a thousand years. With linguistic uniformity comes intellectual constraints; whether or not it was previously possible to articulate dissent or rebellion in the language of the peoples, over time their linguistic armory is depleted by the acceptance and use of the language—and the philosophy it reflects—of the conquering culture. The longer an empire has control over the political, social, and religious institutions of the areas it has conquered, the less able the conquered people are able to sustain an intellectual culture distinct from that of their conquerors—thus increasing the likelihood that such an empire will continue because no one can conceive of another way of making things work.

Colonialism—a practice that existed long before the European powers made it an art in the nineteenth century—was the means by which the empire was not only propagated but also sustained, through the use of military, political, economic, religious, and intellectual tools.

This is a coercive model of globalization, but it tends to be the one first thought of when discussing how to overcome the various geographical, social, and cultural barriers that divide various groups. It is also the model that is reflected most obviously in history, which tends to be a record of the various conquests of one people or nation by another.

Is it possible, however, for there to be an impulse to "one planet" that is not inherently coercive? Is it possible for these kinds of boundaries to be overcome through mutual goodwill, or a collective self-interest, in which all parties cooperate because it is to the advantage of all players that they do so? This is the million-dollar question because in the absence of some way in which such cooperation might take place, all that remains is a coercive model, however well the coercion is disguised.

Of the current models for breaking down regional boundaries, most of them are economic and arguably coercive in nature. There is the International Monetary Fund (IMF), coupled with the World Bank, both operating within the framework approved (if not designed) by the countries of the G8 (and now G9, if one includes China). Within that framework, although countries identified as "developing" are offered financial assistance, the assistance is tied to certain monetary and trade policies in such a way that they are, in effect, coerced into compliance. Where countries—including members of the G9—try to go their own way, it is still within the framework of international trade agreements (such as the GATT, the General Agreement on Tariffs and Trade) and under the watchful eyes of global currency markets whose volatility is legendary. In the absence of a global gold standard, certain economies set a global economic standard through their national currency; for example, the value of other currencies used to be measured primarily against the U.S. dollar, though increasingly it is measured as well by the Japanese yen and by the euro from the European Union.

It would be one thing if this approach to globalization were successful, but for too many people, it is not, and the number of critics from all perspectives grows. Oswaldo de Rivero, the head of the Peruvian delegation to a round of the GATT talks, lays out very clearly in *The Myth of Development: The Non-Viable Economies of the 21st Century* why the current structure not only favors the wealthy but also entails the failure of the economies of developing countries in the South. Similarly, Joseph Stiglitz, 2001 Nobel Prize winner in economics, reached the same conclusions about the unequivocal failures of the IMF and the World Bank, from the perspective of an insider (*Globalization and its Discontents*). For those who wonder why and how such a situation came about, in *The Wealth and Poverty of Nations: Why Some Are So Rich and Some So Poor,* historian of technology David Landes set out the historical development of industrial economies through to the present and makes it clear why there are winner and losers.

There is a difference, however, between the macroeconomic globalization that organizations such as the IMF and the World Bank promote and what can be termed commercial globalization. Commercial globalization, through the merchandising of certain products worldwide, promotes an economic model of consumption that is not restricted by national boundaries. Because the objects sold through such global channels are always value-laden, this reflects a globalization, if not of the commercial culture itself that produced the items, at least of some of its values and mores. For example, it is not possible for McDonald's restaurants to be found worldwide without there also being an element of the American burger culture that is found wherever there are golden arches, regardless of what food is actually served (even the McLobsters that seasonally grace the menu in Prince Edward Island). Given the worldwide availability—albeit at a higher price—of virtually any item to be found on the shelves of a North American supermarket or department store, and the capacity of advertising to be beamed simultaneously to multiple audiences watching television from the four corners of the globe, it becomes understandable how and why commercial globalization has become a potent economic, political, social, and cultural force in the twenty-first century.

Thus, the material aspirations of a 21-year-old in Beijing may well be parallel to someone of the same age in Kuala Lumpur, or Mumbai or Dallas or Moose Jaw. Exposed to the same images and advertising, their material desires in response are likely to be the same; regardless of their culture of origin, their culture of aspiration is likely to include cars, computers, iPods and fast food.

One might say the primary implication of commercial globalization is the globalization of consumer culture, specifically Western consumer culture. Whether such a culture is good or bad in and of itself, its implications are arguably negative in terms of what it does to the local culture through supplanting local values and replacing them with (usually) more alluring and exciting values from far away.

In addition, the diversity of local cultural values—reflected in everything from forms of government to traditions around medicine and healing to cultural practices related to agriculture, cooking, and eating to religious belief systems and habits of dress—is endangered by the monoculture of mass consumerism as it is represented in the venues of mass media.

There is a difference, however, between globalization and standardization. It is important to distinguish the two, especially in light of the social and cultural requirements of industrial (and postindustrial) society. A very strong case can be made that the impulse to globalize is an effort to regularize and systematize the messy world of human relations into something that fits a mass-production, mass-consumption model. From the introduction of the factory system (1750) onward, industrial processes have become more and more efficient, systematizing and standardizing the elements of production, including the human ones. Ursula Franklin refers to the emergence of "a culture of compliance" in which the activities of humans outside the manufacturing process become subject to the same terms and conditions as are required in the process of mass production. This culture of compliance requires individuals to submit to systems; it requires them to behave in socially expected as well as socially accepted ways, thus removing the uncertainties and vagaries of human behavior from the operations of society. Although in the mechanical sphere of production, such habits of compliance are essential for the smooth operation of the system, taken outside into the social and cultural spheres in which people live, the antihuman effects of such standardization—treating people in effect like machines to be controlled and regulated—are unpleasant, if not soul-destroying.

Thus, in any discussion of globalization, it needs to be established from the outset what the benefit is, both to individuals and to societies, of some kind of uniformity or standardization in the social or cultural spheres. What is lost, and what is gained by such changes, and by whom? Much has been made of the comment by Marshall McLuhan that humans now live in a "global village," thanks to the advent of mass communication devices such as the radio, the television, the telephone, and now the Internet. Yet studies were done of what television programs were being watched by the most people around the world and therefore had the greatest influence on the development of this new "global" culture that was replacing local and traditional cultures. Imagine the consternation when it was discovered that the two most watched programs were reruns of *Wagon Train* and *I Love Lucy!* Globalization and the cultural standardization that mass-production, mass-consumption consumption society assumes to be necessary may mean that the sun never sets on the fast food empires of McDonald's or Pizza Hut, just as 150 years ago it was said to never set on the British Empire. Yet if the dietary habits of local cultures, in terms of both the food that is grown or produced and the ways in which the food is eaten, are merely replaced by standardized pizzas or burgers (or McLobsters, instead of the homemade variety), one cannot help but think something has been lost.

In the same way as colonies were encouraged to supply raw materials to the homeland and be captive consumers of the manufactured goods it produced (along with the culture and mores that the homeland dictated), so too the commercial colonization of mass-production/consumption society requires the same of its cultural colonies. The irony, of course, is that the "homeland" is much less identifiable now than it was in the days of political empires; although corporate America is often vilified as the source of the evils of globalization, the reality is that corporate enterprises are much less centralized and less entrenched than any nation state. Certainly the burgeoning economic growth of the European Union (with its large corporate entities that not only survived two world wars and a Cold War but even thrived on them), along with Japan, and the emergence of China and India as economic superpowers indicate that the capital of empire today is entirely portable. The reality that some corporations have larger budgets and net worth than many of the smaller nations in the world also indicates that borders are neither the boundaries nor the advantages that they used to be.

Although the economic examples of globalization today are arguably coercive (despite the inevitable objections that no one is forcing us to buy things), it is possible at least to conceive of other ways in which globalization might be noncoercive, incorporating mutually beneficial models instead. In a subsequent book, *Making Globalization Work*, Joseph Stiglitz works through the ways in which the current problems he and others identify with economic globalization could be overcome; while he proposes solutions to the major problems, he does not effectively address the motivational change that would be required for decision makers to make choices reflecting social responsibility on a global scale.

In the political realm, the United Nations (UN) has, in theory, the potential to be a body that—while respecting the national boundaries of its member states—works to find constructive ways of collectively responding to regional and global issues. Whether its first 60 years reflects such an ideal, or whether instead the UN has been a facade behind which coercion has been wielded by one group against another, is a subject for debate; in the absence of a clear global mandate for intervention or the effective economic and military means to intervene, moreover, even within a coercive framework, it is hard to see the UN as a model for good global government.

(In terms of any other models of globalization, one might point to the Olympic movement, but because it has always been a stage for personal and national self-aggrandizement, it is hard to see how it could become a step to some positive global culture.)

In the larger scope history provides, there are positive signs for political organizations that transcend the boundaries of the nation-state and in which participation is voluntary, benefits accrue to all, and the elements of coercion become less significant over time. No one who witnessed the aftermath of the Napoleonic era, the revolutions of 1848, the Franco-Prussian War, the Great War, World War II and the Iron Curtain, ever would have expected either the peaceful reunification of Germany or the formation (and success) of the European Union. Begun first as an economic union, it has continued to grow and mature into a union that has lowered many of the barriers to social, cultural, and political interaction that hundreds of years of nationalism had created.

Whether the EU model is exportable to other parts of the world raises some serious questions about how political globalization might succeed. The EU is regional, involving countries with a long and similar history, even if it was one in which they were frequently at war. The export of its rationale to other areas and cultures, with a different range of historical relations, is unlikely to meet with the same success. There should be some considerable doubt that democracy—as a Western cultural institution—will be valued in the same way in countries that do not have a similar cultural heritage or as desirable to the people who are expected to exercise their franchise. William Easterly is quite scathing in his account of why such cultural colonialism has done so little good, however well-meaning the actors or how noble their intentions (*The White Man's Burden: Why the West's Efforts to Aid the Rest Have Done So Much Ill and So Little Good*).

Certainly the effects of globalization are far from being only positive in nature; globalization in the absence of political and economic justice that is prosecuted through military and economic coercion creates not only more problems than it solves but also arguably bigger, even global, ones. Whatever the potential benefits of a global perspective, they are undercut by what globalization has come to mean in practical terms for many people (as the articles in *Implicating Empire: Globalization & Resistance in the 21st Century World Order* so clearly represent). After the events of September 11, 2001 (9/11), one might easily argue against globalization of any sort given that previously localized violence has been extended worldwide as a consequence of what is now the "global war on terror."

All of these issues combine to ensure what John Ralston Saul describes as "the collapse of globalism." He sees recent events as sounding the death knell for the free-market idealisms of the post–World War II period, noting that the promised lands of milk and honey that were to emerge from the spread of global markets and the demise of the nation-state have simply failed to materialize. In fact, the current reality is so far from the economic mythology that, in retrospect, it perhaps would not be unfair to regard the architects of this plan as delusional and their disciples as blind.

Saul does add a subtitle to his book, however in which the collapse of globalism is succeeded by "the reinvention of the world." Out of the ashes of this kind of economic globalism, in other words, and the unmitigated disaster it has spawned, it might be possible to reinvent a shared perspective on global problems that seeks to find a way other than those that have failed. Although Saul is rather bleak in his outlook and much more effective in describing the collapse of globalism than in setting out the character of such a reinvention, he makes a useful point. The failures of economic globalism are so painfully obvious that there can be no reasonable doubt that some other means of working together must be found.

If there is a perspective that has potential to be a positive rationale for globalization, it might be an environmental or ecological one. One of the most significant issues pushing some cooperative means of globalization is the environment, as we consider the ecological effects of human activities on a planetary scale. Global warming, ozone depletion, and the myriad means of industrial pollution whose effects are felt worldwide make it clear that, in the absence of a global response, we will all individually suffer serious consequences.

As much as we like to divide up the planet in human terms, laying out the grid lines of political boundaries and economic relationships, the fundamental limitations of the planet itself establish inescapable conditions for what the future holds. Although this may seem just as counterintuitive as Saul's analysis of the failure of global economic systems reinventing the world, the global spread of pollution, combined with catastrophic climate change, may catalyze changes that overcome local self-interest in favor of something bigger than ourselves. The artificial boundaries that humans create, everything from the notion that one can possess the land to the idea that one can control a part of the planet, are seen through even a crude ecological lens to be nonsensical and even dangerous. If the idea that people have the right to do what they please with the land, water, or air that they "own" is replaced by some more ecologically responsible understanding, then there may be a common ground for cooperation on a planetary scale that does not as yet exist. Whether such global cooperation will be in response to some global disaster or whether it will be the result of some new and more positive understanding remains to be seen.

It may seem like pie in the sky, but there are noncoercive ways of conceiving of a global community in which globalization consists of the universal acceptance of ideals and values. If justice, human rights, and respect were tied to the provision of the necessities of life to people in all areas of the planet, and peaceful means were used to settle whatever disputes might arise, then a global culture that reflected these things would be good for everyone.

This is not a new idea, but it is one that Albert Schweitzer elaborated on in his book *The Philosophy of Civilization*. The first two sections were written "in the primeval forest of Equatorial Africa" between 1914 and 1917. The first section of the book, "The Decay and Restoration of Civilization," locates the global problem not in economic forces but in a philosophical worldview that has undermined civilization itself; for Schweitzer, the Great War was a symptom of the spiritual collapse of civilization, not its cause. He asserts that society has lost sight of the character of civilization and, having lost sight of it, has degenerated as a result. That degeneration is primarily ethical; civilization is founded on ethics, but we are no longer aware of a consistent ethical foundation on which we can build a life together. The second section, not surprisingly, is titled "Civilization and Ethics"; in it, Schweitzer explores this ethical (and spiritual) problem. Schweitzer's answer, reached in the third section published after the War, was to found ethical action on a principle Schweitzer called "the reverence for life." By doing this, he said, it would be possible to make decisions that were more fair, just, and life-giving than society at the present time was making; he noted that the principle was a general one, for it was not only human life, but all living things, for which people were to have reverence.

The idea of "reverence for life" entailed not only an ecological view of life but also one in which a spiritual dimension in all living things was acknowledged and respected. Moreover, it was not merely a Christian spirituality that Schweitzer said must underpin ethics in civilization, but it was a spirituality in general terms that—across religious boundaries, as well as cultural and political ones—had not just a respect for life, but a reverence for it.

In the search for some noncoercive means of uniting people across social, political, cultural, and economic as well as geographic boundaries, working out some vague consequentialist environmentalism to guide the activities and choices of individuals in the global community is not likely going to be enough. There does, however, need to be some ethical framework within which to consider options that, in some form and in the service of some greater, global good, will not have negative effects on people, places, and human institutions. Such a framework will be difficult to find, to articulate, and to accept. Perhaps Schweitzer's idea of reverence for life might turn out to be as useful an ethical touchstone for global decision making today as he thought it would be nearly a century ago.

See also Technology; Technology and Progress.

Further Reading: Aronowitz, Stanley, and Heather Gautney, eds. *Implicating Empire: Globalization & Resistance in the 21st Century World Order.* New York: Basic Books, 2003; De Rivero, Oswaldo. *The Myth of Development: The Non-Viable Economies of the 21st Century.* New York: Zed Books, 2001; Easterly, William. *The White Man's Burden: Why the West's Efforts to Aid the Rest Have Done So Much Ill and So Little Good.* New York: Penguin, 2006; Franklin, Ursula. *The Real World of Technology.* 2nd ed. Toronto: Anansi, 1999; Landes, David S. *The Wealth and Poverty of Nations: Why Some Are So Rich and Some So Poor.* New York: Norton, 1999; Saul, John Ralston. *The Collapse of Globalism and the Reinvention of the World.* Toronto: Viking, 2005; Schweitzer, Albert. *The Philosophy of Civilization.* Trans. C. T. Campion. New York: Macmillan, 1949; Stiglitz, Joseph. *Globalization and Its Discontents.* New York: Norton, 2003; Stiglitz, Joseph. *Making Globalization Work.* New York: Norton, 2007.

Peter H. Denton

GREEN BUILDING DESIGN

Professionals in the building industry are becoming more aware of the enormous effects that buildings have on the environment. This awareness has brought forward many ideas, discussions, and solutions on how we can reduce the harmful environmental effects of the buildings we construct. Green building design and practices are all based on the fundamental respect and management of the resources we use to create and use spaces.

Green building design thus takes into consideration the following: energy use and generation, water use and conservation, material and resource selection, land use and selection, waste management, environmental management, built versus natural environments, and occupant comfort and well-being.

There are a number of debates surrounding green building design. The key areas of concern deal with the shift in the design methods, the vision of its value beyond the higher initial costs of design and construction, and the superficial focus on the marketing and status that come with a green building.

Green designers must look at the different ways that buildings interact with the outside environment and make choices to minimize or at least reduce the effects of that interaction. Buildings use raw materials for construction, renovations, and operations. They also use natural resources such as water, energy, and land. They generate waste and emissions that pollute our air, water, and land.

For most, green building is considered a building approach that includes environmental concerns in the design, construction, and operation of a building. The environmental impact or footprint of the green building is smaller than that of a typical building. Typically, its energy consumption is the first measure of a building's level of green but there is much more to green (or sustainable) design principles.

Buildings have a significant effect on the environment, creating a complex series of considerations for a green approach. One must look at the different phases of any building's existence, from the resource consumption in the production of the building materials used in its construction to its daily operation and maintenance to its decommissioning at the end of its usefulness. At each phase the building affects the environment either by consuming natural resources or by emitting pollution. Such effects can be either direct or indirect; a direct effect is the greenhouse gas emission from the building's heating system, whereas an indirect effect is the resource consumption or pollution caused by the utilities creating the energy that lights and provides power to the building.

Proper green designers, builders, owners, and users look at all these environmental aspects when dealing with their building. In some cases, they might be dealing with an isolated system, but increasingly the systems and technologies used for a building depend on other systems or technologies to be effective.

Green building is thus a systemic approach to building. Those involved must be knowledgeable about many areas of a building, not just their own specific area. This is one of the first hurdles in green building. An architect always must have an understanding of all of the systems and intricacies of a building's design, construction, and function. Now, with green building, so must everyone else involved in the building project. The contractor must understand the needs and concerns of the mechanical engineer. Both of them must understand the plans of landscape architect and the plumbing and electrical engineers. The building's systems work together; they are not isolated as past designs would have made them. The landscaping can shade the building in the summer, reducing the air conditioning and allowing sunlight into the space in the winter when light and solar heating are needed. With everyone understanding each other, the contractor will know that he cannot cut down the existing trees on the property because they affect the building's design. The mechanical engineer knows that the heating and cooling systems can be smaller. The plumbing and electrical engineers can consider rainwater collection and solar power arranged around the landscaping and the modified mechanical system.

Studies have found that the most effective green buildings require having the project start with what is called an Integrated Design Process (IDP). This process requires that all of those who are involved with designing the building participate together in a series of design meetings to discuss the design of the building. The building owner, the builder/contractor, the architect, the structural and mechanical engineers, and any other individuals with a stake in the building project must attend all of the meetings. Everyone shares ideas and design suggestions. Everyone discusses the possibilities for this project. Those with experience in this process have found that many inefficiencies are removed from the design before the plans are even drawn up because everyone is at the table from the beginning to point out what will or will not work.

When it comes to green building and the interdependence of the green systems and technologies, everyone needs to be part of the design from the beginning. This new approach is difficult for many who are accustomed to the old, more linear approach to design and construction. Many design professionals are comfortable with their standard methods of design and do not adjust well to the shift to a new style. Building owners need to see immediate progress for the money that they are spending and thus are uncomfortable with the outlay of time and money involved in a longer IDP design process. It is difficult for them to see the money that they will save from numerous oversights caught in an IDP meeting that would have been overlooked following the older methods.

After the design process and the construction of the building are completed, the critics are there to share their opinions on the "greenness" of the project. There is constant criticism that a particular building project was not built green enough. In fact, every decision in green design is continually tallied. From a material choice to the construction method, the questions are asked: What is the greenest choice? What is reasonable? Just as there are many options for building a conventional building, there are many more for green building. The designers must weigh many criteria to determine the best solution. In the example of material selections, the designers must consider the durability of the material, the amount of renewable resources for the raw materials used to produce the material, the embodied energy of manufacturing and delivering the material, and the toxicity involved in the material's installation, use, and maintenance. In a specific example such as hardwood flooring, bamboo is durable and highly renewable, and the installation has little in volatile organic compounds (VOCs). The other end of the scale of consideration shows that there is a high level of embodied energy when this product is shipped from another continent. There are also the finishing choices to protect the flooring once it is in place and to be reapplied throughout the flooring's life. Some of those choices are high in VOCs. Local hardwood as another option has the same durability but less embodied energy in its delivery. New finishing choices provide a low- or no-VOC option. The designer must weigh the flooring options, with bamboo's renewable quality versus the locally available hardwood. Whichever choice the designer makes, critics will point out that there were other options.

Money is obviously also a factor in the decision making, either in the cost of innovative products and technologies or in the rewards of status and reputation created by being part of a significant green building project.

As a result of purchasing and installation expenses of the newer technologies or construction methods, the initial costs of a green building may be higher. Once in place, however, the operation costs of the building may be much lower. For example, energy and water usage are reduced, so utilities costs are lower. Green building designers first reduce consumption of energy and water with efficient fixtures and operations. If they want to go further, they add technology to generate power or reuse their water to reduce dependency on the utility providers. Unfortunately, building owners and their financial advisors are hesitant to pay higher costs upfront without proof of the future savings.

There are hidden savings in the human resources area, as well. The people working inside these buildings are not tired or sick as often, so they take fewer breaks or sick days. These conditions differ for each project, so the savings are not easily calculated, but research is being done that will eventually quantify such positive outcomes of green building design.

There are many other reasons a building owner should build green, not all of them environmental. The status of a green building is a big draw. The competition for elite building owners used to be to build the tallest tower. Now, almost every building owner may participate in the competition of green building—the first green building in this locale, the first green building of that type, or the first green building to have a particular new innovative technology. Building design teams can be driven into these competitions, ironically losing the original vision of the green building itself as they compete.

Design teams that seek status with their work sometimes get caught in what has been called "greenwashing" or "green ornamentation." Greenwashing and ornamental green buildings have confused the public about what green building intended to do, which was to build buildings that reduce their environmental footprint in construction and operation.

Greenwashing is making the claim of being a green building on the basis of merely having installed one or two resource-reducing systems or products, without incorporating the green philosophy into all the building design considerations. The building design industry has developed systems to attempt to measure the level of green a building has achieved and to prevent greenwashing. These green systems, using labeling techniques similar to nutritional labels designed to ensure the integrity of low-fat and high-fiber claims, were established to eliminate the false claims of green. These systems have also helped to increase understanding and process in green design but have created a new competition in building. Designers and building owners now use these systems to seek the highest level of green measured by the system, decided by a system of points. If a technology or method that promotes sustainability does not provide a point, it is not included in the project. This point-chasing approach leads to the other status building, the ornamental green building.

The ornamental green buildings seek grand reputations and status based on their building by adding the systems and technologies that create the buzz of green building with the public and the design and building industry. These systems, though green in their intention, lose their green effectiveness in relation to the complete building design. These buildings have their green technologies on display not for public education purposes but for marketing and status. The technologies and systems put into place are selected based on their appearance and potential marketability more than their resource reduction and ecological purpose. Although these technologies are assisting with sustainability, the neglected areas of the building project may as a result be undercutting any good things the rest of the design accomplishes.

Green building practices are working their way into governmental and corporate policies. Leading industry groups are working to incorporate some of these practices into the official building codes. As these codes and policies are being developed, the fundamental questions must still be answered: Is this the greenest approach? Is this reasonable? Much of our future may depend on our ability to answer the questions effectively.

See also Ecology; Global Warming; Sustainability.

Further Reading: Edwards, Brian. *Green Buildings Pay.* London: Routledge, 2003; Fox, Warwick. *Ethics & the Built Environment.* London: Routledge, 2001; Harrison, Rob, Paul Harrison, Tom Woolly, and Sam Kimmins. *Green Building Handbook: A Guide to Building Products & Their Impacts on the Environment.* Vol. 1. London: Taylor & Francis, 1997; Johnston, David, and Kim Master. *Green Remodeling: Changing the World One Room at a Time.* Gabriola Island, British Columbia: New Society Publishers, 2004; Woolly, Tom, and Sam Kimmins. *Green Building Handbook: A Guide to Building Products & Their Impacts on the Environment.* Vol. 2. London: Taylor & Francis, 2000; Zeiher, Laura. *The Ecology of Architecture: A Complete Guide to Creating the Environmentally Conscious Building.* New York: Watsun-Guptill, 1996.

Shari Bielert

HEALING TOUCH

Healing Touch is a noninvasive energy-based approach to improving one's health and well-being that complements traditional methods of medical intervention. Janet Mentgen (1938–2005) was a nurse who noticed the healing aspect of touch during her nursing career. Mentgen studied how energy therapies influenced the healing progress of her patients. She went on to develop the Healing Touch program, which now is taught worldwide. In conjunction with other healers and Mentgen's own practice, Healing Touch evolved and developed throughout her lifetime.

Healing Touch is a complementary alternative medicine (CAM) based on the principle that a person's energy extends beyond the parameters of his or her physical skin and that there are layers of energy, physical, mental, emotional, and spiritual in nature, surrounding each individual. Proponents of Healing Touch believe that healing can occur in these domains by modulating the energy fields of a person to aid in the relief and sometimes the elimination of an ailment in the physical, mental, emotional, or spiritual realm. Opponents of Healing Touch question the validity of such claims of recovery and ask for scientific proof to back up the anecdotal stories of healing. Some go as far as to say that Healing Touch has about as much efficacy as the snake-oil approach to treatment in days gone by.

People have gravitated toward Healing Touch when traditional solutions to healing have been unable to deliver the results for which they had hoped. Ongoing physical pains and discomfort can become tedious, and a more educated consumer culture has become wary of addressing issues with medication. Given the increased frequency of news stories reporting on the overuse of over-the-counter medications, it is no wonder that people are concerned about the effects on their systems of overusing and abusing drugs.

Consider the overuse of antibiotics, enabling bacteria to become "super bugs" that are no longer susceptible to antibiotics. Television ads promote a variety of drugs to solve sleeping problems, headaches, stress, and a host of other ailments.

Healing Touch can complement medical intervention and sometimes be used as an alternative to medication. Cancer patients have found a decrease in physical pain when Healing Touch therapy has been provided. Parents have used Healing Touch extensively with their children. In the everyday play of running and jumping, falls that incur bumps and bruises are not uncommon. When Healing Touch has been applied, children have calmed down quickly, and the swelling from bruises has significantly decreased—even the color of the bruise has rapidly changed to the yellow-green color of a bruise that is almost healed. The bruise has disappeared much more quickly with a Healing Touch treatment than would a bruise that is left untreated.

According to the Healing Touch philosophy, we are each surrounded by an energy field and a mental energy field. People talk in terms of "energy" in their everyday language in reference to how they feel. A person may describe his or her weakened mental capacity by saying, "I don't have the energy to do this," which can refer to the person's inability to complete a cognitive task. Or the ability to focus on one task may be compromised because the person's mental energy at the time is not "balanced." Another typical expression is "my head's not clear." Healing Touch practitioners pay attention to these words and translate them into a Healing Touch assessment. A depletion of mental energy can be attended to by means of a variety of forms of Healing Touch treatments, such as the "mind clearing technique" or a "chakra connection" in which the client's energy field is modulated to address his or her condition. People who have experienced the mind-clearing treatment in which the practitioner places his or her hands in a sequence of positions on the person's head, forehead, and face over the course of approximately 20 minutes, have described feeling much more relaxed and calm after the treatment. It is not uncommon for the client to fall asleep. The healing touch practitioner would say that the smoothing of the mental energy field has facilitated the client's self-healing.

People suffering from emotional distress whose feelings of well-being are compromised by a variety of life-event stressors often enter into therapy to find relief, answers, and solutions to their personal problems. Therapists or counselors are trained to deal with emotional health issues via talk therapy. Many who have accessed this form of help have improved their emotional health. Medication for depression, anxiety, and other types of problems has been useful to people as well. Sometimes talk therapy and medication are not effective, however, and again people look for alternate solutions. People remain "stuck" and unable to move forward in their lives. Therapists and counselors have referred these clients to healing touch practitioners for additional support.

According to Healing Touch theory, an emotional trauma can be stuck in a person's emotional energy field. "I'm feeling drained" describes what it feels like when the energy reserves of a person are at a low point, and the person is finding it difficult to deal with personal problems. By modulating their energy, individuals can feel relief without having to actually talk about the indescribable emotional suffering they are experiencing. The practitioner may teach clients a technique that the they can use on themselves in order to feel better emotionally. The chakra connection is one such technique commonly taught to participants in Healing Touch sessions (chakra, from the Sanskrit, meaning a wheel of light—energy centers within the body that help to generate an electromagnetic or "auric" field around the body). The major chakras of the person are all connected, beginning with the chakras in the feet and moving up to the knees, hips, solar plexus, spleen, heart, arms, neck, forehead, and crown (top of forehead) and then connecting to the energy of the universe. Once a person's chakras are connected, the individual often feels less stressed and calmer and has a renewed capability to handle his or her fears, worries, and emotional problems.

Questions of a spiritual nature often get addressed within the context of a person's spiritual beliefs. Eastern and Western societies have numerous theologies or philosophies.

In Canada, Healing Touch has taken place in some religious institutions such as the United Church of Canada. Christian physical therapist Rochelle Graham founded the Healing Pathway Program, which incorporates Healing Touch theory within the Christian tradition. This program is designed to train people who want to develop their healing skills in accordance with their Christian beliefs. Healing Touch has been reported to provide a spiritually peaceful tranquility to the person receiving the treatment. One anecdote tells of a man who was involved in a horrendous car accident on the 401 highway outside of Toronto. A doctor asked this man how he had managed to live, because all of the doctor's experience and the surely life-threatening injuries of the accident seemed to preclude survival. He told the doctor, "It was as if I was being held in a bubble," in an attempt to explain how the healing touch treatment had made him feel.

Opponents of Healing Touch within the Christian sphere would argue that Healing Touch is unbiblical. They might say there are only two sources of power: God and Satan, good and evil. When illness or tragedy strikes, sometimes there is healing, and sometimes there is not; whether healing happens is up to God. A person may want and pray for healing, but whether or not healing occurs is God's will.

Touch is something to which humans are drawn. A hug or embrace can be congratulatory, comforting, or consoling; the impact of such is frequently immeasurable. A person can relieve stress by gazing out a window away from the business of work. A person's spirits can be raised by an unexpected phone call from a loved one. Research has demonstrated that one of the most effective pills is the placebo. So whether an individual uses biomedicine or Healing Touch or both, what ends up being the determining factor in the success of the treatment may be that person's belief system.

See also Health and Medicine; Health Care.

Further Reading: Bruyere, R.L.O. *Wheels of Light*. Ed. Jeanne Farrens. New York: Simon & Schuster, 1989; Healing Touch Canada. http://healingtouchcanada.net; Hover-Kramer, D. *Healing Touch: A Resource Guide for Health Care Professionals*. New York: Delmar, 1996.

Susan Claire Johnson

Healing Touch: Editors' Comments

It is easy enough to make touching sound mysterious, mystical, and ineffable. And yet it goes hand in hand (so to speak) with the fact that humans are radically social, the most social of all living things. Touching is a basic feature of becoming human, not just during infancy and early childhood, but throughout one's life. The significance of touching has been long known among social scientists and is the basis for much of the not-always-rigorous talk about energy. To be social is to be connected, and the connection that goes along with being human is the basis for our feelings and emotions. We are human in the best and fullest sense of that term only to the extent that we have the opportunity to experience the touching aspects of social life. When any two people come into each other's presence, it is as if two coils of wire with electricity coursing through them were brought into proximity. Electrically, this phenomenon generates a field linking the two coils. Socially, something like an energy field is generated between people. It is this field that provides the pathway for emotions and ultimately for communication. The anthropologist Ashley Montague was one of the strongest advocates of touching as a fundamental and necessary feature of the human condition. And in his work on interaction ritual chains, Randall Collins has provided solid social science grounding for our common-sense ideas about touch and energy.

Further Reading: Collins, Randall, *Interaction Ritual Chains*. Princeton, NJ: Princeton University Press, 2005; Montague, Ashley. *Touching: The Human Significance of the Skin*. 3rd ed. New York: Harper Paperbacks, 1986.

HEALTH AND MEDICINE

Many dimensions of health and medicine are controversial. The issues range from conflict within medicine over explanations of disease processes to public policy issues surrounding research priorities and the allocation of health care resources. Few of these issues divide neatly into sides, or into sides that hold across more than one case, because people's everyday experiences may put them on one side or the other in unexpected ways. For example, a person who holds an antiabortion or pro-life position regarding reproduction may nonetheless think that it is important for people to be able to refuse medically invasive care in the case of a terminal illness. Some might find these two positions to be in conflict, and it is difficult to decide in the abstract what the right medical choices might be for concrete cases based on real people's lives and values.

Health and medicine are sources of public controversy because there are points of detailed scientific controversy at many levels. Scientists can differ in theorizing the causality of illness, and many conditions and illnesses have complex and multiple causes. This creates difficulties for researchers, health care practitioners, and patients. For example, it is difficult to determine the cause of breast cancer in general and for any specific patient. Despite the existence of several genetic markers (e.g., BRCA1 and BRCA2), the presence of those genes in a woman does not determine if she may get cancer or when, although these genes strongly correlate with higher risks for cancer. But further, many individual cases of breast cancer are not attributable to known genetic factors or specific environmental causes. In only a few cases can any specific cancer be attached to a known cause. This confronts researchers, physicians and medical professionals, and of course patients with a variety of complex issues, questions, and problems. The lack of certainty makes it difficult for a patient to determine the best course of action to avoid cancer or to select cancer treatments. The different sources of causation also lead to questions about research directions and policies. For example, some breast cancer activists argue that the focus on genetics in breast cancer means that there is not sufficient attention paid to environmental causes, such as exposure to toxic chemicals, in setting the research agenda and awarding funding.

Besides the complexity of causation creating controversy in health and medicine, there are sometimes more specific controversies over causation per se. For example, although largely discredited now, Dr. Peter Duesberg developed the hypothesis that AIDS is not caused by the HIV virus, but by exposure to recreational drugs and other lifestyle choices that wear down the immune system. (The millions of people in nonindustrial nations who do not do the recreational drugs in question yet do have HIV/AIDS should be seen as fairly strong evidence against the Duesberg hypothesis.) Sorting out this controversy slowed down the research process and produced confusion for patients and also contributed to policies that were unjust. Scientists had to spend time and energy repeatedly refuting competing claims, and people often put themselves or partners at risk by denying the importance of HIV test results. Some part of the controversy signals concerns with the pharmaceutical industry and the way that it makes enormous sums of money from the current definition of the HIV/AIDS link. Some part of the controversy also reflects a support for a punitive approach to HIV/AIDS by people who think that those who get HIV/AIDS somehow deserve their illness. In international contexts, the resistance to the HIV/AIDS link is both a resistance to the pharmaceutical "colonization" of a country that must import expensive drugs and an attempt to deny a problem on the international level that produces feelings of shame or inferiority. Thus, many controversies in medicine are about more than just the science; they are about the association of different facts with values and policy implications.

Controversies in medicine also arise in places where there are not yet frameworks for understanding conditions. For example, chronic illnesses and chronic pain are difficult for many physicians and researchers to conceptualize because in mainstream medicine there is a general expectation of getting over an illness or getting better. So conditions that do not have a clear trajectory of improvement are of course difficult for patients, but often difficult for professionals (and insurance companies) to understand as well. Conditions that are relatively new can also produce controversy. For example, fibromyalgia and related chronic fatigue diseases are sometimes discounted by medical professionals as not really illnesses or as signs of depression or other medical or mental disorders. Although there are many professionals who take these conditions seriously, the lack of clear diagnostic criteria and causal mechanisms means that even the existence of the condition as a treatable disease is sometimes called into question, creating hardships for those who experience the symptoms associated with those conditions.

There are other kinds of controversy that emerge around health and medicine, for example in the definition of something as an illness or medical condition. This is part of the process of medicalization and can be part of the jurisdictional contests between different professional groups. For example, physicians worked hard in the eighteenth and nineteenth centuries to discredit midwifery, to exclude traditional women practitioners from the domain of childbirth. Certainly at the beginning of the professionalization of medicine, the "male midwives" and professional obstetricians did not produce successful outcomes by intervening in the birth process. One reason was that the concept of asepsis had not taken hold in medicine. Furthermore, many traditional midwives were very experienced at taking care of the health of women and infants. But the new medical men portrayed midwives as dirty, superstitious, and backward and emphasized the dangers of the childbirth process as well as their professional qualifications to treat women. The medicalization of childbirth helped exclude women from this realm of action they had traditionally controlled and eventually led to a backlash. American women are particularly subject to unnecessary caesarean section yet still suffer high rates of birth-related mortality. Their general lack of control over the birth process led to a countermovement that focuses on childbirth as a natural process, not a medical procedure. The result has been more natural and home-birth experiences and in many states a reemergence of midwives as alternatives to medicalized birthing.

Medicalization can be thought of as the professional capture of the definition of a condition. In particular, forms of social deviance are candidates for medicalization. For example, over the past several decades, alcoholism has been increasingly interpreted as a medical condition (including the newest genetic explanations), in contrast to older interpretations of alcoholism as a moral failing or weakness of will. This, of course, greatly expands the research and treatment possibilities for the condition. Medicalization also contributes to removing moral stigma from a condition, perhaps making it easier for people to admit that they have a problem and seek treatment. Some would argue, however, that medicalization occurs at the expense of holding people responsible at least in part for their behavior and condition. Similarly, attention deficit hyperactivity disorder (ADD/ADHD), premenstrual syndrome (PMS), obesity, and some forms of mental illness have undergone medicalization processes in recent years. This validates peoples' struggles with the conditions to some extent and removes moral stigmas. This is at the expense, however, of finding nonmedical solutions or alternative interpretations or treatments of physical and mental processes and cedes control and definitions of experience to a small and relatively powerful homogeneous group: physicians. On occasion, there are also demedicalizations of conditions, such as homosexuality, which was removed from the major psychiatric diagnostic encyclopedia in the 1980s because of pressure from gay and lesbian activists and allies and increased understanding by medical professionals themselves.

Controversies in health and medicine are both the cause of and a result of the emergence of alternatives for health care and understanding disease and illness. Alternative medicine is both the continuation of traditional medical practices not recognized by formal medical authorities and the emergence of new perspectives in part engendered by resistance to medicalization processes. Alternative medicines take many forms and have varied greatly over time, and one question that should always be considered is, what is alternative to what?

The category of traditional or indigenous medicine refers to the long-standing practices of groups not part of the Western scientific medical trajectory. Sometimes these practices are part of long legacies of local medical practices, predating Western medicine, such as the herbal and practical medicine of European cultures before 1800. Sometimes parts of these practices have been maintained, although with changing circumstances, by groups who are still connected to older traditions. Similarly, Chinese and other Asian cultures have maintained strong legacies of traditional medical practice, and many people who do not have access to conventional Western medicine have medical systems that do not conform to Western medicine in either theory or practice. It is easy to dismiss these practices as somehow primitive or inferior, and certainly many older practices, such as bloodletting in Europe through the eighteenth century, often did more harm than good. More than a billion people in China, and nearly as many in other cultures, however, have medical traditions that seem to have been effective for promoting the health of many people over many generations.

Within Western contexts, practices such as homeopathy (treatment with very small amounts of medicines that produce specific counter-symptoms), naturopathy (focusing on the whole patient in the context of only naturally available substances and diet), osteopathy (focusing on musculoskeletal alignment and preventive medicine), and chiropractics (using physical manipulation to treat pain and illness) are among many alternatives to the usual routines of conventional Western medicine. Faith healing and "scientology" can be considered alternatives to conventional medicine as well. Mainstream scientific Western medicine is labeled allopathic or orthodox medicine in relation to these other traditions.

The alternative medical practitioners have launched powerful criticisms against conventional Western medicine, including the lack of attention to preventive health care. They have also criticized the capture of medicine by the pharmaceutical industry, which puts profits before people, and the dire harms often experienced by patients as a result of medical error or iatrogenesis, which is the harm caused by medical treatment itself. For example, some debate whether chemotherapy and its horrible side effects improve the quality of life of many cancer patients, when there are potentially other less invasive and less destructive models for treating cancer. Philosophically, some within and outside of conventional medicine critique Western medicine's approach to death as a failure, rather than as a part of the life process, leading to use of invasive and expensive medical technologies at the end of life and invasive and painful treatments of illnesses such as cancer. Also, rigorous models of scientific practice, discussed in the next paragraph, often impinge on some professionals' ideas about the art of medical practice and their creativity and autonomy as practitioners.

In part to protect patients from quackery and harm, and in part to protect professional jurisdiction, a specific model of scientific experimentation is considered the gold standard to which medical treatments are held. The double-blind experiment, where neither patients nor researchers know if a patient is in a control group receiving (1) no treatment or an already established treatment or (2) the experimental treatment, is considered a key way of sorting out the effects of a potential medical intervention. Any medical experimentation is complicated by the placebo effect, where people feel better just because they are getting some sort of medical attention, and also confounded by the problem of people sometimes, even if rarely, spontaneously getting better from all but the most dire medical conditions. It is also unclear in advance whether a treatment that is effective for 85 percent of patients will necessarily be the right treatment for any specific patient who might be in the 15 percent for whom the treatment does not work. The need for rigorous medical experimentation is critiqued by those who advocate for alternative medicines, for it is nearly impossible to test the very individualized treatments of many of these practices. And medical experimentation also creates controversy over problems of informed consent and risk. For example, medical researchers must present the anticipated risks of medical experimentation and not scare patients away from consenting to participate in clinical trials of new medicines or procedures. This leads, some would argue, to systematically understating the risks of treatments, contributing to problems such as the need for drug recalls after products are on the market and people have used medicines that have harmed them.

Scientific medicine is also critiqued for its use of nonhuman animals as experimental models. To the extent that other species are like humans and are scientifically good models, their similarities with humans could be argued to require that we extend many human rights to them, including the right to not be used in medical experiments against their best interests. To the extent that other species are unlike humans (and thus not deserving rights), then the question as to how accurate they can be as models for humans in the experiments can lead one to entirely reject the use of animals in experimentation. Although some do hold the position of preventing all use of animals in medical experimentation, most who see the necessity of animal experimentation also believe in avoiding unnecessary suffering for species that are used.

Even with the problems of scientific medicine, many proponents of alternative medicines sometimes strive for the recognition of their practices as scientifically valid medicine. Because science has the power to culturally legitimate medical practices, there are numerous new research programs examining alternative and complementary therapies, from herbal and behavioral therapies for cancer or chronic illnesses to experiments on the power of prayer or other faith-based interventions in pain relief and mental health. Even the existence of these attempts at scientific experimentation are sometimes scoffed at by the most scientifically minded orthodox practitioners, who see these experiments as trying to put a veneer of respectability on quackery. More activist critics of scientific medicine see these as capitulating to the legitimacy of scientific reductionism and defeating the entire purpose of an alternative medicine truly focused on individuals. Still, many more persons find it reasonable to try to sort out whether different medical systems, sometimes viewed as radical, can safely improve human health.

Even within contemporary conventional medicine, controversies frequently emerge. For example, anti-vaccination movements arise sporadically, either because of side effects or potential links of vaccination to illness outbreaks. For example, the live polio vaccine has been rejected by several African nations in part because the live vaccine can produce the polio illness, although at far lower rates than in an unvaccinated population. Preservatives such as thimerosal have been removed from most childhood vaccines because some believe they are implicated in autism, despite the lack of concrete evidence. Many dismiss critics of vaccination as irrational, but that does not recognize the real risks of vaccinations, which sometimes do produce illnesses and reactions, and also the fact that vaccination programs seem to be an infringement on people's rights to control their own health. However small the risks might be, people are generally opposed to risks that they perceive they have no control over, in comparison to risks they might voluntarily take.

Because health is fundamentally a cultural and value-based phenomenon, the emergence of controversy in health and medicine should be expected and if not embraced, not dismissed as somehow irrational either. Many controversies surrounding health and medicine are of course focused at the beginning and end of human life, where values and emotion are most explicitly engaged. For example, the use of embryonic stem cells is seen by some people as the destruction of potential human beings, clearly a value judgment that shapes the perception of possible research. To pronounce a human embryo as not a person having rights because it cannot yet think or live independently, however, is also a value judgment. Controversies at the end of life (such as over organ donation or the use of feeding and respiration tubes) are similarly inseparable from value decisions about the quality of life.

These value questions intersect with political processes and become public debates about health care research priorities and questions about the distribution of health resources. Research agencies such as the Medical Research Council of Canada (MRC), National Institutes of Health (NIH), and the National Science Foundation (NSF) as well as the military and space agencies all fund or conduct health and health-related research, and there is an extensive federal bureaucracy and regulatory schema for allocating medical resources to the poor, the disabled, children, and the elderly and (in the United States) for putting restrictions on what private insurance companies must and must not do for their subscribers.

The U.S. federal ban on stem cell research is an example of a controversy over a cultural definition of what life is intersecting with a narrative of scientific progress and the search for future benefits from medical experimentation. California passed its own enabling legislation, and the research can go on in the private sector (where it will be far less regulated), but this ban constrains the opportunities for research and some approaches to solving medical problems. Although less controversial, per se, other issues, such as the lack of attention to breast cancer and women's experiences with disease, also reflect how priorities and opportunities are shaped by values in public policy. This neglect has been addressed by women activists and other leaders. The distribution of wealth and power in society clearly shapes the attention to medical needs. For example, international commentators have noticed that although the United States and European countries have the most resourceful and powerful medical research systems, because the populations do not suffer from infectious diseases such as malaria or cholera, little research into prevention and treatment is available for the poorer populations in the world that do not have the wealth or access to the medical research enterprise. People in the United States and Canada, however, know a lot about heart disease, erectile dysfunction, and other conditions of an affluent elite.

Access to health care more generally is a matter of both national and international inequality. Although access to clean drinking water and adequate food supplies goes a long way to ensuring the health of populations (and some would argue is more important than specific medical interventions such as antibiotics in improving human health), nonetheless, access to medical intervention is considered a human right by many observers. However, because health care is distributed by market-based mechanisms, this produces inequalities. The pharmaceutical industry is a multibillion-dollar enterprise of global scope, and the insurance industries are immensely profitable national organizations. Critics argue that the pharmaceutical industry shapes research in ways that promote its interest in profits, not a more general interest in human health. For example, prevention of illness generally makes less profit than treating people after they become sick, so prevention is not emphasized in research policy. Although industry representatives argue that drugs are expensive because they must undergo such rigorous research and testing, much of that testing is supported in part by public research grants and has already been paid for, again in part, by the public. So people who need drugs are often the least able to pay for them.

The United States, in particular, values individualism and trusts in the market to meet social needs. This leads to an ideology of healthism, which converges with neoliberalism as a model of social relations and government intervention. Healthism is the application of sanctions, moral and economic and both positive and negative, to the health of individuals. For example, physical beauty is assumed to be healthy, and health, beauty, and physical fitness represent goodness in the media. An individual struggling with his or her health or fitness is assumed to be at fault in some way, often despite medicalization processes that have a tendency to remove blame from the individual. As a political philosophy, neoliberalism in general mandates that an individual is responsible for his or her own success, and an individual's participation in the free market will guarantee his or her happiness. State intervention in health care or the market is seen as a distortion or inefficiency in the market or an impingement on personal liberty. Of course with regard to health, if you are not healthy, you cannot participate in the market to gather wages to pay for health care, producing a vicious cycle. Each of these ideologies is unable to explain the role of context and social power in shaping the opportunities of individuals to take care of themselves either financially or physically. Insurance discrimination is an example of this: if a person has a preexisting medical condition, it is extremely difficult to get insurance, and if it is available, it will be very expensive. It is in the best interests of an insurance company to avoid people who might be expensive, whether because of a known preexisting condition or because of anticipated risks measured through behavior (such as smoking) or medical tests (such as DNA testing). Healthism and neoliberalism both assume an equality of access to opportunity, which is not possible in a highly stratified society.

Bioethics and more specifically medical ethics are fields of study that attempt to help sort out some of the value-based controversies that surround medical issues. The approaches of bioethics have both strengths and weaknesses. The systematic discussion of values and principles can help to clarify key issues and examine ideologies and assumptions that are shaping people's responses to ethical problems. But bioethics can also serve as a form of rationalization for pursuing a specific course of action. For example, some commentators on the U.S. federal stem cell ban posit that economic competitiveness is an ethical good (because it produces jobs and money) and should be considered in the evaluation of the benefits and risks of medical innovations (such as stem cell research). Because bioethicists take a professional position of neutrality, they can often illuminate the sides and possibilities of an ethical debate without adding much clarity as to what actually ought to be done, whether as a matter of public policy or as a matter of personal decision making.

There are many more specific health and medical controversies, and more are expected to come as scientific and technological innovation shape what is possible and what is desirable for human health. Because U.S. mainstream culture places a high value on innovation, change is often assumed to be good without detailed examination of short- and long-term consequences and consequences that are a matter of scale—from the personal to the political. Individuals and social groups need to be more informed as to the real (rather than the fantastic or imaginary) potential of medicine and medical research and to the processes that shape the setting of research priorities and the institutions that shape the ethical conduct of research and health care delivery.

See also Health Care; Immunology; Medical Ethics; Reproductive Technology.

Further Reading: Birke, Linda, Arnold Arluke, and Mike Michael. *The Sacrifice: How Scientific Experiments Transforms Animals and People.* West Lafayette, IN: Purdue University Press, 2006; Conrad, Peter, and Joseph W. Schneider. *Deviance and Medicalization: From Badness to Sickness.* Philadelphia, PA: Temple University Press, 1992; Hess, David. *Can Bacteria Cause Cancer? Alternative Medicine Confronts Big Science.* New York: New York

University Press, 2000; Lafleur, William, Gernot Bohme, and Shimazono Susumu, eds. *Dark Medicine: Rationalizing Unethical Medical Research.* Bloomington: Indiana University Press, 2007; Scheper Hughes, Nancy. "Parts Unknown: Undercover Ethnography of the Organs-Trafficking Underworld." *Ethnography* 5, no. 1 (2004): 29–73; Star, Paul. *The Social Transformation of American Medicine.* New York: Basic Books, 1984; Timmermans, Stephan, and Marc Berg. *The Gold Standard: The Challenge of Evidence-Based Medicine.* Philadelphia, PA: Temple University Press, 2003.

Jennifer Croissant

HEALTH CARE

Health care is a term laden with political implications, particularly in Western societies. Although there are many conflictual issues relating to science and technology in medicine, when it comes to health care, those issues are woven together with a series of equally complex social and political considerations. If medicine is the system by which human diseases and physical infirmities are addressed, then health care is about how that medicine is delivered, by whom, to whom, for what reason, and at what costs.

Health care needs to be understood in terms of three primary divisions: geographic, political, and social. Where someone lives, globally, affects the nature and kind of health care received; in North America or Europe, for example, more patients are closer to advanced health care (such as hospitals, diagnostic equipment, and medical specialists) than in Africa. Within all societies, patients tend to be closer to advanced health care if they live in urban areas; because of the economic costs of advanced health care, such facilities tend to be located in densely populated urban areas and not in the countryside. In the areas of the world identified as "developed," the accumulation of wealth leads to a multiplication of health care options, from diagnostics to treatment; in "developing" countries, such options (for economic reasons) simply do not exist. Mandatory vaccination programs covered by governments will reduce or eliminate certain diseases in some regions of the world, whereas, in other regions, because governments cannot afford such vaccination programs, there is a high child mortality rate from these same diseases. Similarly, in areas of the world where there is good nutrition and clean water, the incidence of disease tends to be lower, requiring less cost for health care; in areas of the world where there is little food or the water is not safe to drink, health care costs can increase dramatically—in a society that is impoverished and under threat to begin with.

Political divisions that affect health care tend to be national boundaries; different countries have different health care systems that reflect the political choices made by the government or its citizens. Although Cuba, for example, does not have the money to deliver advanced health care to many of its citizens, there are more doctors per capita than in developed countries, such as Canada and the United States, where a large percentage of the population do not have a doctor at all. Canada has a universal health care system funded largely by the government; in the United States, health care is dominated by private health management organizations (HMOs) and is funded primarily by the private sector. In both countries, critics of the health care system make it a political hot potato for candidates in every election at both local and national levels.

Social divisions that affect the delivery of health care relate to the socioeconomic status of different patient groups. In a privately funded health care system, people who do not have adequate medical insurance do not receive medical treatment, except perhaps in emergency circumstances. Needed diagnostic procedures or treatments may be deferred or not happen at all because of the economic costs to the individual; doctors are, in effect, not able to provide medical treatments they know are necessary unless the health management authority authorizes the expenditure. Critics bluntly state that people die because they cannot afford the health care that would save their lives; illnesses, accidents, or operations can mean financial catastrophe for low-income families whose incomes are too high to be covered by programs (such as Medicaid in the United States) but not high enough to afford health insurance. In a publicly funded health care system, everyone may be eligible for diagnostic procedures and treatments, but because of insufficient resources, patients must wait their turn, often to the point that treatable medical conditions become untreatable, or minor operations become major ones. For certain diseases, such as cancer, where early diagnosis and treatment significantly affect outcomes, such delays may have serious, even fatal, consequences. These queues are lengthened by the number of people who, because health care is "free" (at least on the surface), inappropriately seek medical attention; some attempts have been made to impose a user pay system to address this problem, but these have not made a significant difference.

Although economic costs tend to be the determining factor in what health care is delivered—and debates about the respective systems, private or public, tend to be handled in terms of their financial implications—focusing only on the economics of health care sidesteps or ignores other, more fundamental issues.

Health care in the context of Western medicine (often termed "biomedicine" to distinguish its scientific content) is acceptably delivered in a series of very specific ways. Biomedicine is focused on institutional delivery, where the patient comes to a central location to receive health care. Whether this is a clinic, for primary care; a diagnostic facility for tests; or a treatment center (such as a hospital), health care is delivered through institutions by health care practitioners with a defined set of skills, competencies, and accreditations. Whereas many individuals in Western society alive today will remember the doctor making house calls, delivering babies, or performing operations on the kitchen table, these avenues of health care delivery are outside the norms of current biomedicine. This is why efforts to increase "home care" (providing health care in the home, not in institutional contexts) are met with opposition, as are programs for everything from palliative care to midwifery. The lack of access to biomedical specialists or specialized equipment, especially in the event of an emergency, is seen as unnecessarily risky and—in the context of liability—not something that insurance companies or governments in Canada and the United States want to promote. Similarly, even if the diagnosis or treatment someone receives from an "unqualified" practitioner is exactly the same as from a practitioner with official credentials, there is an institutional bias in the system, one that inflicts significant penalties on those not permitted to practice or who practice in a place or manner that has not been vetted and approved.

The result of the institutional delivery of health care is a huge administrative cost, whether it is borne by the patients themselves or by the society through government funding. As these administrative costs increase, often exponentially, health care funding must increase, even if less and less money is actually being spent on doctor–patient interactions. Growing criticism of the institutional delivery model of health care is aimed at everything from its costs to the burden of risk borne by patients who may emerge from a treatment facility with more diseases than they had initially (like superbug antibiotic-resistant bacteria), to the long waits in emergency wards, to inappropriate or unnecessary tests and treatments, to the lack of personal care felt by patients, and so on.

Another distinct area of debate focuses on who delivers the health care. If medicine is understood as biomedicine, and the care consists of operations and prescription drugs, then care is delivered by surgeons and pharmacists. If health care consisted of advice about diet, focusing on preventive rather than curative procedures, then it could better be delivered by a dietician. If the patient presents with a psychological disorder, whether the care is chemical (back to the pharmacist) or psychological (a psychologist or psychiatrist) or spiritual (a chaplain), how the problem is presented and understood directs the care to the appropriate staff or system.

The philosophers' mind–body problem has become an issue for the new institutional medicine. In particular, insurance companies are not as prepared to insure for mental illnesses as they are for physical illnesses. There has even been some debate about whether an illness that does not or cannot show up in an autopsy is an (insurable) illness at all. This debate may be resolved as we develop post-Cartesian (after Descartes) models that unify the body–mind system. New research on the plasticity of the brain demonstrating the brain's potential for functions in the wake of brain damage will contribute to the development of such a unified model.

Biomedicine is very much a descendant of the medieval guilds system—in fact, modern biomedicine shares many features with medical practice back to the time of Hippocrates in ancient Greece (the father of Western medicine, to whom is attributed the Hippocratic Oath that—in various versions—is still offered to new doctors as a code of medical practice). What is understood by "medicine" leads directly to choices about the persons who might appropriately deliver health care to someone who is sick.

In essence, because medicine is itself a cultural system, health care too is a cultural system; in different cultures, there are differing roles for health care practitioners and differing standards for how they are trained, how they are recognized by the community, and how they practice their medical arts. The growing acceptance of alternative therapies—and alternative understandings of health and disease, as well as the nature of medicine itself—in societies dominated by biomedical culture suggests a sea change in what health care is, who is entitled to it, and how it is delivered. Thus, there are now schools of midwifery; naturopathy and homeopathy have adopted an institutional structure like that of biomedicine to train "doctors" who practice these methods; nurse practitioners are starting to take a formal role on the front lines of health care delivery, to substitute for nonexistent general practitioners or to provide a cheaper alternative; and acupuncture is increasingly accepted as a therapy, even if explanations from Chinese medicine of how and why it works are not. Critics of biomedicine promote these and other alternatives in reaction to what they see as the protection of privilege by the medical profession, especially physicians, to maintain both their monopoly on medical practice and the financial rewards associated with it.

If the boundaries of "health care practitioner" start to blur, and the preferred status of the health care institution is replaced by a more decentralized system of health care delivery, then who receives this health care also starts to change. If some of the barriers—perhaps most of the barriers—to health care delivery are economic, the result of urban-focused and expensive facilities emphasizing the machinery of biomedicine and the virtues of better living through prescription drugs, then removing these barriers will increase access to health care in North America, regardless of whether the system is private or public. Shifting the focus away from treatment to prevention, moreover, requires far less of an institutional framework for the delivery of preventive medicine that, in its turn, should lower the social burden of providing biomedical care for the people who need the operations or medications.

Looking at the global context, there is no way that developing countries (in "the South") will be able to establish or afford the kind of medical infrastructure found in developed countries (in "the North"); attempting to do so will waste health care dollars on an administrative structure, leaving less for actual health care. In fact, even developed countries find this medical infrastructure unsustainable, which is why there is an ongoing shortage of medical professionals; provinces in Canada poach doctors and nurses from each other, offering money and other incentives, and American hospitals lure Canadians south for the same reasons. (Both countries poach medical personnel from developing countries, leaving them with fewer and fewer medical professionals to provide basic care to more and more people.) Different models of health care delivery, based on different understandings of health and disease, are essential if medical crises are to be averted.

At least in part, the global spread of Western biomedicine—and the culture that spawned and supports it—needs to be countered with a realistic assessment of the values and attitudes it embodies. Critics have a field day describing Western biomedicine as death-denying, a product of a culture that denies the reality of death through everything from Botox to plastic surgery, one that pretends the world will go on like this forever. (Close-ups of the ravaged faces of geriatric rock stars on tour give the lie to this fantasy!) The point of biomedicine is to increase longevity, to focus on the quantity of life, it seems, not its quality—in effect, to regard death as a failure. Studies have shown that, in Canada and the United States, the overwhelming majority of the health care dollars a person consumes in his or her lifetime (from 50% to 80%) is spent in the last six months of life; in other words, the system does not work! Ask any person traveling on the bus with you for stories, and you will hear many anecdotes of unnecessary treatments and diagnostic procedures performed on the dying and nearly dead. The dead, when witnessed firsthand by the living at all, are embalmed, casketed, and rendered into a semblance of what they were like when alive; gone are the days when a corpse was washed and dressed by the family, laid out on the dining table, and carried to the cemetery by his or her friends.

With a more realistic view of the inevitability of death, the critics argue, comes a revaluing of life; accepting its inevitable end may well prove to be the means by which scarce medical resources are freed up to provide health care for those whose lives are not yet over, but who do not have the money or means to secure the treatment they need.

See also Death and Dying; Health and Medicine.

Further Reading: Barr, Donald, A. *Health Disparities in the United States: Social Class, Race, Ethnicity and Health.* Baltimore: John Hopkins, 2008; Illich, Ivan. *Limits to Medicine.* London: Penguin, 1976; Konner, Melving. *Medicine at the Crossroads.* New York: Vintage, 1994.

Peter H. Denton

HIV/AIDS

Human Immunodeficiency Virus (HIV), a virus affecting the human body and organs, impairs the immune system and the body's ability to resist infections, leading to Acquired Immune Deficiency Syndrome or Acquired Immunodeficiency Syndrome (AIDS), a collection of symptoms and infections resulting from damage to the immune system. Medical confusion and prolonged government indifference to the AIDS epidemic was detrimental to early risk reduction and health education efforts. Initial facts about AIDS targeted unusual circumstances and unusual individuals, thereby situating the cause of AIDS in stigmatized populations and "at risk" individuals. Although current efforts to curb the spread of HIV/AIDS are based on a more realistic understanding of transmission and infections, government policies and educational campaigns still do not fully acknowledge the socioeconomics, drug-use practices, cultural attitudes, and sexual behaviors of populations. Global HIV prevalence stabilized with improvements in identification and surveillance techniques, but reversing the epidemic remains difficult. The pervasive spread of HIV in particular populations and geographic areas continues as economic realities influence infection rates.

In 2007, 33.2 million people were estimated to be living with HIV, 2.5 million people became newly infected, and 2.1 million people died of AIDS. Since the first recognized and reported death on June 5, 1981, AIDS has killed more than 25 million people, making HIV/AIDS one of the most destructive epidemics in history. The number of new HIV infections per year peaked in the late

1990s with over 3 million new infections, but the infection rate never plummeted. Although the percentage of people infected with HIV leveled off in 2007, the number of people living with HIV continues to increase; the combination of HIV acquisition and longer survival times creates a continuously growing general population. Treatments to decelerate the virus's progression are available, but there is no known cure for HIV/AIDS.

Two strains of HIV, HIV-1 and HIV-2, infect humans through the same routes of transmission, but HIV-1 is more easily conveyed and widespread. Transmission of HIV occurs primarily through direct contact with bodily fluids, for example, blood, semen, vaginal fluid, breast milk, and preseminal fluid. Blood transfusions, contaminated hypodermic needles, pregnancy, childbirth, breastfeeding, and anal, vaginal, and oral sex are the primary forms of transmission. There is currently some speculation that saliva is an avenue for transmission, as evidenced by children contracting HIV through pre-chewed food, but research is ongoing to determine if the hypothesis is correct.

Labeling a person HIV-positive or diagnosing AIDS is not always consistent. HIV is a retrovirus that primarily affects the human immune system by directly and indirectly destroying $CD4^+$ T cells, a subset of T cells responsible for fighting infections in the human body. AIDS is the severe acceleration of an HIV infection. When fewer than 200 $CD4^+$ T cells per microliter of blood are present, cellular immunity is compromised, and in the United States, a diagnosis of AIDS results. In Canada and other countries, a diagnosis of AIDS occurs only if an HIV-infected person has one or more AIDS-related opportunistic infections or cancers. The World Health Organization (WHO) grouped infections and conditions together in 1990 by introducing a "stage system" for classifying the presence of opportunistic infections in HIV-positive individuals. The four stages of an HIV infection were updated in 2005, with stage 4 as the indicator for AIDS. The symptoms of AIDS do not normally develop in individuals with healthy immune systems; bacteria, viruses, fungi, and parasites are often controlled by immune systems not damaged by HIV.

HIV affects almost every organ system in the body and increases the risk for developing opportunistic infections. Pneumocystis pneumonia (PCP) and tuberculosis (TB) are the most common pulmonary illnesses in HIV-infected individuals, and in developing countries, PCP and TB are among the first indications of AIDS in untested individuals. Esophagitis, the inflammation of the lining of the lower end of the esophagus, often results from fungal (candidiasis) or viral (herpes simplex-1) infections. Unexplained chronic diarrhea, caused by bacterial and parasitic infections, is another common gastrointestinal illness affecting HIV-positive people. Brain infections and dementia are neurological illnesses that affect individuals in the late stages of AIDS. Kaposi's sarcoma, one of several malignant cancers, is the most common tumor in HIV-infected patients. Purplish nodules often appear on the skin, but malignancies also affect the mouth, gastrointestinal tract, and lungs. Nonspecific symptoms such as low-grade fevers, weight loss, swollen glands, sweating, chills, and physical weakness accompany infections and are often early indications that an individual has contracted HIV.

There is currently no known cure or vaccine for HIV/AIDS. Avoiding exposure to the virus is the primary technique for preventing an HIV infection. Antiretroviral therapies, which stop HIV from replicating, have limited effectiveness. Post-exposure prophylaxis (PEP), an antiretroviral treatment, can be administered directly after exposure to HIV. The four-week dosage causes numerous side effects, however, and is not 100 percent effective. For HIV-positive individuals, the current treatment is "cocktails," a combination of drugs and antiretroviral agents administered throughout a person's life span. Highly Active Antiretroviral Therapy (HAART) stabilizes a patient's symptoms and viremia (the presence of viruses in the blood), but the treatment does not alleviate the symptoms of HIV/AIDS. Without drug intervention, typical progression from HIV to AIDS occurs in 9 to 10 years: HAART extends a person's life span and increases survival time 4 to 12 years. Based on the administration of cocktails and the increase in the number of people living with HIV/AIDS, the prevailing medical opinion is that AIDS is a manageable, chronic disease. Initial optimism surrounding HAART, however, is tempered by recent research on the complex health problems of AIDS-related longevity and the costs of antiretroviral drugs. HAART is expensive, aging AIDS populations have more severe illnesses, and the majority of the world's HIV-positive population do not have access to medications and treatments.

In 1981 the U.S. Centers for Disease Control and Prevention (CDC) first reported AIDS in a cluster of five homosexual men who had rare cases of pneumonia. The CDC compiled four "identified risk factors" in 1981: male homosexuality, IV drug use, Haitian origin, and hemophilia. The "inherent" link between homosexuality and HIV was the primary focus for many health care officials and the media, with drug use a close second. The media labeled the disease gay-related immune deficiency (GRID), even though AIDS was not isolated to the homosexual community. GRID was misleading, and at a July 1982 meeting, "AIDS" was proposed. By September 1982 the CDC had defined the illness and implemented the acronym AIDS to reference the disease. Despite scientific knowledge of the routes and probabilities of transmission, the U.S. government implemented no official, nationwide effort to clearly explain HIV mechanics or promote risk reduction until the surgeon general's 1988 campaign. Unwillingness to recognize HIV's pervasiveness or to fund solutions produced both a national fantasy about the AIDS epidemic and sensationalized public health campaigns in the mass media.

Prevention advice reinforced ideas of safety and distance; the citizenry was expected to avoid "risky" behavior by avoiding "at risk" populations. Strategies to prevent HIV/AIDS were directed at particular types of people who were thought to engage in dangerous behaviors. Homosexual sex and drug use were perceived to be the most risky behaviors, and thus heterosexual intercourse and not doing drugs were constructed as safe. Disease prevention programs targeted primarily gay populations but were merely health precautions for everyone else—individuals not at risk.

Citizens rarely considered how prevention literature and advice applied to individual lives because the public was relatively uninformed about the routes of

HIV transmission. Subcultures were socially stigmatized as deviant, and at-risk populations were considered obscene and immoral. "Risk behavior" became socially constructed as "risk group," which promoted a limited understanding of how HIV was contracted. The passage of the Helms Amendment solidified both public perceptions and government legislation about AIDS and AIDS education. Federal funding for health campaigns could be renewed each year with additional amounts of money as long as they did not "promote" homosexuality and promiscuity. Despite lack of funding, much of the risk-reduction information that later became available to the public was generated by advocates within the homosexual community.

Although avoidance tactics were promoted by the national government to the citizenry, precautionary strategies were adopted and utilized by gay communities. Distributing information through newspapers, pamphlets, and talks, the community-based campaigns emphasized safe sex and safe practices. Using condoms regardless of HIV status, communication between sexual partners, and simply avoiding intercourse were universal precautions emphasized in both American and European gay health campaigns. With a nontransmission focus, safe sex knowledge was designed and presented in simple language, not medical terminology, so the information was easy to understand. Although "don't ask, don't tell" strategies were still adopted by many gay men, the universal safe-sex strategy employed by the gay community promoted discussions about sex without necessarily the need for private conversations. The visibility and accessibility to information helped gay men understand HIV and promoted individual responsibility. The national pedagogy, by contrast, banned sexually explicit discussions in the public sphere. Individuals were encouraged to interrogate their partners in private without truly comprehending either the questions asked or the answers received. Lack of detailed information and the inability to successfully investigate a partner's sexual past facilitated a need for an organized method of identifying HIV-positive individuals.

With the intention of stemming HIV, the CDC's Donald Francis proposed, at the 1985 International Conference on AIDS in Atlanta, that gay men have sex only with other men who had the same HIV antibody status and presented a mathematical model for testing. Shortly thereafter, HIV testing centers were established, and the national campaign, centered on avoiding HIV-positive individuals, was implemented. Instead of adopting safe-sex education and behaviors, the government merely inserted technology into existing avoidance paradigms. HIV tests were only valid if the last sexual exchange or possible exposure had occurred within six months to a year earlier. Many people misinterpreted negative test results as an indicator of who was "uninfected," however, merely reinforcing educated guesses. With the test viewed as an ultimate assessment for determining a sexual partner's safety, many individuals relied on HIV test results to confirm individual theories of who was and was not infected. Unrealistic discussions about sexual practices and behaviors were detrimental to the American population, especially adolescents and young adults.

In 1990 epidemiologists confirmed that a wide cross-section of American youth were HIV-positive. Minority and runaway youth were particularly affected, but millions of young people had initiated sexual interactions and drug use in the previous decade. Because health campaigns focused on prevention, there was little and often no help for individuals who were infected. Diagnosing the onset of symptoms and tactics to delay AIDS progression were almost nonexistent. Instead of recognizing the sexual and drug practices of middle-class white kids, society classified young people into categories of "deviance": deviant individuals contracted HIV; innocent children did not. Refusing to acknowledge that young people were becoming infected, many parents and government officials impeded risk-reduction information. Consequently, few young people perceived themselves as targets of HIV infection, and much of the media attention focused on "tolerance" for individuals living with AIDS.

Under the false assumption that infections among youth occurred through nonsexual transmission, HIV-positive elementary school children and teenagers were grouped together and treated as innocent victims. Although drug use and needle sharing were prevalent behaviors in teenage initiation interactions, the public agenda focused on sexuality as the primary transmission route. Knowing about or practicing safe sex was dangerous; ignorance would prevent HIV. Representations of youth in the media reinforced the naiveté and stereotypes that initially contextualized AIDS in the adult population; *New York Times* articles suggested HIV infections in gay youth were the result of liaisons with gay adults or experimentation among themselves. In a manner reminiscent of the initial constructions of AIDS in the 1980s, HIV-infected youth were effectively reduced to deviant, unsafe populations. Securing heterosexuality became, yet again, a form of safe sex and the primary prevention tactic for HIV. Refusal to acknowledge non-intercourse activities as routes for HIV transmission pervaded government policies of the twentieth and twenty-first centuries.

Recommendations for avoiding HIV infections were limited in both scope and funding. Because heterosexual women were increasingly becoming infected, the Federal Drug Administration (FDA) approved the sale of female condoms in 1993. However, female condoms were largely unavailable, and the price was prohibitive for many women. Approved in 1996 by the FDA, the viral load test measured the level of HIV in the body. As with the female condom, the test was expensive and continues to be cost-prohibitive. Needle exchange programs demonstrated great effectiveness in reducing HIV infections via blood transmission. Although the U.S. Department of Health and Human Services recommended needle exchange programs in 1998, the Clinton Administration did not lift the ban on use of federal funds for such purposes. Needle exchange remains stigmatized, and the primary source of funding continues to come from community-based efforts. In 1998 the first large-scale human trials for an HIV vaccine began, but no vaccine has been discovered. Despite community and government efforts, people continue to become infected with HIV/AIDS.

With growing numbers of individuals contracting HIV, the government implemented some treatment strategies. The AIDS Drug Assistance Program (ADAP) was established to pay for HIV treatments for low-income individuals. In 1987 azidothymidine (AZT)/zidovudine (ZDV) became the first HIV/AIDS drug to receive the FDA's approval. AZT's toxicity was well documented, but the effectiveness of the long-term monotherapy was questionable. Regardless, AZT was administered to the population, and the FDA approved three generic formulations of ZDV on September 19, 2005. AZT continues to be the primary treatment in reducing the risk of mother-to-child transmission (MTCT), especially in developing countries. There were few effective treatments for children until August 13, 2007, when the FDA approved a fixed-dose, three-drug combo pill for children younger than 12 years old. Treatments are improvements for developing awareness of HIV/AIDS, but the realities of transmission and the costs associated with HIV infection remain largely ignored.

People Living With AIDS (PWA, coined in 1983) became the faces of HIV infections, and individuals were the impetus for increased attention to the AIDS epidemic. Rock Hudson, an actor world-renowned for his romantic, heterosexual love scenes, appeared on ABC World News Tonight and announced he had AIDS. He died shortly after his October 1985 appearance. President Ronald Reagan, a close friend of Hudson, mentioned AIDS in a public address in 1986, the first time a prominent politician specifically used the words HIV and AIDS. In 1987, the same year the CDC added HIV to the exclusion list, banning HIV-positive immigrants from entering the United States, Liberace, a musician and entertainer, died from AIDS. *Newsweek* published a cover story titled "The Face of Aids" on October 10, 1987, but the 16-page special report failed to truly dispense with the stereotypes of HIV infection. With the growing number of PWAs, government policies toward HIV changed somewhat. In 1988 the Department of Justice reversed the discrimination policy, stating that HIV/AIDS status could not be used to prevent individuals from working and interacting with the population. December 1, 1988, was recognized as the first World AIDS Day. However, even with such social demonstrations of goodwill, recognizable faces remained aloof; the public "saw" HIV but did not associate HIV with the general population until a so-called normal person grabbed community attention.

One of the most public and media-spotlighted individuals was Ryan White, a middle-class, HIV-positive child. White contracted HIV through a blood transfusion. His blood-clotting disorder fit existing innocence paradigms and, thus, provided opportunities for discussions about HIV, intervention, and government aid. At age 13, White was banned from attending school, prevented from associating with his classmates, and limited to classroom interactions via the telephone. The discrimination White endured throughout his lifetime highlighted how "normal" people were affected by public reactions and government policies. In 1990, the year White died at age 18, the Ryan White Comprehensive AIDS Resources Emergency (CARE) Act was passed. With 150,000 reported AIDS cases in the United States, CARE directed attention to the growing incidences of HIV and increased public compassion.

The teen culture of the 1990s continued to be affected as additional celebrities were added to the seropositive list. Earvin "Magic" Johnson, an idolized basketball player and all-time National Basketball Association (NBA) star, announced his HIV-positive status in 1991. The perversion labels normally associated with HIV were momentarily suspended as the public discourse tried to fit Johnson's wholesome role-model status into the existing risk paradigm. Much of the public, including individuals in methadone clinics, referred to positive HIV serostatus as "what Magic's got" and avoided the stigmatized label of AIDS. The compassion and understanding for HIV-positive individuals was short-lived, however. Freddie Mercury, lead singer of the rock band Queen, died in 1991 from AIDS. Because he was a gay man, Mercury's life was quickly demonized, and he did not receive the same "clean living" recognition from the press. Preaching compassion was good in rhetoric, but not in practice.

Limited public empathy did not quell the diversity of individuals affected by AIDS. In 1992 tennis star Arthur Ashe announced his HIV status, and teenager Rick Ray's house was torched (Ray, a hemophiliac, and his siblings were HIV-positive). During 1993, Katrina Haslip, a leading advocate for women with AIDS in prison, died from AIDS, and a young gay man living with HIV, Pedro Zamorn, appeared as a cast member on MTV's *The Real World*. Zamorn died in 1994 at age 22. Olympic Gold Medal diver Greg Louganis disclosed his HIV status in 1995, which sent shockwaves into the Olympic community. Louganis had cut his head while diving during the 1988 Olympics, and concern quickly entered scientific and media discussions about HIV transmission. The discrimination Louganis endured affected athletic policies and issues of participation in sports for HIV-positive athletes. Even though HIV/AIDS was the leading cause of death among African Americans in the United States in 1996, the public continued to focus on individuals, whose faces, displayed in the media, informed much of the understanding of HIV in the United States.

During the June 2006 General Assembly High-Level Meeting on AIDS, the United Nations member states reaffirmed their commitment to the 2001 Declaration of Commitment. Efforts to reduce the spread of AIDS focused on eight key areas, including reducing poverty and child mortality, increasing access to education, and improving maternal health. Universal access to comprehensive prevention programs, treatment, care, and support were projected outcomes for 2010. Strategies to improve HIV testing and counseling, prevent HIV infections, accelerate HIV/AIDS treatment and care, and expand health systems were four of the five suggestions WHO expected to implement. The sheer numbers of people infected with HIV, however, tempered the hope and optimism surrounding intervention techniques.

Globally, between 33.2 and 36.1 million people currently live with HIV. Predictions for India in 2006 estimated 2.5 million people (.02% of the population) were infected with HIV; India ranked third in the world for HIV rates (UNAIDS 2006). Using improved analysis techniques, however, the 2007 revised statistics for India indicate the HIV epidemic is less prevalent than initially predicted. Indonesia has the fastest-growing epidemic, and HIV prevalence among men has increased in Thailand. Eastern Europe and central Asia have more than 1.6 million people living with HIV, a 150 percent increase from 2001 estimates (estimated between 490,000 and 1.1 million). Sub-Saharan Africa continues to be the most affected region, with 20.9 to 24.3 million people (68% of the global total) living with HIV. In 1988 women living with HIV/AIDS exceeded men, and current infection rates continue to be disproportionately high for women in sub-Saharan Africa. Women are more susceptible to HIV-1 infections, but their partners (usually men) are often the carriers and transmitters of HIV. For women as mothers, MTCT can occur in utero during the last weeks of pregnancy, during childbirth, and from breastfeeding. Approximately 90 percent of all children with HIV worldwide (two million) live in sub-Saharan Africa. Although risk behavior has changed among young people in some African nations, the mortality rates from AIDS is high because of unmet treatment needs. Delivery of health service and monetary funding remain inadequate for prevention efforts and HIV treatments.

The majority of the world's population does not have access to health care settings or medical techniques that prevent HIV infections. Universal precautions, such as avoiding needle sharing and sterilizing medical equipment, are not often followed because of inadequate health care worker training and a shortage of supplies. Blood transfusions account for 5 to 15 percent of HIV transmissions because the standard donor selection and HIV screening completed in industrial nations are not performed in developing countries. Health care worker's behaviors and patient interactions are impacted by the lack of medical supplies, including latex gloves and disinfectants. Approximately 2.5% of all HIV infections in sub-Saharan Africa occur through unsafe health care injections. Implementing universal precautions is difficult when economic funding is severely restricted or outright absent.

Education efforts are also constrained by the lack of monetary support. HIV prevalence has remained high among injecting drug users, especially in Thailand, where HIV rates are 30 to 50 percent. AIDS-prevention organizations advocate clean needles and equipment for preparing and taking drugs (syringes, cotton balls, spoons, water for dilution, straws, pipes, etc.). Cleaning needles with bleach and decriminalizing needle possession are education efforts advocated at "safe injection sites" (places where information about safe techniques were distributed to drug users). When needle exchanges and safe injection sites were established, there was a reduction in HIV infection rates. Individuals, especially young people, engaged in high-risk practices with drugs and sex, often because of a lack of disease comprehension. Although aware of HIV, young people continue to underestimate their personal risk. HIV/AIDS knowledge increases with clear communication and unambiguous information.

Questions surrounding HIV/AIDS have stemmed from both a lack of understanding and a desire to understand the complexities of the disease. Early misconceptions about transmission—casual contact (e.g., touching someone's skin), and engaging in any form of anal intercourse—created fear and folklore. Certain populations—homosexual men and drug users—were incorrectly identified as the only people susceptible to HIV. National pedagogy mistakenly proclaimed that open discussions about HIV or homosexuality would increase rates of AIDS and homosexuality in schools. The false belief that sexual intercourse with a virgin would "cure" HIV was particularly detrimental to many young women. Although much of the early fictional rhetoric was rectified through the distribution of scientific knowledge, denial and delusion continue to influence individuals' perceptions of HIV/AIDS.

A small group of scientists and activists questioned the testing and treatment methods of HIV/AIDS, which influenced government policies in South Africa. Established in the early 1990s, the Group for the Scientific Re-Appraisal of the HIV/AIDS Hypothesis launched the Web site virusmyth.net and included a collection of literature from various supporters, including Peter Duesber, David Rasnick, Eleni Papadopulos-Eleopulos, and Nobel Prize winner Karry Mullis. As a result, South Africa's president, Thabo Mbeki, suspended AZT use in the public health sector. At issue was whether AZT was a medicine or a poison and whether the benefits of AZT in MTCT outweighed the toxicity of the treatment. Retrospective analyses have raised criticisms about Mbeki's interference. The expert consensus is that the risks of AZT for MTCT were small compared to the reduction of HIV infection in children. The South African AZT controversy demonstrated how science could be interpreted in different ways and how politics influenced public health decisions.

The biological ideology of most scientific inquiry has influenced HIV investigations, with much research focused on understanding the molecular structure of HIV. During the late 1980s Canadian infectious-disease expert Frank Plummer noticed that, despite high-risk sexual behavior, some prostitutes did not contract HIV. In spite of being sick, weak from malnourishment, and having unprotected sex with men who were known to have HIV, the women did not develop a seropositive test result. The scientific community became highly interested in the women (known as "the Nairobi prostitutes") and hypothesized that the women's immune systems defended the body from HIV. Of the 80 women exposed to HIV-1 and determined to be uninfected and seronegative, 24 individuals were selected for immunological evaluation. Cellular immune responses, like T cells, control the infection of HIV; helper T cells seem to recognize HIV-1 antigens. The small group of prostitutes in Nairobi remained uninfected, even though their profession exposed them to prolonged and continued exposure to HIV-1. Cellular immunity prevented HIV from infecting the body, not systemic humoral immunity (i.e., defective virus or HIV-antigens). The Nairobi prostitutes' naturally occurring protective immunity from the most virulent strain of HIV was a model for the increased focus and development of vaccines.

Historically, vaccine production has concentrated on antibodies and how the human body can be tricked into fighting an infection. A benign form of the virus infects the body, and the immune system's white blood cells respond; antibodies attack the virus in the bloodstream and cytotoxic T lymphocytes (T cells) detect infected cells and destroy them. Vaccines for measles, yellow fever, and pertussis operate within this scientific paradigm. HIV mutates rapidly, however, and different strains exist within the population. A vaccine for one subtype would not provide immunity to another HIV strain. The unpredictability of HIV requires a scientific transition in the research paradigm and a willingness to use human beings as test subjects.

For the effectiveness of a vaccine to be gauged, thousands of people will have to be part of the research trials. The ethical problems associated with human subjects and the costs of long-term investigations prohibit many researchers from committing to vaccine research. Additionally, the economic market mentality provides more incentive for making a product for mass consumption. The costs and risks of a vaccine limit financial gains for companies; inoculation against HIV reduces the number of consumers who need the product. Instead, antiretroviral medications and treatments are the primary focus for research funding. An AIDS vaccine would benefit the entire world, but no company or country is willing to devote the economic and scientific resources to the research. The International AIDS Vaccine Initiative, a philanthropic venture-capital firm dedicated to finding a vaccine, has received funding from private and public donations, including significant contributions from the Bill and Melinda Gates Foundation. Researchers, however, continue to reduce the AIDS virus to its genetic components instead of approaching HIV vaccines from new perspectives.

The complexity of HIV creates difficulties in finding a single, permanent solution. Education and prevention have had limited success, and antiretroviral therapies cannot cure the vast number of people infected with HIV/AIDS. A partially effective vaccine or a vaccine that targets only one mutation of HIV is not a solution. Ignoring a population's behaviors, economic situations, and beliefs has proven detrimental to the AIDS epidemic. The difficulties of the disease make HIV/AIDS a formidable problem.

See also Health and Medicine; Health Care; Immunology.

Further Reading: Barton-Knott, Sophie. "Global HIV Prevalence Has Leveled Off." *UNAIDS.* http://www.unaids.org; Fowke, Keith, Rupert Kaul, Kenneth Rosenthal, Julius Oyugi, Joshua Kimani, John W. Rutherford, Nico Nagelkerke, et al. "HIV-1-Specific Cellular Immune Responses among HIV-1-Resistant Sex Workers." *Immunology and Cell Biology* 78 (2000): 586–95; Jakobsen, Janet, and Ann Pellegrini. *Love the Sin: Sexual Regulation and the Limits of Religious Tolerance.* Boston: Beacon Press, 2004; Kallings, L. O. "The First Postmodern Pandemic: Twenty-Five Years of HIV/AIDS." *Journal of Internal Medicine* 263 (2008): 218–43; Laumann, Edward, John Gagnon, Robert Michael, and Stuart Michaels. *The Social Organization of Sexuality: Sexual Practices in the United States.* Chicago: University of Chicago Press, 1994; Patton, Cindy. *Fatal Advice: How Safe-Sex Education Went Wrong.* Durham, NC: Duke University Press, 1996; UNAIDS. "Overview of the Global AIDS Epidemic." *Report on the Global AIDS Epidemic.* http://www.unaids.org; U.S. Department of Health and Human Services HIV/AIDS Web Site. http://www.aids.gov; Weinel, Martin. "Primary Source Knowledge and Technical Decision-Making: Mbeki and the AZT debate." *Studies in History and Philosophy of Science* 38 (2007): 748–60; World Health Organization Web site. http://www.who.int/en.

Laura Fry

HUMAN GENOME PROJECT

Who would not want humanity to be able to read the until-now-hidden genetic code that runs through every cell in the human body? This code contains all the instructions that operate the incredibly complex processes of the human organism. It determines physical details from an individual's eye color to his or her height to whether someone will suffer male-pattern baldness to whether a person has a predisposition to develop breast cancer.

The Human Genome Project was an international scientific effort, coordinated by the U.S. National Institutes of Health and the U.S. Department of Energy, to decode the string of tens of thousands of genes, made up of about three billion DNA (deoxyribonucleic acid) pieces, and help find out what they do. In 2003 the project announced that it had successfully mapped out the entire human genome, 2 years before expected and 13 years after the project had been formed in 1990.

Certainly most biological scientists were elated when the Human Genome Project announced in 2003 that it had successfully been able to record every gene and its structure along the chain that makes up the human genetic code. Scientists saw a golden age of biomedical science that might be able to cure human diseases, extend human life, and allow humans to reach much more of their full genetic potential than ever before. Simply understanding why the human organism works the way it does might help humans to understand who and what they are at a fundamental level.

Something about the galloping pace and huge strides being made in human genetic knowledge sent shivers down the spines of critics, however. They worried about a future in which "designer" people would be purposely created with genetic alterations that would never have naturally occurred. They saw grave dangers in genetic knowledge being used to deny people with potential problems everything from medical insurance to jobs, thus affecting even their ability to survive. They were concerned that new abilities to understand and alter human genes would allow the wealthy and influential to give themselves and their children advantages over poorer and less influential people who could not afford to pay for the manipulation of their genetic makeup.

One of the possible gains of decoding the human genome is that knowledge of which genes, or combinations of genes, can cause certain physical diseases, deformations, or other problems should allow scientists over time to produce tests that can reveal whether individuals have certain predispositions or the potential to develop these problems. Many genetic predispositions do not mean that an individual will definitely develop a condition, but knowledge that an individual has that gene or those genes could give the individual a heads-up or a way of trying to avoid or minimize the damage from the condition developing.

A more advanced application of the new knowledge of the human genetic code is to use it to find ways to deal with "bad" genes or bits of DNA, by somehow removing the problem component, by replacing it, or by "turning it off." The purposeful manipulation of human genes to eliminate or minimize the effects of "bad" genes or DNA is often called gene therapy. Early attempts at using gene therapy to improve human individuals' situations have been disappointing, but many scientists have high hopes for eventual, revolutionary success in controlling many human diseases.

Overall, the revelation of the human genetic code, even if the function of all the many thousands of genes and the way they interact to produce effects is not yet known, is a stunning opening up of the human species' understanding of the basic rules that govern the biochemical structure of human existence.

Humans stand to gain new knowledge and powers of manipulation from the unraveling of the human genetic code, but what will people do with the knowledge? Certainly some scientists will use it to develop new therapies, drugs, and treatments for human diseases. Few would want to prevent that. What if an insurance company required all applicants for medical or life insurance to submit to a genetic test that would reveal what potential genetically based diseases or disorders they carry within their cells, however? What if all insurance companies began requiring these sorts of tests? Would this create a class of citizens who could not obtain insurance, not because they actually had a condition, but because they might develop one in the future? Could employers, some of whom offer medical insurance coverage and deal with the effects of diseases, demand such tests and refuse to hire people with potential problems? This potential alarms many people.

The U.S. Congress has wrestled with this issue in pieces of legislation that have attempted to balance the rights of individual workers and citizens with those of employers and insurance companies. Around the world, legislators have wrestled with the issue of how to protect individuals' rights to employment, insurance, and privacy in an age in which knowledge of someone's genetic makeup can say much about their possible future.

Some also worry about the possibility of wealthy and influential people being able to change or manipulate their own genes and DNA in order to minimize weaknesses or produce new strengths, while poorer and less influential people who could not afford to manipulate their genetic structure would be stuck with their natural-born abilities and weaknesses. Would this become a new basis for a class system, with "supergeniacs" attaining an enduring supremacy over weaker and poorer people? Could wealthy and influential people customize their children's genetic structure to give them advantages over less privileged people? Would these new abilities, if not made equally available, exacerbate the social inequalities that already exist?

The revelation of the human genetic code does not yet provide the ability to create Robocop-type or Cylon-like human–machine hybrids or part-human, part-animal hybrids, but what if that became possible? Should that ability be left to scientists and research companies to develop, or should there some sort of governmental legal structure put in place first before disturbing possibilities become realities?

The basic list of genes and their components is now available for humans to ponder and consider. What humans will do with this knowledge remains to be seen.

See also Cloning; Eugenics; Genetic Engineering; Nature versus Nurture.

Further Reading: The Human Genome Project. http://genome.gsc.riken.go.jp/hgmis/project/hgp.html.

Edward White

Human Genome Project: Editors' Comments

It is important that readers of this entry recognize the extent to which ideas about what genes are and what they can do are biased, even among scientists, by the emphasis

American culture places on individualism and individual responsibility. American culture is biased toward psychological, genetic, and neurological explanations for all aspects of human life, including why we behave the way we do. At the same time, the culture is resistant to social and cultural explanations. Readers are urged to explore the texts in the following further-reading list in order to assess the power of genes and whether they are as significant as many front-page stories suggest. They will discover among other things that no single gene determines eye color and that the shadow of racism darkens many efforts to reduce human behavior to genes.

Further Reading: Hubbard, Ruth, and Elijh Wald. *Exploding the Gene Myth: How Genetic Information Is Produced and Manipulated by Scientists, Physicians, Employers, Insurance Companies, Educators, and Law Enforcers.* Boston: Beacon Press, 1999; Lewontin, R. C., Steven Rose, and Leon J. Kamin. *Not in Our Genes: Biology, Ideology, and Human Nature.* New York: Pantheon, 1984; Moore, David S. *The Dependent Gene: The Fallacy of "Nature vs. Nurture."* New York: Henry Holt, 2002; Ridley, Matt. *Nature via Nurture: Genes, Experience, & What Makes Us Human.* New York: Harper/Collins, 2003.

IMMUNOLOGY

Immunology, a branch of the medical sciences, is the scientific study of the immune system. This system or set of physiological interactions was first identified in the early 1900s as constituting a system parallel in importance to those of digestion and blood circulation. Like other systems, the immune system became the basis for a series of interventions and therapies known as vaccinations that have often become issues of public debate. The medical establishment and public health agencies have used immunology to argue strongly for the requirement of vaccinations; individuals and groups have opposed vaccinations with appeals to religious, naturalist, and civil rights arguments and on occasion have questioned the science of immunology.

In 1796 Edward Jenner discovered that immunity to smallpox, a highly contagious virus that causes blisters to form on the skin, could be acquired by individuals exposed to cowpox, a similar virus that afflicts cattle. Jenner used the term *vaccination* to describe his discovery because his vaccine originated from a virus affecting cows (*vacca* is the Latin word for cow). Louis Pasteur later used the term to describe immunization for any disease. In 1900 Paul Ehrlich introduced his theory of antibody formation (side-chain theory) to explain how the immune system identifies pathogens. His research in immunology led to the Nobel Prize in Medicine in 1908. This discovery by Ehrlich opened the door to the modern study of the human immune response.

The immune system includes key organs within the human body: the lymph nodes, lymph vessels, thymus, spleen, and bone marrow, as well as accessory organs such as the skin and mucous membranes. The human body is surrounded by a variety of agents that can cause disease under certain circumstances. A disease-causing agent is called a pathogen and can include viruses, bacteria, fungi, protozoa, and parasites.

To protect against pathogens, the human body exhibits three levels of defense. Skin and mucous membranes form the body's nonspecific first line of defense. The skin provides a keratin barrier that prevents organisms from entering the body. Mucous membranes, found in areas with no skin, trap and stop the action of many types of microorganisms.

The second line of defense includes the nonspecific inflammatory response induced by histamines. Histamine causes vasodilation (enlargement of blood vessels), which increases blood flow to the infected area; it also causes an increase in temperature that will result in the destruction of some pathogens. Another mechanism is the production of interferon (protein) by cells infected with viruses. Interferon is transmitted to healthy cells and attaches to receptors on the cell surface, signaling these cells to produce antiviral enzymes that inhibit viral reproduction. Macrophages, white blood cells that ingest pathogens, also have a role in the second line of defense.

The third line of defense, working simultaneously with the second line, targets specific pathogens. The immune response is classified into two categories: cell-mediated and humoral. The cell-mediated immune response acts directly against pathogens by activating specific leukocytes called T cells ("T" stands for thymus, the organ where these cells mature). Macrophages initiate the cell-mediated response by displaying parts of pathogens, called antigens, on their surface. An antigen is any molecule that can be identified by the body as foreign. In the lymph nodes, T cells that recognize specific antigens produce helper T cells and cytotoxic T cells. Cytotoxic T cells destroy the pathogen directly, and helper T cells produce chemicals that stimulate the production of other leukocytes.

The humoral immune response begins when the chemicals secreted by helper T cells activate another type of leukocyte called B cells ("B" stands for bursa of Fabricius, an organ in birds where these cells mature). The humoral response acts indirectly against pathogens. Clones of B cells differentiate into (1) plasma cells that make antibodies to mark pathogens for destruction and (2) memory cells that help the body react faster to subsequent invasion by the same pathogen. Antibodies are specialized proteins that bind to antigens on the surface of pathogens and inactivate them. B cells use this method to mark pathogens for destruction by macrophages or stimulate the production of proteins that cause pathogens to lyse.

Progress in the field of immunology as well as advances in technology have provided opportunities for humans to enhance the immune system response. One approach is the use of antibiotics, substances that target organelles of microorganisms, such as bacteria or parasites. Antibiotics boost the immune response by helping with the destruction of specific pathogens. Other approaches involve prevention of infection. Individuals may obtain passive immunity from the acquisition of antibodies from another organism, such as a newborn from breast milk. Artificial immunity, requiring the inoculation of an organism with a vaccine, has received the most attention in recent years and is a source of some controversy in society.

Vaccines contain either weakened pathogens or antigens (obtained from parts of the pathogen). Vaccines cause a primary immune response in an organism, resulting in the production of plasma cells that make antibodies to ultimately destroy the weakened pathogen and memory cells that will enhance the immune response when the pathogen is encountered a second time. The individual receiving the immunization is unlikely to have symptoms of illness because the pathogen is severely weakened, and the immune response is sufficient to destroy the pathogen before it causes full-blown disease. The person's next response to a live pathogen will be a much quicker secondary immune response that begins with memory and plasma B cells.

Even with all the technological advances in immunology, our knowledge of the immune system response is still limited in that we cannot predict when the immune system may overreact to a common substance or act against one's own body cells. An overreaction of the immune system to a common substance, such as dust, pollen, or mold, is considered an allergic reaction. An allergic reaction can result in a full-blown immune response that, if not treated, can result in death. Similarly, autoimmune diseases result when the immune system fails to recognize cells in the body as "self." Rheumatoid arthritis (leucocytes attack joint cells) and multiple sclerosis (leucocytes attack the covering of neurons) are examples of autoimmune diseases.

The medical community and public health agencies, such as the World Health Organization (WHO) and the Centers for Disease Control and Prevention (CDC), strongly advocate for the use of vaccines in society. Currently, vaccines are seen by these organizations as the most effective way to prevent and eventually eliminate infectious disease. Although a list of recommended vaccinations and a routine vaccination schedule for infants is supported by international governments, international health regulations require only the yellow fever or meningococcal vaccine for travelers to certain countries where these diseases are prevalent.

Although the medical community and public health agencies admit that no vaccine is completely safe for every individual, they advocate that the advantages to international disease prevention outweigh the risks to individual health. Risk-cost-benefit analysis weighs the chance that an individual will experience suffering against the suffering of the larger population if they remain unvaccinated. An example of the impact an immunization can have worldwide is the 10-year, WHO vaccination campaign against smallpox that eventually led to its eradication in the late 1970s. Public agencies and the medical community would argue that the success of several vaccinations against common childhood disease led to an increase in overall public health in the last century. This raises the question of whether governments can or should require routine vaccinations for all their citizens to dramatically reduce or eradicate other common infectious agents.

Further support for vaccination has come from government and public health agencies since September 11, 2001. The current threat of terrorist attacks includes the use of biological weapons such as smallpox or anthrax. Government and public health agencies argue for the need to maintain a supply of necessary vaccines to contain the spread of infectious disease if biological attacks are carried out.

Ethical arguments against vaccination revolve around the following question: if the implementation of vaccine technology causes harm to some individuals, how much harm is acceptable to achieve a benefit to public health? This question addresses not only injury caused by routine vaccination but also injury caused to individuals as a result of the initial testing of a vaccine in the population. The threshold of harm may be different for different groups in society based on their values and beliefs. Although the arguments of the medical community and public health agencies have already been established, religious groups would argue that every life is valued by God and that only God has the power to give life and take it away. Similarly, human rights organizations would argue for the protection of the Third World populations who are often participants in vaccine research, specifically AIDS vaccine investigations that are currently being conducted on the African continent.

Public concerns with safety extend to the medical side effects of vaccination. Parent groups have argued that some vaccinations, specifically the measles-mumps-rubella (MMR) vaccine, have links to autism as a result of a harmful mercury additive, thimerosal. Thimerosal was used as a preservative in many multi-dose vaccines before July 1999. Although research findings have provided no conclusive link between vaccinations and autism, the Food and Drug Administration (FDA) mandated that childhood vaccinations created after 1999 contain no thimerosal preservative.

Other ethical issues include the argument by naturalists and some religious fundamentalists that vaccines are unnatural, foreign substances to the human body. Vaccines can contain preservatives and other chemicals that are seen as contaminants. Religious fundamentalist groups might cite a section of Scripture such as "your body is a temple" (1 Corinthians 7:19) to defend physical purity. This leads to a more recent controversy over the use of vaccines to prevent sexually transmitted diseases. A human papillomavirus (HPV) vaccine was licensed in 2006 by the FDA for use in preventing HPV in young girls in order to decrease their risk of cervical cancer later on. A case has been made by religious and parent groups that these vaccines may increase risky sexual behavior in young people. They argue that as the risk of infectious disease decreases because of immunization, risky behavior increases.

Civil rights organizations argue that decision making should be left in the hands of individuals; parents should be allowed to make vaccination choices for their families without government interference. Advances in reproductive immunology, specifically the anti-hCG (human chorionic gonadotropin) vaccine, raise civil rights and ethical issues as we move into the field of immunological contraception. Proponents of anti-fertility vaccines advocate that these vaccines would give families more choices in family planning methods as well as help in population control in locations where conventional birth control methods are unpopular or unavailable. Women's rights advocates question the safety of the vaccine and its long-term effects on fertility. Similarly, questions of whether this new technology gives women more autonomy over reproductive decisions have also been raised.

One economic argument surrounding vaccinations involves population growth and sustainability. Critics argue that increasing vaccinations in developing countries causes an increase in the worldwide population beyond what its resources can sustain. When the human population reaches its carrying capacity, the initial decrease in infant mortality as a result of vaccination will eventually lead to death by starvation in early adulthood.

The expense of vaccination production has led to some controversy in how funding should be distributed in immunological research. The licensing and safety requirements involved in the production of vaccines often make the cost of vaccines prohibitive to developing countries. Developing countries that are too poor to buy the needed vaccines for their populations often rely on Western countries or international agencies to provide these vaccines free of charge. Lack of demand and money for vaccinations has caused pharmaceutical companies to put less money into vaccine development and more money into research for treatments and antibiotics for which individuals in wealthy countries will pay. At issue is whether more money should be put into developing prevention methods as opposed to finding cures and treatments for infectious disease. On one side of the issue are public health agencies and the medical establishment that place the utmost importance on vaccine development; on the other side are pharmaceutical companies that are interested in financial gains from selling their product. Governments have traditionally offered funding to drug companies for research and development of specific vaccines; however, critics argue that a better plan would be for governments to promise to pay for vaccines that actually work, thereby increasing the initial market for the vaccine and competition among drug companies. Research on development of a human immunodeficiency virus (HIV) vaccine illustrates these economic issues. Developing and selling costly treatments for patients diagnosed with HIV is currently more lucrative for pharmaceutical companies than research and development of a vaccine that may be impracticable given the mutation rates of the HIV virus.

See also Health and Medicine; HIV/AIDS; Vaccines.

Further Reading: Campbell, Neil A., and Jane B. Reece. *Biology.* 7th ed. San Francisco: Pearson Education, 2005; Fenner, Frank, Donald A. Henderson, Isao Arita, Zdenek Jezek, and Ivan D. Ladnyi. *Smallpox and Its Eradication.* Geneva: World Health Organization, 1988; Immunization Safety Review Committee. *Immunization Safety Review: Vaccines and Autism.* Washington, DC: National Academies Press, 2004; Offit, Paul A., and Louis M. Bell. *Vaccines: What You Should Know.* 3rd ed. Hoboken, NJ: Wiley, 2003; Sprenger, Ute. "The Development of Anti-Fertility Vaccines: Challenging the Immune System." *Biotechnology and Development Monitor* 25 (1995): 2–5; Surowiecki, James. "Push and Pull." *The New Yorker,* December 20, 2004.

Betsy A. Frazer

INDIGENOUS KNOWLEDGE

Indigenous knowledge is the knowledge that indigenous peoples all over the world have gathered over generations of living in harmony with nature.

Indigenous knowledge includes all knowledge that is needed for an indigenous society and culture to survive; it includes traditions, ceremonies, culture, environment, forestry, farming, artwork, prayers, and dancing. People have existed for centuries in a way that is respectful to animals and the environment. For centuries, they did not take more than they needed, and in return they were given the ability to live from the land. Spiritual and traditional ceremonies are involved in indigenous knowledge, as they have helped people pass on their knowledge from generation to generation through dancing, singing, and storytelling and to keep active and alive.

A few years ago, if you had asked a scientist about indigenous knowledge, you likely would have been told that it does not exist; it is just part of the folklore Native peoples rely on to explain where and how they live. Scientists felt that the medicines Native peoples used and the ceremonies held were all part of the traditions of indigenous people, and they did not assign any scientific value to the people's knowledge. On the other hand, if you spoke with an indigenous person, he or she would tell you that it is oral knowledge that has been passed through generations through medicine men, elders, and healers. Knowledge gleaned from their environment is used to predict the weather, heal sickness, and live within a specific ecosystem. Without such knowledge, indigenous people would not have survived in their environments; without their knowledge, early Europeans likely would not have survived in the countries they explored. Through accumulating and passing along knowledge from generation to generation, indigenous peoples have found answers to the questions of what foods are safe to eat, what medicines are needed to heal sickness, what the weather patterns are, and whether there are predator animals nearby. If indigenous people did not take the time to observe and learn from their environment, they did not survive.

Although indigenous knowledge and what we would recognize as scientific knowledge are both based on observation, there are some very real differences. Indigenous knowledge has been passed on orally for centuries, whereas scientific knowledge has been written down with all of the proofs or evidence that support it. Indigenous people had elders and healers who would carry the knowledge and teach younger people their knowledge in order to help the rest of their community survive. Their education was based on talking, observing, listening, and trying. The longer the students worked with their teacher, the more they learned. If a student did not learn from an elder or a healer, the lives of the community members would be jeopardized. Scientific knowledge is learned by going to school, learning what your instructor teaches you, and taking tests to prove you know what you are doing. Scientists who discover new knowledge continually have to test and retest their theories in order to prove they are accurate.

Indigenous knowledge is based on years of observations and learning. Indigenous people spend their entire lifetime studying with their teacher to ensure they have the skills and knowledge to pass on to their own students, whereas scientific knowledge has to be demonstrated, over and over. Indigenous knowledge is very holistic and spiritual because indigenous people see all of the things in the world as interconnected. The sky provides sunshine and rain, which nourishes plants to grow and thrive; these plants grow and (with the help of insects) reproduce. Insects are food and nourishment for small animals, which thrive and grow to become food and nourishment for larger animals. Larger animals become food for humans, and humans also utilize the animals' muscle fibers and fur. Anything they do not use fully is eaten by insects and other animals or returned to Mother Earth. When people pass away, their bodies are returned to the ground and become soil. On the other hand, scientific knowledge is based on different categories and their theories. In biology, people and animals are broken down to their body parts, organs, cells, and DNA. In chemistry, everything is broken down to chemical properties, elements, molecules, and atoms and then further to protons, neutrons, and electrons. Physics is the study of how things work and move. People's minds are studied in psychology, their culture in anthropology, and their society in sociology. Every aspect is separate and individual.

In contrast, indigenous knowledge is based on a holistic view of the world. Everything is connected, and everything has a place. Indigenous people did not own the land, but they were caretakers of it. They did not take more from the land than they needed, and they thanked Mother Earth for what she offered them for survival. They were also aware of weather patterns because a storm, hurricane, drought, or other adverse weather event would affect their life and their well-being. People survived based on their knowledge of their surrounding environment. Hunters would follow the tracks of animals to find food. While following those tracks, they would also look for signs of how long it had been since the animals had passed through the area. Based on the animal's droppings and footprints, they could identify the animal, the size of the animal, and the size of the herd. If they needed to find water, they looked for the signs they had learned that told them if water was nearby. If a plant was growing in the desert, that meant there was water underground; because wildlife would survive only in an area where food and water were available, if there were no plants or animals or insects around, there probably was no water either. Indigenous people needed to be aware of the weather and had to look for signs that told them if the weather was changing. Was the wind picking up; was the ground trembling; were animals looking for shelter; was the sky growing dark? Indigenous people all over the world looked for signs based on where they were living. Various signs would tell them if it was a snowstorm, thunderstorm, tidal wave, or tornado that was coming. Each of these weather conditions would be life-threatening to the indigenous people who would need to survive them. If people were sick, medicine men and healers were called in to assist in their healing. Their traditions and practices were based on years of knowledge and studying to find which plants were the best medicines to heal people. If something was discovered to soothe a stomach ache, it would be used, and that knowledge would be passed on; plants that soothed a sore throat or healed a wound were also identified and used. The methods that were used were passed across generations.

For centuries indigenous people relied on observation and trial and error, learning to live off the land. They knew what berries were safe to eat and which ones not to eat based on observation and experience. If a berry that was poisonous was eaten, the person either got sick or died. Other members of their family groups would know not to eat those same berries again. When picking fruit caused stains on their skin, they knew that those plants could be used to help dye fabrics and skin. By observing animal behaviors, they learned what was good to eat, what watering holes were safe to drink from, and which animals were dangerous to them. Animal behavior would signal them if dangerous animals were in the area because an animal's life depends on always keeping an eye out for predators. Animals communicate with each other by making sounds and gestures that warn others of danger or predators in the area; indigenous people would learn the differences between the sounds and gestures to ensure they were alerted as well.

Today, scientists such as David Suzuki are writing and speaking about the need to return to indigenous knowledge in order to help stop the destruction of our planet and to heal sickness without relying on man-made chemicals. You cannot turn on your television without seeing advertisements to purchase ancient remedies known by indigenous peoples to cure arthritis or rheumatism or heartburn. Indigenous people are being looked to for their knowledge of plants that heal. Scientists are interested in how they lived in their environment without destroying the area in which they lived. Educational institutes are studying the way they teach and learn because it works with the student in a holistic way and does not just teach the mind. Doctors are interested in the drugs they used to cure their sick. The world is changing, and people are beginning to realize that indigenous knowledge is a distinct area of knowledge and needs to be accepted as a way of life.

See also Ecology; Globalization; Science Wars.

Further Reading: Battiste, M., and J. Y. Henderson. *Protecting Indigenous Knowledge and Heritage.* Saskatoon: Purich Publishing, 2000; Knudtson, P., and D. Suzuki. *Wisdom of the Elders.* Toronto: Douglas & McIntyre, 1992; Sefa Dei, G. J., B. L. Hall, and D. G. Rosenberg, eds. *Indigenous Knowledge in Global Contexts.* Toronto: University of Toronto Press, 2002.

Marti Ford

Indigenous Knowledge: Editors' Comments

The indigenous, traditional, or local knowledge battleground focuses on the differences and conflicts between these forms of knowledge and modern scientific knowledge. Immediately, one can argue that modern scientific knowledge is just another form of indigenous, traditional, or local knowledge. That is, science in this sense is the local knowledge of modern Western industrial societies, writ large. Furthermore, it is easy to forget that some indigenous peoples have destroyed their environments in their pursuit of survival strategies. Mesolithic peoples were already engaged in activities that led to deforestation. Slash and burn strategies common throughout history can be part of a "shifting cultivation" strategy that sustains fertile land, but it can also result in irreparable damage to an ecology. One must therefore be careful in attributing sustainability values universally to indigenous peoples. What is more, when used to drive a culture that uses different technology, these same sustainability values can result in environmental degradation. In fact, culture itself may be an environmentally degrading addition to the world's evolutionary history. Social necessities leading to the growth and development of cultures are by definition environmentally exploitive. Even with an awareness of the ways in which our activities degrade, corrupt, and homogenize ecologies, we may be powerless to do little more than postpone the inevitable disasters and catastrophes.

INFLUENZA

The term *influenza* is derived from the Italian word for "influence" and dates from 1357. Italian astrologers of that time believed influenza was the result of the influence of celestial bodies. Influenza is commonly known today as the flu. It is an infectious disease that affects both birds and mammals. Influenza has been at the center of many debates between private and government scientists and within the government itself, and these debates have become an obstacle to medical scientists and physicians seeking to discover an effective treatment and vaccine.

There are many different strains of influenza, some more dangerous than others, but all are caused by an RNA virus from the Orthomyxoviridae family. Influenza is not a disease natural to humans, and it is believed to have originated in birds and spread to humans during the last ice age. There are three types of influenza viruses, classified as A, B, and C. Type C rarely causes disease in humans, and type B causes illness, but not epidemics. Only type A is capable of producing an epidemic or pandemic. Individuals suffering from seasonal influenza generally recover in two weeks, with 20,000 to 50,000 individuals dying of influenza viral infections annually within the United States.

Influenza can weaken the body's immune system, leaving an individual susceptible to secondary infections. Although influenza has been known for centuries, it became infamous during the Great Influenza pandemic of 1918–19, also known as the Spanish flu (type A, H1N1). Interestingly, it received the name Spanish flu simply because the Spanish newspapers were the first to report it, even though it had appeared in the United States months before. This strain of influenza was particularly lethal and is thought to have originated in Haskell County, Kansas. Although this influenza might have died out, the political state of the country at the time helped to spread it worldwide. America had just entered the Great War (1914–18) and was preparing to ship thousands of soldiers to France. Before this could be done, the soldiers needed to be trained. This training took place in cantonments throughout the country, with each cantonment holding tens of thousands of young men in cramped quarters, and influenza spread rapidly among the soldiers and support staff on the base. The movement of troops between U.S. bases, forts, and cantonments ensured that almost no American community went untouched by the disease.

Shipping men overseas helped to promote the spread of influenza throughout Europe and eventually the world, with cases appearing as far as the Arctic and on remote islands in the South Pacific. Nearly all residents of Western Samoa contracted influenza, and 7,500 were killed—roughly 20 percent of the total population. As surgeon general of the army, William Gorgas was responsible for ensuring the effective and successful performance of military medicine. But although Gorgas was known internationally as an expert on public health, in reality he was given little authority by the U.S. government. Gorgas recommended that drafts be postponed and that the movement of soldiers between cantonments and overseas cease. President Wilson, however, continued to transfer soldiers from bases throughout the country and to ship them overseas, creating strained relations between the president and his military medical advisers.

Because the natural home of influenza is birds, and because influenza can survive in pigs, the survival of humans is not necessary in order for influenza to survive. As a result, mortality rates in humans can reach extremely high numbers. Contemporary estimates suggest that 50 to 100 million individuals were killed worldwide during the Great Influenza—2.5 to 5 percent of the world's population—and 65 percent of those infected in the United States died.

A second battle was being fought during the Great War, this one between the scientists and influenza itself. It was no mystery that disease followed war, and on the eve of the United States' entrance into this war the military recruited the top medical minds in the United States. These included William Welch, founder of Johns Hopkins University; Victor Vaughan, dean of the Michigan medical school; Simon Flexner, Welch's protégé; Paul Lewis from Penn; Milton Rosenau from Harvard; and Eugene Opie at Washington University. Eventually the entire Rockefeller Institute was incorporated into the army as Army Auxiliary Laboratory Number One by Surgeon General of the Army William Gorgas. As the pandemic raged on, scientists found themselves in a race against time. Scientists worked night and day, at times around the clock, in an attempt to develop a treatment and a vaccine or antiserum for influenza. The risk was great, as more than one scientist was struck down by the disease itself.

The cause of influenza was not known at this time, and two camps emerged: those who believed influenza to be a virus and those who believed that the bacterium *B. influenzae* caused the disease. During this time a number of medical discoveries were made, such as a treatment for three different types of pneumonia. Unfortunately, no true progress toward creating an influenza vaccine occurred until 1944, when Thomas Francis Jr. was able to develop a killed-virus vaccine. His work was expanded on by Frank MacFarlane Burnet, who, with U.S. Army support, created the first influenza vaccine.

The American Red Cross was another principal player. Given the tremendous number of both civilian and military deaths as a result of influenza, and the cost of the War overseas, the government could not put together the necessary funds and personnel to care for matters on the home front. Assistance was needed, and when it became apparent that the influenza had reached the scale of a pandemic, the Red Cross created the Red Cross National Committee on Influenza to coordinate a national response. The Red Cross proved invaluable. The Red Cross National Committee took charge of recruiting, supplying, and paying all nursing personnel and was responsible for providing emergency hospital supplies when local authorities were unable to do so and for distributing doctors through the U.S. Public Health Service to wherever they were needed. The shortage of medical personnel created by the War meant that the Red Cross was more or less single-handedly responsible for coordinating the movement of medical personnel throughout the country. Between September 14 and November 7, 1918, the Red Cross recruited over 15,000 women with varying degrees of medical training to serve in military and civilian posts. By spring of the following year, the Red Cross had spent more than two million dollars in services.

The severity of the 1918–19 epidemic was not forgotten, and since then, influenza has been a concern for physicians, scientists, and policy makers. With the exclusion of recent avian viruses passed directly from bird to human, all type A influenza viruses globally have originated from the 1918 H1N1 virus. In the early 1930s, scientist Richard Shope proved that the feared H1N1 virus was alive and thriving in the country's pig population. This is particularly feared because pigs can act as an intermediary animal, allowing avian flu strains to adapt to mammals and then be passed onto humans. This strain of the H1N1 virus in the pig population is often referred to as Swine Flu. In 1957 the threat of another pandemic appeared. Government and medical officials feared the return of the H1N1 virus, or swine flu. That was not the case. Although the virus killed upward of one million individuals, it was not the H1N1 virus and instead became known as the Asian Flu, an H2N2 virus. An earlier, and much less documented, influenza virus had occurred between 1889 and 1890. This pandemic was known as the Asiatic (Russian) flu. The Asiatic flu killed roughly one million individuals, and it is suspected that it too was an H2N2 virus. The most recent pandemic occurred from 1968 to 1969. Known as the Hong Kong virus (H3N2), it infected many, but the mortality rate was low. It was responsible for 750,000 to 1,000,000 deaths. Although there has not been a pandemic since the Hong Kong flu, public officials, hypersensitive to the threat of a flu epidemic, were concerned for the potential of a swine flu epidemic in 1976 and Asiatic flu pandemic in 1977.

In 1976, at Fort Dix, New Jersey, an 18-year-old private, feeling the symptoms of influenza, decided to join his platoon on a night march anyway. A few hours into the hike, he collapsed. He was dead by the time he reached the base hospital. Although the young private's death was the only suspicious death to occur, it was a reminder of the 1918–19 virus's ability to kill young adults quickly, and officials feared another epidemic was at hand. Simultaneously, a young boy living on a Wisconsin farm did contract swine flu, surviving thanks to the antibodies produced by handling pigs, which were infected with Shope's swine flu virus. Overwhelmed by the potential consequences of being wrong, medical and government officials chose to prepare themselves for the worst and declared the potential for an epidemic. Dr. David J. Sencer, director of the Centers for Disease Control, requested a $134 million congressional allocation for developing and distributing a vaccine. Following a dramatic televised speech give by the President, Congress granted $135 million toward vaccine development and distribution in a last-minute vote. The President signed Public Law 94-266, allocating funds for the flu campaign on national television, stating that the Fort Dix virus was the cause of the 1918–19 pandemic. The epidemic never surfaced. The American flu campaign was criticized on both a national and an international level, and Sencer was removed from his position at the CDC in 1977.

The most recent influenza scares have centered on avian flu (H5N1) and have most often been located in Hong Kong and other Asian countries. Avian influenza, also known as bird flu, is an extremely virulent virus that generally infects only birds. In recent years, however, it has been documented as infecting pigs and most recently, although rarely, humans. It spreads rapidly though animal populations and can produce a mortality rate of 100 percent within 48 hours. In 1997 the H5N1 virus spread directly from chickens to humans, and it killed 16 out of 18 infected. It is this particular virus that the term *avian influenza* most commonly refers to. After this incident, all chickens in Hong Kong (1.2 million) were slaughtered in an effort to contain the virus. This protective measure failed because the virus had been able to spread to the wild bird population. In 2003 two more people were infected with avian flu, and one died. When scientists first tried to develop a vaccine for avian flu using the traditional vaccine growth medium, chicken eggs, they found that the virus was too lethal; the virus was killing the eggs in which it was being grown. A vaccine for avian flu now exists, but it took more than a year to develop, and it has not been stockpiled should a pandemic arise. All of those who caught the virus were infected directly by chickens, and the virus did not develop the ability to spread human-to-human.

The potential for creation of a new, lethal virus exists, however. If one of the individuals who caught the avian flu had simultaneously been infected with a human influenza strain, it would have been possible for the two different strains of influenza to separate and recombine, using the human individual as an incubator to create a new strain of avian flu capable of being spread through human-to-human contact. It took a year to develop an avian flu vaccine. Should the virus mutate once more, it would have done the majority of its damage by the time a new vaccine could be developed by scientists. In an effort to stem this possibility, the World Health Organization (WHO) established a formal monitoring system for influenza viruses in 1948. Eighty-two countries and 110 laboratories participate by collecting information, which is then processed by four collaborating WHO laboratories. Any mutations in existing viruses are documented and are then used to adjust the next year's vaccine. The surveillance system also actively searches for any signs of a new influenza strain, especially one with the potential to mutate into the next pandemic.

See also Epidemics and Pandemics; Vaccines.

Further Reading: Barry, John. *The Great Influenza: The Epic Story of the Deadliest Plague in History.* New York: Viking Adult, 2004; Garrett, Laurie. *The Coming Plague: Newly Emerging Diseases in a World Out of Balance.* New York: Penguin, 1995; Taubenberger, Jeffery, and David M. Morens. "1918 Influenza: The Mother of All Pandemics." *Emerging Infectious Diseases* 12, no. 1 (2006), http://www.cdc.gov/ncidod/EID/vol12no01/05-0979.htm; World Health Organization. "Influenza." *Epidemic and Pandemic Alert and Response (EPR).* http://www.who.int/csr/disease/influenza/en.

Jessica Lyons

INFORMATION TECHNOLOGY

In our contemporary world, information technology (IT) has gained a pervasive significance in practically every sphere of life. Together with its technological innovations, IT has contributed to profound transformations in management practices, work processes, planning and administration, political action, scientific theories and methods, communication, and culture. It is not unusual to hear talk today about an "information revolution," conflated with hype and enthusiasm for a new age of digital technologies. Nevertheless, this optimistic technological determinism recurrently emerges historically as unsustainable. Such optimism is readily deterred in the face of comprehensive social criticism. The intersection of technological optimism and social criticism is the focal point of the continuing debate over the definition, extent, scope, and direction of IT.

Materially and fundamentally, IT refers to methods, apparatus, and infrastructures that operate through and with information, as codes, images, texts, and messages, stored and transmitted by databases, telecommunications networks, satellites, cable, television, telephones, and computers. Assuming a technical and cybernetic definition regardless of its semantic content, information is now understood as a quantitative measure of communicative connections, translated into a binary code (bit, either 0 or 1) and carried by a channel to a receiver. This gives rise to another common IT expression, information and communication technology (ICT). From the first computers in the 1950s, enhanced in the next two decades with the transistor and integrated circuit, coupled with the digitization of telecommunications, up to the first networks that led to the Internet in the 1990s, we seem to have achieved a digital convergence and integration of media, information, and communication in an interconnected computerized world.

Different approaches and perspectives have emerged concerning the importance and extension of these trends. Some observers of these trends endorse the idea of an "information society" or an "information age" as a new postindustrial society; others regard it rather as an extension of preestablished social relations and structures and of unfulfilled and unfulfillable promises of freedom, well-being, democracy, community, and progress. Similar promises have accompanied earlier technologies, from radio and television to nuclear energy. These promises are part of the ideologies of technological determinism and technological progress.

The more enthusiastic visionaries see technology in general, and IT in particular, as decisive and neutral agents of social change, largely focusing on technology's "social impacts" in a natural, desirable, beneficial, and inevitable course of technological development. This radical evolutionist technological determinism tends to stand opposite perspectives on "social shaping" or "social construction" of technology. Social shaping and construction critics and theorists are more concerned with the continuous social, cultural, economic, and political processes of technology. As David Lyon emphasizes, if societies are technological products, conversely technologies are social products because they are shaped by diverse social factors, adapted to certain necessities and choices, and appropriated in different ways by users who can accept, refuse, or reformulate them, and they entail varying positive and negative effects.

Daniel Bell was one of the most notable authors to announce a new type of society, discontinuous by a quantitative and qualitative shift based on information and knowledge. This postindustrial society would increasingly see the replacement of the manufacturing economy by a service economy, the rise of scientific research and development in production and innovation, and the creation of new dominant professional, scientific, and technical groups, ranging from scientists, teachers, librarians, journalists, managers, secretaries, clerks, and lawyers to computer programmers, engineers, and analysts.

These economic transformations are currently being understood under an "information economy" or "new economy" model where productivity, competitiveness, and power depend on the capacity to generate, process, and efficiently apply knowledge-based information. Manuel Castells, a leading student of IT trends, speaks of a new mode of production and development, "informationalism." In a "network society," constant fluxes of financial, technical, and cultural information tend to increase globalization and dominance of finance capital, diminish the role of nation-states, promote flexible production and horizontal corporations, and lead to new business models based on innovation, flexibility, and global coordination.

The fragilities of deterministic approaches and high expectations surrounding the "new economy," however, can probably be exposed using the example of the dot-com crash of 2000. A huge investments' peak in Internet-based companies finally burst its "bubble," showing the significance of more continuous approaches in identifying traditional market and management constraints. Decentralized structures and flows of information do not necessarily change hierarchical systems, traditional class differences, or power relations. We witnessed an increase in multinational vertical concentrations intertwining multiple areas of information, media, and entertainment. So-called new forms of stratifications between the "informational labor" of well-educated and connected workers, service, and industrial workers and an excluded underclass, poorly educated and unskilled, can be traced to established forms of social inequality. Supposedly fragmented nation-states may still assert some planning, administration, surveillance, and control over global networks of finance, technologies, production, merchandises, and work forces run by powerful corporations.

These rearranged power equilibriums have been largely discussed in controversies over the political roles and actions of nation-states, institutions, movements, and citizens in an interconnected world. Nation-states have partly extended the scope and diversity of their power by using computerized systems of information collecting and recording and also by developing a media network of public relations, spin doctors, opinion polling, news management, image production and advocacy, and political advertising. Nonetheless, the coupling of digital technologies with political strategies has put more and more stress on concerns about spectacle politics, centralization of databases with personal information, privacy and electronic surveillance of citizens, social movements, and even political opponents.

Yet, IT is considered by the more optimistic as playing an important role in revitalizing democratic processes in an "electronic democracy." The basic assumption is that access to information and dissemination of one's points of view are essential in democracy, so networks allow people to be more informed and thus more interested and able to intervene in more direct forms of participation and dialogue. Digital technologies are believed to facilitate alternative channels of up-to-date information, electronic voting, online petitions, political Web blogs and chat rooms, mobilization of interest groups, and networks of social movements (witness, for example the antiglobalization protests in Seattle and Genoa). From a more skeptical perspective, however, there are serious doubts about whether IT can radically change powerful traditional political frameworks, visible for example in trends that those who engage more directly in electronic media are activists already involved in political actions.

Concerning science and technology, information has been employed primarily in developing electronic circuits, computers, networks, and artificial intelligence. These technologies have changed scientific practices through the use of computing devices in research, as for example in modeling and simulation. Furthermore, information has also been used as a concept in human and social sciences such as psychology and economics and in natural sciences such as biology and genetics, leading to an "informational" convergence particularly visible in new areas of biotechnology and nanotechnology. But the information metaphor is still extremely controversial in issues such as cloning, ownership of genetic information, use of human or live models, manipulation of organic characteristics, military investments and applications, and even our conceptions of nature, human, and technique.

Significant changes brought by IT can be felt maybe more intensely in our social and cultural experience. Perhaps we do not live closely together in the "global village" of Marshall McLuhan, but we do live more connected in a media society with televisions, videos, satellites, telephones, computers, radios, cinema, cell phones, Internet, portable audio, cameras, books, newspapers, magazines, and so on. In particular, the vitality and vastness of online interactions have entailed many theories, cultures, and ways of living in the realm of cyberspace (a term coined by science fiction author William Gibson).

Today there are many different instant messaging services, such as ICQ, Yahoo!, and MSN; chat rooms; Web forums; newsgroups; mailing lists; and wireless devices such as the Blackberry. Online interactions can be also structured in "virtual reality" environments, such as MUDs, MOOs, and MMORPGs, where you can experiment anonymously with gender, age, race, sex, and violence, or the virtual world "Second Life" where "residents" can socialize, work, play, trade, and even exchange their earnings "outside." Networking interactivity, participation, and openness are now seen to be enhanced in Web 2.0. This refers to Web-based services that allow users to create, change, and customize contents to a great extent, emphasizing online collaboration such as Web logs, wikis, podcasts, social networking on MySpace, and free video, photo, and music sharing Web sites such as Flickr and YouTube.

These immense computerized networks of information and communication are nevertheless subject to various critiques. Many see virtual worlds as superseding the real world, leading to a decline of face-to-face communication and to addiction, social isolation, or multiple and unstable identities, although IT certainly has increased channels of communication between formerly unrelated people and parts of the world.

But on the other hand, we can find some puzzling clues in Albert Borgmann's account of blurring differences between real and technological worlds. Information seems to be losing its cultural, social, and political references to the "reality" of things, and "technological information" is thus presented as a reality in itself, often in situations of uncontrollable overload, saturation, misinformation, and disinformation. In his words, "in detaching facets of reality from their actual context and setting them afloat in cyberspace, information technology not only allows for trivialization and glamorization but also for the blurring of the line between fact and fiction" (p. 192).

Other central issues also trouble cyberspace domains of freedom and expression, especially with regard to copyright and intellectual property. Music downloading through P2P file-sharing programs such as Napster and Audiogalaxy has been heavily prosecuted by the music industry. Responses range from copyright treaties to open source and "copyleft" movements, which seek to protect freedom of information, although it is increasingly an asset to commercial interests as seen in paid content, narrowcasting, online meters, and cookies or more recently in the question of "net neutrality." Attempts by ISPs to prioritize data from their own sponsors or associated companies have led to controversies over restrictions based on content, Web sites, services, or protocols.

IT's rapid and exciting developments are undeniable, making information a powerful resource, commodity, currency, cultural framework, and theory. Considering our economic, political, and social dependence on computerized systems, however, it is essential to determine how technologies are built and by whom, for whom, and how they are used everyday. Critical perspectives are needed to engage in such analyses of values and priorities of technology construction, design, and use.

One of the main questions is equal and democratic access to information. The digital divide between industrialized and developing societies, and between information-rich and information-poor in each nation, is already reinforcing preexisting social disparities based on income, education, skills, resources, and infrastructures. Nicholas Negroponte announced in 2005 the $100 laptop initiative to developing countries, but the fact remains that "e-learning" doesn't necessarily change the balance of power. Equal access must also mean democratic choice in deliberation, planning, decision making, testing, and evaluation of IT, for example through public consultations to ensure adequate technological systems.

Another fundamental question concerns the status and definition of information. As a quantitative and homogeneous measure, information seems to disregard the quality, the sense or character, of what is being communicated—whether it is significant, accurate, absurd, entertaining, interesting, adequate, or helpful. So it is pertinent to ask which type of information is being produced; what is its finality, function, or content; and who decides its price, ownership, and applicability. Maybe then it will be possible to disentangle present confusions between information and knowledge, a difference between, on one hand, a supply of information mainly valued as a commodity within electronic systems that gather, organize, analyze, use, and transmit data and, on the other hand, the ability to gain knowledge and understanding to act more freely and consciously.

See also Computers; Internet; Privacy; Search Engines.

Further Reading: Bell, Daniel. *The Coming of Post-Industrial Society.* New York: Basic Books, 1976; Borgmann, Albert. *Holding on to Reality: The Nature of Information at the Turn of the Millennium.* Chicago: University of Chicago Press, 1999; Castells, Manuel. *The Information Age: Economy, Society and Culture.* Vol. 1, *The Rise of the Network Society.* Vol. 2, *The Power of Identity.* Vol. 3, *End of Millennium.* Oxford and Cambridge: Blackwell, 1996–98; Lyon, David. *The Information Society: Issues and Illusion.* Cambridge: Polity Press/Blackwell, 1988; Negroponte, Nicholas. *Being Digital.* New York: Knopf, 1995.

Susana Nascimento

INTELLECTUAL PROPERTY

Intellectual property is at the center of several controversies in science and technology. The two primary forms of intellectual property are copyright and patent. Trade secrets and trademarks are also considered forms of intellectual property. Copyright refers primarily to written and visual forms of expression, whereas patents are meant to protect inventions, whether devices or processes. A patent guarantees a monopoly for an inventor for a fixed period of time (up to 20 years), but on the condition that the invention is disclosed. Patented inventions must be new and "non-obvious." Copyright protects the exact expression of an artist or author, but not the core ideas. Both copyrights and patents are exclusionary in that they prevent others from using a new technology, but they do not guarantee the rights of the creator or inventor to implement the new technology, which may be based on intellectual property held by others.

"Fair use" copyright controversies are centered on the reproduction of a creator's images or text by others. Fair-use exemptions include documentary or scholarly work on the image or text. The exact use of text by others is known as plagiarism. The reproduction of images for satire or commentary has been contested in the courts. For example, Mattel, the owner of the Barbie doll, has unsuccessfully sued artists who have used the doll in satires and other media productions. They have sued on the basis of both copyright and trademark infringement. To date, however, the right to parody has been protected by judges referring to the First Amendment of the U.S. Constitution.

Electronic file sharing of music and film have, obviously, come under scrutiny as intellectual property cases. Many people think that once they have purchased a form of media, whether book, computer program, or media file, they are free to do what they want with it. It is clear that, to date, you can read, reread, sell, or donate a book when you have finished reading it. Electronic media have been categorized separately from traditional texts, however, and so far the courts have sided with the original manufacturers or producers of these media in arguing that consumers do not have the right to use or reuse the intellectual property expressed in the media. This in part reflects the identification of software as protected by patent law rather than copyright law. Because computer software is the implementation of a process, rather than a form of self-expression, it is protected as intellectual property by patents, which limit the use of the technology through licensing.

Patenting has produced its own set of intellectual property controversies. The most recent include the patenting of whole organisms and genes. Although it is fairly unambiguous that a test that can detect a virus or genetic sequence might be patentable, given that it is an invented, useful process, it is not clear whether the virus or genetic sequences themselves are patentable. The U.S. Patent and Trademark Office and the courts so far have said that organisms and genes are patentable. Critics argue that discovery is not the same as invention. Because the genes are not modified by their detection, they should not be considered intellectual property. The patenting of biological products includes human tissues. To date, if a patient has cells removed from his or her body that are then patented by researchers, the researchers, and not the patient, own the cell lines and information about those cells as intellectual property. The case for patenting organisms is perhaps more robust. Specially bred or genetically modified whole organisms, which include mice and other laboratory animals, are changed, and the modifications for the germ line are not discoveries, but innovations. Because they are not available in nature, they have been protected by patent law.

The patenting of biological products, from genetic information to whole organisms, is controversial because it drives up the costs of research. People must pay for, rather than freely share, research materials. This leads to disincentives to replicate and verify work, which may allow for the perpetuation of errors. Increased secrecy is considered bad for science, although the patent itself is a kind of disclosure of information.

In international contexts, the desire to capture information has led to what some call bioprospecting or even biopiracy, which is patenting genes or organisms in other parts of the globe. Cases include an attempt to patent the neem tree for its antifungal properties and to patent the antibiotic properties of the spice turmeric. The patents were thrown out because the uses of these products were known to local populations, and there was no inventive process. (In U.S. terms, the existence of traditional knowledge means that the patents were not for "non-obvious" uses.) Other plants, genes, and extracts have received patents, however. Activists are concerned that the information is taken to First World pharmaceutical companies, and great profits are made that do not come back to the locales that provided the materials. In fact, it may be that traditional peoples would have to pay royalties for using a process that they initiated.

Related controversies have also occurred in plant biotechnology. Monsanto has sued several farmers for patent infringement because they saved seeds, a traditional farming practice, that contained patented genetic sequences. Their purchase of Monsanto seeds included an explicit license that prevented seed saving, requiring farmers to purchase new seeds every year instead of saving a small part of any given year's crop for future plantings. This pushes up farmer costs, and the risks of crop failure fall squarely on the farmer. There have also been suits because corn plants were fertilized with pollen that blew in from nearby fields planted with patented strains of corn. Other cases involve soybean seeds from a previous year that volunteered to grow a year later or that grew from seeds that fell off plants as they were being harvested the previous year. Monsanto has very aggressively protected its intellectual property rights in these cases, bankrupting some farmers.

There are also critiques of the intellectual property system per se. In the United States, the patent and copyright system was designed in the Constitution for individuals, not for corporations. The courts have continually upheld the rights of a company to own the intellectual property of its employees, however. This is especially the case when the intellectual property is work related. The company's rights can apply, depending on the nature of the employment contract, for a specific time in the employee's future even if he or she should leave the company. This can take the form of trade secrets protection or noncompetition clauses that prevent a person from taking a job with a competitor or that limit practice in both place and time for a fixed interval. It does not matter if an employee-inventor comes up with an idea in the shower (or other private space). If it is related to the employee's expertise and role in the company, the company or organization has the first right to determine ownership and dispensation. If the individual's idea is clearly separated from his or her expertise and work role (say a chemical engineer has an idea for a wooden child's toy at home on the engineer's own time), then he or she can probably claim ownership of this idea, but it will depend on the exact nature of the employment contract and the context of the invention.

The most important but as yet invisible critiques of patenting and copyrighting have to do with the question of whether patents and copyrights are necessary at all to protect intellectual property. For example, patents are not useful in fields with rapid technological turnover. In the computer components industry, the life span of the technology may be shorter than the time necessary to file the patent, and thus being first to market, not legal protections, will be the guarantor of market success. In fact, keeping the intellectual property open may allow more people to adopt the component and integrate it into their own inventions, expanding market share and profits. The long development time for drugs, by comparison, may mean that the patent is necessary to protect the invention at least for a while until some profits can be made. Products with high barriers to industry entry may not need patents because simply having access to a new idea does not provide the infrastructure to develop it.

Because people believe that patents are necessary for protecting and providing an incentive for invention, they act in ways that reinforce the seeming necessity of the patent system. This has also led to excesses such as patent "trolls"—companies that buy up or formulate patents and extort money from people who actually develop the technology or service, although the trolls are not selling the goods or services themselves. Patents can also be held to prevent others from providing an innovation, although the European patent system prevents nonproductive exclusion. Others think that the term *troll* should be replaced by NPE, for *nonpracticing entity,* to recognize the valuable services that might be provided by patent-holding companies.

Finally, the cultural specificity of copyrights and patents is often overlooked. In many cultures, copying and sharing information is considered good and considered a spur to innovation and invention. For example, blues and hip hop are musical traditions where borrowing and pastiche are expected and valued, providing continuity and a sense of legacy to the art forms. There are cultural objections to the patenting of living organisms as a debasement of the intrinsic worth of living things. There are those who question whether ideas can or should be turned into property at all. Knowledge, in this perspective, is not a "thing" in the way that material goods are and thus cannot be effectively contained, and attempts to contain intellectual property are thus impractical as well as probably immoral—because knowledge is power, and limiting the flow of knowledge can be a source and signal of oppression. It is odd that even the staunchest libertarians who decry any involvement of the state in the free market seek the protections of conventional intellectual property regimes.

See also Information Technology; Privacy.

Further Reading: Center for Food Safety. *Monsanto vs. U.S. Farmers.* http://www.centerfor foodsafety.org/pubs/CFSMOnsantovsFarmerReport1.13.05.pdf; Martin, Brian. *Information Liberation: Challenging the Corruptions of Information Power.* London: Freedom Press, 1998; Viadhyanathan, Siva. *Copyrights and Copywrongs: The Rise of Intellectual Property and How It Stifles Creativity.* New York: New York University Press, 2001.

Jennifer Croissant

INTERNET

The Internet is a worldwide system of computers, a network of networks in which someone with one computer can potentially share information with any other computer. With the number of such linked computers around a billion, the Internet is often called the most significant technology advance in a generation.

Understanding the Internet is not simply a matter of describing how it works, however. It also requires looking at the consequences of using the World Wide Web (WWW). The amazing ability of the Internet to hide its complex technologies leads some to think it is easy to understand. Anyone can point and click and traverse the globe. Fewer can speak sensibly about the way modern culture has changed for better and worse in the Internet age.

Today the terms Internet and World Wide Web mean the same thing for most people. Strictly speaking, they are different. The World Wide Web is the collection of documents, files, and media people access through the Internet. The Internet is the network technology that transports World Wide Web content. Put another way, the Internet makes the World Wide Web possible; it is the World Wide Web that makes the Internet essential.

The two terms are a useful way to talk about "the Internet," as most people call it. The first part of the story is the quiet building of the Internet among academics over 25 years. They had no idea of the eventual significance of their inventions. The second part of the story is the rise of the World Wide Web in popular culture, when it seemed everyone knew they had a revolution on their hands.

Before either story began to emerge, one of the elements of the Cold War between the United States and the Soviet Union was the significance of science. A few years earlier, the United States had established its superiority in science with the development and detonation of the atomic bomb (1945). Each side knew that scientists could win wars, and the A-bomb seemed indisputable proof of this truth at the time. The Soviets raced to develop their own nuclear weapons and then surpassed the United States by launching the first satellite in 1957. Was Soviet science now better than American science? Did the advantage of space mean victory for the Soviets? A shocked U.S. military responded by forming the Advanced Research Project Agency (ARPA), bringing together the best minds in the nation to regain the technological lead. But how could they work together and communicate across the country? In particular, how could their computers talk to each other and share research? The Internet began simply as the answer to that question.

Dozens of innovations mark the way to the Internet wave of the 1990s, but three building blocks stand out, all beginning with the letter *p:* packets, protocols, and the PC (personal computer). None were created with today's Internet in mind, but all three were used to build today's World Wide Web.

"Packets" were designed for a time of war. Planners needed a way to ensure command and control in the event of a nuclear attack. Regular telephone connections would be useless in an attack, and radio broadcasts were too easily intercepted or jammed. ARPA scientists struck on a way to break up all information into packets, each carrying its destination address and enough instructions to reassemble thousands of packets like itself into original information at the end. Breaking down information into thousands of packets meant messages were hard to intercept and useless on their own. Because they were small, they were capable of traveling to their destination through any available route, even by many routes if one was blocked or busy.

The Internet still works this way. Packets transfer all information, whether that information is Web pages, e-mails, file downloads, or instant messages. Trillions of packets flood through any available network and are routed to their destination by powerful gateway computers. These computers do not examine, filter, or store the packets. They simply send them on to a destination computer that reassembles them perfectly. Imagine a trillion postcards sent out every hour to millions of addresses everywhere in the world and arriving accurately in under a second. This is how the Internet functions, and it works amazingly well. During the 9/11 attack on New York City, regular phone service broke down almost immediately. Cell phone networks were overwhelmed. But e-mails continued to get through because they relied on a method of communication intended to function during a nuclear war.

All elements considered, however, the Internet most certainly would not withstand a real nuclear attack. Although the network itself and the packet method of communication would not fail, the electromagnetic pulse (EMP) of a nuclear explosion would incapacitate 95 percent of the computer chips around the blast zone. The network might continue to work, but the computers hooked up to it would not.

Interestingly, the original military point packets also make it extraordinarily hard to block, filter, or censor Internet content. What was simply a design feature for a time of war has now defined the Internet for those who resist all attempts to censor or to control information. It is ironic that technology for command and control now inspires those refusing any command and control at all over the Internet.

It is not surprising that the efficient method of letting computers talk together through packets caught the attention of university researchers in the 1960s. By the end of the decade, what might be recognizable as an Internet went online under the name ARPANET (Advanced Research Project Agency Network). It only linked a few computers used strictly for research. Private, personal, and commercial uses were not permitted. What was needed for the scientists was simply a way to yoke together multiple computers for solving complex problems. Packet communication was quickly adopted by universities as an excellent way to send large amounts of data through a single network.

The common protocol is the second building block of the Internet (a protocol is an agreed-upon way of doing things). Computer networks spoke the same way (packets); now they needed a common language in which to communicate. Because networks of the day were built for diverse purposes, many languages were invented. Imagine talking in the United Nations lobby. Vinton Cerf, an ARPA scientist, proposed in 1974 a common protocol for inter-network exchange of information. His invention, called TCP/IP (Transmission Control Protocol/Internet Protocol), meant local computers always communicate with outside networks in a common language. The protocol did not achieve immediate adoption, but the benefit of using a common protocol spurred adoption. With it any computer network could access any other network anywhere in the world, and today TCP/IP is called the glue that holds the Internet together. It was at this time Cerf coined the word *inter-net* as a short form of *inter-network*.

The 1970s and 1980s saw steady growth in Internet connections, but things were still in the hands of researchers. Using the Internet required expensive equipment and mastery of arcane commands for each request. There was little popular awareness of the Internet, and few saw any particular use for it outside academic and military activity. A few small events, in hindsight, provided a catalyst for the eventual explosion of public Internet use in the 1990s.

One was the first e-mail, in 1972. Scientists needed a way to send instructions back and forth. Though officially frowned upon, messages soon involved birthday greetings, weekend plans, and jokes. Soon, the number of e-mails far exceeded the number of research files being exchanged. Another sign of things to come was the first online games played across the network. As early as 1972, administrators started noticing unusually high network traffic on Friday nights after someone uploaded a Star Trek game. People used the network to blast Klingons and compete with friends at other universities. These may have been the first computer nerds, and the significance of their gaming to the development of the Internet today should not be overlooked.

Another tool that in hindsight paved the way for the World Wide Web was USENET (this 1979 term is a contraction of *user network*). Large numbers of users "subscribed" to a special interest topic and were able to conduct two-way discussions. Soon the "news groups," as they were called, went far beyond research and even news and became online communities. They were the precursors of today's discussion forums, chat rooms, and RSS feeds. USENET groups were the watershed development for the shift to having users pull what they wanted personally from the network and then use the medium for the composition of popular content. The first Internet communities thus were born, giving a glimpse of how the World Wide Web would eventually work. USENET also introduced the first spam (unwanted communications), the first flame wars (often vicious online disputes), and the first online pornography.

Two more small events had important consequences for the Internet. One was the introduction of the Domain Name System (DNS) in 1984. In place of hard-to-remember numbers such as 74.14.207.99 for network addresses, simple names such as google.com were enough. Now the network was far easier to use, and a name on the network took on potential value. The smallest but most significant event was the lifting of the prohibition against commercial use of the Internet in 1987.

The third building block for today's Internet was the PC (personal computer) introduced by Apple in 1976 and the widespread marketing of business versions by IBM in 1980. The key word here is *personal*. Until then computers were expensive tools for researchers or for the geeks who could build them. The personal computer was aimed at the general public. Soon companies developed graphical user interfaces (GUIs) to replace arcane command languages, and thus simple-to-use software was developed for the novice. The mouse, the icons, and the WYSIWYG (what you see is what you get) interface brought everyday computer use into mainstream society. Anyone could do it. By the end of the decade, personal computers numbered in the millions and were affordable and in the hands of people who played with them in addition to using them at work. With millions of computers in the hands of the utterly uninitiated, everything was ready for an Internet revolution 25 years in the making.

The unintentional revolutionary was Tim Berners-Lee, yet another researcher using the Internet in the late 1980s at CERN (European Laboratory for Particle Physics) in Switzerland. He relied on the network to collaborate with colleagues around the world. Though the network was fine, the documents and files were not in the same format or easily found. He thought it would be much easier if everybody asking him questions all the time could just read what they wanted to know in his database, and it would be so much nicer if he could find out what these guys were doing by jumping into a similar database of information for them. He needed a simple way to format documents and describe their location and some common way to ask for them. It had to be decentralized so that anyone anywhere could get information without asking someone. Ideally the requests could come from inside the documents as links to other documents so that a researcher did not need to use some other application. Most of all, it had to be easy.

Berners-Lee sat down in 1990 and penned the specifications for a global hypermedia system with now-universal acronyms: HTTP (HyperText Transfer Protocol), HTML (HyperText Mark-up Language), and URL (Uniform Resource Locator). Though originally designed for far-flung researchers to collaborate on projects without bothering each other, the resulting universal information space set in place the keystone of today's Internet. For good measure Berners-Lee even gave his creation a name: the World Wide Web (WWW). He capped off these innovations with a small piece of software called a browser. He intended it only to make it easier for his peers to retrieve and read documents. He did not know it would touch off the modern Internet revolution.

For 25 years the word *Internet* was little known outside of academic circles. As the 1990s unfolded, however, everyone was talking about the Internet, also known as the Information Superhighway, Cyberspace, Infobahn, or simply the Web or the Net, as the technology took hold of popular culture. Everyone wanted to be on the Web, and users who hardly topped 100,000 at the beginning of the decade were on course to surpass 200 million by the end.

Why the sudden growth? In part the Internet was cheap and easy to use. Moreover, it was the effect on people's imagination the first time they clicked around the new frontier. Old rules of geography, money, and behavior did not apply. No one was in charge of the Web. Everything was available in this new world for free. Founded in 1993, the magazine *WIRED* trumpeted a techno-utopianism where the Internet would transform the economy, society, and even humanity itself. The experimental layouts and bold use of fluorescent and metallic inks in *WIRED* sum up the personality of the Internet in those early years, and the magazine is still published today.

For example, one 21-year-old innovator, Marc Andreessen, took Tim Berners-Lee's lowly browser made for research papers and added pictures, color, and graphical design. Others would soon add audio, video, animation, and interactive forms. His company (Netscape, formed in 1994) simply gave the browser away for six months and then went to the stock market with an IPO (initial public offering) worth $2.4 billion on the first day.

No wonder people began saying the Internet had started a "new economy." The WWW erased geography and time constraints. Anything digital could be multiplied a million times and distributed worldwide for free. Entrepreneurs lined up for the new gold rush of the information age. Billions poured in to fund every imaginable innovation, the stock market soared, and for years it seemed true that there was more profit in clicks than in a bricks and mortar industry.

What is called the "dot-com bubble" burst in 2000, draining away these billions and delivering the sobering reminder that, even in the New Economy, certain Old Economy values such as profitability, accountability, and customer service still mattered. Nevertheless, the Internet proved a seismic shock to business economics. Even the smallest business, no matter where located, could consider the world its marketplace. Companies that "got the Net" could outmaneuver large corporations. For the most part, traditional businesses did not disappear with the Internet; they adapted their old models to use it. Because most goods and services were physical, traditional business controlled means of production but used the Internet to improve supply management, ordering, and customer service.

Many point to Amazon and eBay, both launched in 1995, as examples of the "new economy." Amazon at first simply sold the old commodity of books. They built success on the frictionless character of Internet access. Books were the same anywhere; the real problem was finding them in a local bookstore. Amazon saw they could let people find a book easily, review what others thought of it, make payments with a single click, and never have to leave the house. It worked, and today every online seller works on the same principle as Amazon. The Internet enables better selection, cheaper prices, and faster delivery. Nevertheless, though Amazon is 100 percent online, this is still the old economy made better using new technology. To this success should be added industries such as banking, travel, and insurance, all transformed by the Internet within a few years. They migrated online with great success but used Internet technology to enhance existing business rather than to fundamentally change it.

eBay introduced an online version of an economic model as old as society itself: person-to-person trading. The now $50 billion company produced nothing. It simply put buyer and seller together using the Internet. By providing a listing service and payment system and taking a commission, eBay makes a good case for being a "new economy" business. Millions of sellers, not just buyers, were now networked. The stroke of genius in eBay was their rating system for buyers and sellers to keep score on the reputation of each user. Anyone could see another's reputation and make a choice about whether or not to do business with a complete stranger. On the seemingly endless anonymity of the WWW, eBay found a way establish old-fashioned reputation as a key economic currency.

It is important to emphasize that the new economy uses information and ease of communication as its currencies. Up to this point, economies were built on the relative scarcity of goods and services. Resources needed to be acquired, marketed, and sold, but they were always finite. The Internet turned this old economic model upside down. Instead of scarcity, it was built on an endless supply. Digital multiplication of information and distribution through the Internet were essentially without limit. What astonished users in the early days of the WWW was that people were giving away everything for free. Who was paying for this? Who could make money this way? When talking about the "new economy," it may be best to say the Internet did not create it; rather, the Internet required a new economy.

Google (started in 1997) was an instant and spectacular success in the new economy. It did not enhance an old business; it created an entirely new one, though few saw it at first. The need for powerful search engines on the WWW was apparent quite early. Once access to information on the network was solved, the next problem was finding it. With the growth of the WWW, finding a page was like finding a needle in a million haystacks. But even with a search engine the results could number tens of thousands. How could someone find good information?

When Google appeared, it looked like simply a better search engine, but the young graduate students who built it also designed human intelligence into the tool. Instead of only words and titles, Google also analyzed the number and quality of links to each page. Millions of humans chose what pages they visited and what pages they built links to. Google tracked this. The more links to a Web page, the more likely that Web page has good information. It was a surprisingly simple way to judge relevance. Google offered not only an index of the WWW, but also a snapshot of what the world was thinking about it. eBay built a way to track the reputation of users; Google discovered ways to track the reputation of information.

What makes Google worthy of being included in the new economy is that it traffics wholly in information and the power to make sense of it. How can searches be given away free and the company be worth $100 billion? By giving away information and in some cases paying people to take their information, Google gathers intelligence about what is on the WWW, what people think about it, and most of all what people are looking for. It is a marketer's dream. Put people and their interests together with products they are looking for, and there is business. The bulk of Google's revenue comes from advertising, which is systematically targeted by demographic, habit, and personal interest. Google does not want only to index the WWW; it intends to analyze its users. The larger the WWW, the greater the use of it, and the more profitable Google's share of the new economy.

Far different is the situation where an old-style business does battle with the "new economy" principles of the Internet. The prime example is media. If the Internet means that anything digital can be reproduced instantly across the whole system, is it possible to copy-protect music, movies, and books? Is it even desirable? The only thing that keeps this book from being copied a million times on the WWW is the minor inconvenience of transferring the paper-based text to a digital format. If all digital media becomes potentially free, how will media conglomerates ever make a profit? How will artists earn a living? Software sales are another example. Copying software and posting it for others to use for free is a time-honored use of Internet technology.

One response to unauthorized copying is increasingly sophisticated Digital Right Management (DRM) software, which makes media impossible to use without payment. In turn, clever coders have always found a way to crack the protection and post the media anyway. Various surveys have discovered that up to 50 percent of Internet users believe there is nothing wrong with taking illegal copies of software and music. It is likely that illegal downloads will never go away and that people will pay for media simply for the convenience of downloading it from one place and having access to support if there are problems. Neither honesty nor technology will have much to do with it. People will pay for not having to root around Warez sites (collections of illegal software) or locate P2P (peer to peer) repositories willing to share.

Another response has been high-profile lawsuits against people and companies with unauthorized media. The purpose is to frighten others into paying for valid copies. Although this works well against business and corporations, it has made barely a dent in the downloading of music and videos by individuals, especially the young. Today sales of CD music are down even as the number of people listening to songs increases, proving the point that the old-style business of media companies is under serious pressure from the "new economy" principles of the Internet.

A third response recognizes that the Internet may have changed the rules. It says that copying is not only allowed but encouraged. It turns the old media economy of control and distribution upside down. Now the artist or programmer wants the widest possible distribution of the media and gives it all away for free. The goal is exposure, increased sales of related items, or simply the desire to create and see others enjoy the creation. Opponents claim the practice will undermine the ability to control and profit from intellectual property. Others point out that art is in good health on today's Internet and that software development has never known such vitality.

The debate over what kind of "new economy" the Internet has helped to spawn leads to no consensus, but there is general agreement that the impact of the Internet on the worldwide economy, whether new or old, cannot be measured. It is somewhere in the trillions of dollars.

There is another dimension of "the new economy" that relates to the economy of ideas on the WWW. Here information and ease of communication are the currencies. The slogan "Knowledge wants to be free" is part ideology and part recognition that in digital knowledge, there is no cost of delivery. What is called the Open Source movement in the software industry insists that free distribution, work on projects by unlimited developers, and complete access to source codes will produce the best product. The WWW makes it possible. Another vivid example of the new economy of ideas is Wikipedia, an online encyclopedia where anyone can improve articles written by anyone else. Its popularity now rivals the famed *Encyclopedia Britannica.*

Discussions of a "new society" built through the Internet follow the same pattern as those on the "new economy." Enthusiasts claim the Internet will inaugurate a golden age of global community. No distance, no border, and no restriction on information will improve education, stimulate communication, spread democracy, benefit rich and poor alike, and level the playing field in a new Internet age. Much of the language about the Internet, from the early years, is strongly utopian and uses the word *revolutionary* more often than is wise!

Critics of the so-called Internet revolution fear the Internet will only take people away from real-world problems and genuine human interaction. Government and corporations will use the technology to snoop on and manipulate citizens. Criminals will invent new high-tech crimes, and at best the world will be no better and at worst much worse.

Neither the dreams nor the nightmares of the Internet age have arrived, but both the enthusiast and the critic have seen hopes and fears realized on the WWW.

For example, education, as old as society itself, finds itself a beneficiary of the WWW and an area of major concern. It is true that students now have access to many times the learning content of a few years ago. Books, images, research tools, multimedia, and simulations have been mainstreamed in Western education. Internet literacy is an accepted competency for the educated person. Web-based learning has opened up higher education to greater numbers. The Internet removes many of the physical and time restrictions to learning.

But is the learning available on the Internet good? Where once a teacher could ensure the quality of resources, now the words "found on the Web" can apply to the latest research or to complete nonsense. How will students fulfill the social dimensions of their experience on the Web? Though content-oriented subjects do well in Web-based learning, how can hands-on skills ever be put on the Web? Students find an abundance of information on the Web but can also copy and paste it, claiming it as their own. Completion rates for Web-based learning are less than half of those in face-to-face learning, however.

As it was with the "new economy," the "new society" has turned out to be mainly a version of the old society operating at Web speed. Few things are actually new on the WWW. People use the Internet to chat, visit, flirt, and play. Dating, cyber sex, marriage, and funerals are all on the Web. Birth still poses a challenge, but in every case there is some version on the WWW of what people have been doing for thousands of years. The WWW is more of a reflection of society than a force shaping society.

More often than not, quite unforeseen consequences have emerged from the Internet. For example, could the early adopters of e-mail have predicted that more than half the eventual traffic would be spam (unwanted email)? For years visionaries have promised the paperless office, but each year paper use goes up. Office productivity was meant to increase dramatically once everyone was wired into the network. Instead the WWW became the number one source for wasting office time. Dreamers announced whole armies of knowledge workers who would commute via the Internet. Little did they foresee that those knowledge workers would come from halfway around the world, outsourcing or displacing the jobs of local workers.

What should be regarded as "new" in the wired world is the speed with which things happen and the vast increase in the numbers of people who can be involved. Technology does not much change the way people live on the Internet as much as it multiplies its effects. An embarrassing video once circulated among friends and family now can be found by millions of strangers and can never be taken back. A pick pocket could steal a few wallets in a day. A good hacker now can steal a million credit cards in a minute. A rumor or a piece of false information falls into a database or search engine, and suddenly it competes on equal footing with the truth.

A new and dangerously ignored consequence of the WWW is the persistence of information. Internet technology not only retrieves data but also keeps it around, perhaps forever. Until now people could trust that their words and written notes simply disappeared or at least could be controlled. This is not so on the WWW. Information is kept, and it may be found by anyone in the world.

Privacy, or the lack of it, is certainly an old issue taking a new form on the WWW. In the early days people reveled in the seeming anonymity of their Web browsing. People could hide behind a billion packets and the complex communications of TCP/IP, but not anymore. Online companies track browsing habits. Local Web servers log every request made from a browser. Chat rooms archive information. Governments routinely listen in on the chatter moving across the networks. Unsuspecting users routinely let tracking programs be installed on their computers and give away personal information in exchange for Web-based baubles. Worse, people publish all manner of personal detail on the WWW, not grasping that Google and other search engines make this information permanent and findable by anyone. Already employers are searching the history of potential employees on social networking sites. Many have lost jobs because of some frivolous post made years before. It will not be long before some political candidate for high office will be undone by the record of some indiscreet posting in a forum or visit to an unsavory Web site.

It is certain that the WWW has not created the new society some of its cheerleaders proposed. It is also doubtful that society itself has changed that much as a result of the introduction of the Internet to mainstream culture. The idea that technology by itself will determine the character of human life is naïve. It is fair to say, however, that society has not kept up with the consequences of Internet technology. In part this is because the technology is young, and people are too close to it. The next wave of the Internet is likely to be the widespread linking not just of personal computers but of things. Phones, media players, and gaming are already widespread online. Someday it could be vehicles, appliances, tools, and parts of the human body linked into a global interactive network.

How then can the significance of the Internet be understood today? First and foremost, neither should it be regarded as something entirely new, nor should one listen too closely to either its fans or its cynics. It is one of many innovations dubbed a revolution by some and a threat to society by others. Compare the Internet to electricity, the telegraph, transatlantic cable, telephone, radio, television, satellites, or computers. All struck awe into their first users but were adopted by the next generation as simply the way things are done. None was a revolution by itself. The social changes that have come with these technologies have as much to do with how people envisioned them, reacted to them, and applied them as they do with the inventions themselves.

Human imagination has a remarkable way of adapting technology in ways its inventors did not consider. Therefore society is less likely to be transformed by the Internet than to transform the Internet into areas not yet conceived.

See also Censorship; Computers; Information Technology; Privacy; Search Engines; Software.

Further Reading: Anderson, Janna Quitney. *Imagining the Internet: Personalities, Predictions, Perspectives.* New York: Rowman & Littlefield, 2005; Buckley, Peter, and Duncan Clark. *The Rough Guide to the Internet.* London: Penguin, 2007; Negroponte, Nicholas. *Being Digital.* New York: Knopf, 1995; Standage, Tom. *The Victorian Internet.* New York, Walker Publishing, 1998; Stoll, Clifford. *Silicon Snake Oil: Second Thoughts on the Information Highway.* New York: Anchor, 1996.

Michael H. Farris

MAD COW DISEASE

Mad cow disease, also known as bovine spongiform encephalopathy (BSE), is one of a number of diseases known to be caused by an abnormal protein known as a prion. The origins and extent of mad cow disease and other prion-related diseases potentially transmitted through large-scale animal and meat production continue to be a concern for agricultural producers and consumers.

Prion diseases in nonhuman animals include chronic wasting disease (CWD) in deer and elk, scrapie in sheep, transmissible mink encephalopathy (TME) in mink, and mad cow disease in cattle. Human prion diseases include Creutzfeldt-Jakob disease (CJD), fatal familial insomnia (FFI), and variant Creutzfeldt-Jakob disease (vCJD). Scientists believe that consuming meat from cows infected with BSE causes vCJD. Unlike other disease-causing agents such as bacteria and viruses, prions do not seem to reproduce themselves by replicating their genetic information. In fact, prions do not seem to contain genetic information. All prion diseases are known as transmissible spongiform encephalopathies, or TSEs. All TSEs are contagious ("transmissible"), cause the brain to become sponge-like, with many tiny holes ("spongiform"), and are confined to the brain ("encephalopathy").

Mad cow disease causes cows to stumble around erratically, drool, lose weight, and act hostile, making them seem insane or "mad." Evidence suggests that humans who consume beef infected with BSE can contract vCJD but exhibit symptoms only after an extended incubation period that can last as long as decades. Mad cow disease, when it infects humans, is known as vCJD because of its similarities to CJD. Creutzfeldt-Jakob Disease was first observed and described in the early years of the twentieth century, although doctors did not know what caused it. Patients were observed to lose control over large motor functions and to then progressively succumb to dementia. Centuries before, sheep farmers in England, Scotland, and Wales had observed a disease in their flocks. Sheep farmers called the disease scrapie, after the sheep's behavior of scraping themselves on stone walls to apparently scratch an itch, but with the result of scraping off their valuable wool. Although mad cow disease was not identified until the 1980s, other prion diseases had been making their presence known for centuries, without the disease-causing agent being discovered or named.

It was not until the mid-twentieth century that these diseases were linked by their common causal agent, prions. The link was first suggested by Carleton Gajdusek, an American doctor stationed in Australia. In the early 1950s, Gajdusek heard of a mysterious disease that was killing women and children in Papua New Guinea. The victims of the mystery disease were members of the ethnic and geographical group called Fore, and they were dying from what was locally termed the "laughing disease," so named because the first symptom sufferers exhibited was a kind of uncontrollable nervous laughter. The disease, eventually officially called kuru, rapidly progressed, leading to the symptoms later exhibited by cows with mad cow disease: jerking movements, mental degeneration, and death. Gajdusek and his team dissected the bodies of those who died of kuru and used samples to infect monkeys and other animals. Every animal and human that died of the disease had a brain that showed the "swiss cheese"–like holes that would come to be associated with all TSEs. In 1959 a veterinarian named William J. Hadlow published a paper connecting kuru with scrapie because of the similarities in the brains of the infected. In 1984 the first cow in the United Kingdom exhibited signs and symptoms of BSE, and shortly thereafter it was determined that this newly observed disease in cattle was a prion disease because the brains of the cows that died of it exhibited the telltale holes of other known TSEs. A few years later, scientists determined that prions, which had already been determined to have to ability to infect across species, had been introduced into the British cattle population through relatively new feeding practices that had introduced sheep neural matter into cattle feed.

Mad cow disease came to the attention of public health officials and the meat-eating public in 1995, when the first death from vCJD was identified. Investigations into the disease revealed that it was a lot like the previously identified CJD, but with significant differences that indicated a new strain of prion disease. Soon after the first case, vCJD was connected to exposure to cows infected with BSE.

BSE's sudden occurrence had already been linked to relatively new industrial agricultural practices of giving cattle feed made in part of the processed parts of other dead cows. These industrial agriculture practices, introduced in the early 1970s, were designed to maximize efficiency in the beef industry. Farmers, or farming companies, realized that letting cattle graze on land with grass took up a lot of space because of the acreage required to feed the cattle. Grazing also took up a lot of time, because a diet of grass meant that cows grew at their normal rate and were not ready to be sent to slaughter and market until they reached a certain weight, which could take as long as four or five years. Farming companies hit upon a solution that would drastically reduce the costs of space and time.

They found that cattle could be crowded together in pens and fed a high-calorie diet rich in protein and fat that would speed up their growth and make them marketable in only a little more than a year. In a further move designed for efficiency and low cost, slaughterhouse and feedlot operators recycled the waste left from the slaughter of previous cattle, such as blood, bone and brains. This practice helped farmers produce many more pounds of meat at much cheaper prices than would have been possible with cows allowed to roam free and graze.

Mad cow disease, or rather the prions that cause it, lives in the nervous system tissue of infected animals. The nervous system tissue is part of what is leftover after the usable parts of the cow have been sent to butchers and grocery stores, fast food companies, and pet food factories. All the matter was ground up together and processed into homogenous feed, thus allowing for the wide distributions of prions among herds of cattle. Most people are now protected from eating contaminated meat because industrialized countries have BSE under control; herds in which the disease is observed are typically destroyed, however, making mad cow disease a significant economic crisis as well as a public health crisis. Unfortunately, because the human variant of mad cow disease (vCJD) has such a long incubation period, it may be many decades before we become aware of its extent.

See also Health and Medicine.

Further Reading: Rhodes, Richard. *Deadly Feasts: Tracking the Secrets of a Terrifying New Plague.* New York: Simon & Schuster, 1998; Yam, Philip. *The Pathological Protein: Mad Cow, Chronic Wasting, and Other Deadly Prion Diseases.* New York: Springer, 2003.

Elizabeth Mazzolini

MATH WARS

Mathematics has been part of formal education for centuries. In the United States, it has been a necessary component of schooling since the public school system was devised. Why is mathematics generally considered essential to a solid education? What are the goals of teaching mathematics? How are these goals determined? Should all students master a predetermined level of mathematics because mathematical understanding swings open the doors to financial success and rewarding lives? Are some areas of mathematics more valuable than others? In this technological age, what are the basics? Once content is determined, is there a best way for mathematics to be taught? What does equal opportunity mean in the context of mathematics education? These are a few of the questions embedded in the Math Wars controversy.

The Math Wars describes an ongoing dispute involving educators, parents, government, and publishers—people and organizations with an interest in who teaches mathematics, who is taught mathematics, and how mathematics is taught and in planning the role of mathematics in modern society. Since its beginning, the United States has defined public education as a right, and citizens have been debating the purpose of education and how the government can best meet its responsibilities. Everyone, it seems, has a stake in this argument. The core issue driving the Math Wars in the United States is why we teach mathematics; the ancillary issue is how we teach mathematics.

Disagreement over the right way to teach mathematics is hardly new, but students have changed, and the subject matter has evolved. A hundred years ago, the U.S. population was not nearly as demographically diverse as it is today; the segment seeking a comprehensive education was more homogeneous, tending to be white, male, and more culturally similar. The aims of education were narrower; lengthy schooling was less readily available, with many families struggling to survive, and only a fraction of the student population was able to graduate from high school. Today's student is not as simple to profile. The U.S. population is swiftly growing, a demographically shifting male/female stew of ethnicities, cultures, abilities, aptitudes, and interests.

When did squabbling over the goals and methods of teaching mathematics change from an educational debate and become identified as the Math Wars? The space race of a half-century ago and the events leading up to it were major factors. Important and well-respected educators had long questioned the effectiveness of traditional methods of mathematics instruction, but finding a better way to teach mathematics became an urgent national priority when the Russians sent Sputnik into orbit. The United States, embarrassed by not being first, perceived the need to dominate in the global competition for economic and political sovereignty. Policy makers saw an unacceptable national deficiency in need of correction. The 1960s, adhering to that premise, saw the birth of New Math, a novel approach to teaching mathematics that focused on deeper theoretical understanding of mathematical concepts than the rote facility associated with the three Rs. Regrettably, many of the teachers expected to teach this new curriculum were neither well trained nor well supported in their professional development. Both students and teachers floundered; New Math met its demise a decade after its introduction.

The backlash after New Math led to its antithesis, "back to basics," a conventional program that stressed computational facility over theoretical insight. Back-to-basics, as flawed as its predecessor, produced graduates weak in both mathematical understanding and genuine interest. Where New Math was too esoteric for most learners, back-to-basics was too superficial to give the learner the necessary insight for decent problem-solving skills. This program was also recognized as not meeting the greater goals of learning on either a practical or a theoretical level.

The next reincarnation of mathematics education simply embellished the back-to-basics texts with cosmetic changes. Responding to the argument that mathematics was unpopular with students because it lacked real-life applications, publishers tacked on a few pages of problem-solving exercises to each chapter of the existing textbooks.

The decades go by; the debate continues. The prevailing philosophy today favors the inclusion of different learning styles for students with different ways of understanding, and most textbooks attempt to recognize the range of student ethnicities and give them an opportunity to "discover" the material for themselves. This showcases another side issue, albeit an important one: the power wielded by the publishing industry in the Math Wars.

In theory, the 50 states are educationally autonomous and empowered to choose their own mathematics curricula. The same is true of provinces in Canada, where curriculum and education are provincial responsibilities. Three states, however—California, Texas, and New York—have specifically stated goals; textbooks are chosen from a list of those meeting the states' goals. It is financial folly for a school in those states to choose a textbook that fails to meet designated guidelines; government funding is based on approval of the texts. The schools in these states are dependent on publishers to offer satisfactory options; the publishers themselves are financially dependent on the orders from these states. Publishers are unlikely to attempt an innovative approach to mathematics pedagogy if the consequence is financial adversity. In the end, although it may appear that schools around the nation have freedom to choose as they see fit, their choices are restricted by the criteria adopted by three states.

Unfortunately, these textbooks are hardly classics of mathematical literature. They tend to be designed as packages rather than separately for each grade, allowing school districts to choose their books for each grade sequentially rather than individually. This idea makes excellent common sense; its downside is that no one author or editorial team can produce a complete set of textbooks. It is just not feasible. Although a single author or team appears to be responsible for the entire series, individual authors are hired to follow the scheme of the series. Consistency is compromised in order to meet demand. Innovation is sacrificed as impractical.

On November 18, 1999, an "open" letter appeared in the *Washington Post,* protesting the federal government's support for the study of new and unconventional mathematics curricula. Signed by dozens of prominent mathematicians and scientists, the letter took a strong position against the National Council of Teachers of Mathematics (NCTM), the National Research Council (NRC), and the American Association for the Advancement of Science (AAAS), organizations that promote making mathematics more accessible to underrepresented populations and adopting different teaching methods in order to do so.

Herein lies one of the essential conflicts embedded in the Math Wars: educational organizations envision a mathematically literate general population, wherein every student is given (and understands!) an introduction to algebra and other components of richer problem-solving skills. Historically, this is an optimistic leap of faith. It is assumed that the general population is both capable and interested enough to achieve this goal. Whether the outcome can support the premise is not a subject that any party wishes to address.

It becomes apparent that the scope of this issue is huge. Behind the question of what every student should learn is the need to identify the purpose of education itself. Some see it as a means for creating a more equitable society, more inclusive of its marginalized members. Others look at the numbers of mathematicians the United States produces and question why so many are foreign students. For still others, the Math Wars is about the pursuit of knowledge. Yet another concern is the technological advancement necessary for participation in the emerging global political economy.

The opposing positions in the Math Wars drama are held by the pro-reform and anti-reform extremists, although it is an oversimplification to suggest that all special interest groups lie at one extreme or the other. At the same time, the most vocal activists do tend to be strident, overstating their positions in order to ensure that their voices are heard. At both ends of the debate are qualified professionals, including mathematics teachers, developers of curriculums, parents and other concerned citizens, professional mathematicians and scientists, and politicians and policy makers.

Pro-reform, the progressive view of mathematics education, argues for intellectual freedom. Student autonomy and creativity are the energies driving education. The more conservative view, anti-reform, argues for a standardized curriculum with an emphasis on drill to ensure a basic level of skill.

One of the arguments focuses on what the basics of a modern mathematical education must include. What should students learn in today's world? Pro-reformers argue that the priorities include good number sense and problem-solving abilities—in other words, a "feel" for math. They see students who need to develop mathematical communication skills and understand the "big ideas" behind what they are learning, to be able to reason mathematically and perform computations easily. The ultimate goal of the reformers is mathematical self-empowerment or the confidence that comes with the ability to make sound judgments. Their stance stresses equivalent opportunity for all learners but does not explain why all cultures should be equally motivated to learn the subject and participate at every level of mathematical sophistication.

At the other extreme, the anti-reformers prefer the methods used to teach mathematics to previous generations, methods that have demonstrated historical success. Their position is that skills and facts taught today should be the same as those taught in earlier years; it worked before, and it still works. Basic computational skills are essential. Mathematics education's priority, from the anti-reform viewpoint, should be reinforcement of standard algorithms and procedures, with less reliance on calculators and other technology. Their rationale is that mastering basic facts leads to understanding and that learning skills comes before studying applications. What this argument lacks is the acknowledgement that population demographics have shifted, as well as the need to address the stated philosophy that underrepresented populations need to be better represented across the educational and professional spectrum. Traditional pedagogical methods tend to further marginalize already-marginalized population groups.

Most of the mathematics being taught in the public schools today tries to acknowledge and incorporate the NCTM platform, which advocates education for demographic equality and social mobility. To achieve that aim, NCTM promotes a pedagogy focused on process, stressing the teaching method, whereas the traditional curriculum is content-oriented. NCTM's process-oriented view is intended to encourage local autonomy, leading to democratic equality and education for social mobility. The traditional content-oriented perspective, in contrast, is efficient, anticipating the best results from the best students. The traditional approach supports an agenda biased toward social efficiency, however inadvertently, and reinforces the existing class structure.

The issues underlying the Math Wars spotlight fundamental philosophical differences between the opposing groups. Those supporting the traditional curriculum point to the inescapable argument that the older methods worked, at least for some (in particular, usually for those making the argument!). If it worked before, they suggest, why should it be upstaged by some so-called "fuzzy math" that encourages learners to construct their own computational algorithms? Why would understanding how the process works be more important than getting the right answer? Traditionalists view the learner as passive, needing only access to the necessary tools of the trade. The drawback is that such an attitude penalizes learners falling outside of the traditionally successful demographic strata, marginalized students who are less likely to become involved in mathematics, thus continuing the existing trends. In order to attract these students, rigor is sacrificed in favor of essential understanding.

Teachers have a significant role in learning as well. Again, traditionalists and reformers hold incompatible images of what teachers should do and how they are expected to do it. Traditionalists stress the importance of content knowledge: a teacher must simply know a lot of math to teach a lot of math; they should be accomplished mathematicians above all. In response, reformers argue that content knowledge alone is insufficient; teachers must be able to convey the knowledge so that students are receptive to learning it. Modern mathematics pedagogy, leaning toward the reform position, advocates a constructivist approach, allowing students the opportunity to make sense of ideas and concepts for themselves. The drawback of embracing this philosophy is that clever algorithms developed over thousands of years are not the object of the lesson plan.

Assessment creates another obstacle. The purpose of assessment is to provide a way of estimating and interpreting what a student has learned. Because so much depends on students' academic performance in this era of high-stakes testing, it is vital to find suitable assessment instruments and techniques in order to better evaluate students' knowledge.

Will these issues ever be resolved? Politics is never simple. Because the core issues of the Math Wars revolve around the very role of public education in our nation, conflict will always be a part of the process. Without compromise, however, the consequences will continue to be borne by the students.

See also Education and Science; Mathematics and Science; Science Wars.

Further Reading: Lott, Johnny W., and Terry A. Souhrada. "As the Century Unfolds: A Perspective on Secondary School Mathematics Content." In *Learning Mathematics for a New Century,* ed. Maurice J. Burke and Frances R. Curcio, pp. 96–111. Reston, VA: National Council of Teachers of Mathematics, 2000; Mathematically Correct Web site. http://www.mathematicallycorrect.com; Mathematically Sane Web site. http://mathematicallysane.com; National Council of Teachers of Mathematics Web site. http://www.nctm.org; Schoenfeld, Alan H. "The Math Wars." *Educational Policy* 18 (2004): 253–86.

Deborah Sloan

MATHEMATICS AND SCIENCE

Philosophers, historians, scientists, science writers, and even sociologists wrote a lot about science from the earliest days of the West's scientific revolution (usually considered to have its origins in the seventeenth century) to the middle of the twentieth century. In the late 1960s, writing about science and answering the question "What is science?" started to change. The change was rooted in a new approach to studying science. Earlier studies had relied on the memories of scientists, the reports of journalists who had interviewed scientists, the hagiographic accounts of historians, and idealistic accounts of science by philosophers. When sociologists first entered this mix in the 1930s, they focused on science as a social institution; they studied norms, age grading, the social system of science, scientific elites, scientific roles in industry and in the academy, and other structural features. They deliberately did not study scientific knowledge, the products of scientific work. During the earlier development of the sociology of knowledge in the 1920s, $2 + 2 = 4$ (a paradigmatic example of a universal truth) was believed to exist outside human time and place. The ancient philosopher Plato claimed that facts such as this were necessarily true and independent of any preliminary construction. The sociologists of science who followed in the wake of the sociology of knowledge accepted this Platonic version of reality. The revolutionary idea that the so-called new sociologists of science put in place in the late 1960s and especially during the 1970s was to look at what scientists actually do when they are doing science. In other words, put sociologists and anthropologists in scientific research settings (e.g., laboratories) and let them observe and report on the actual practices of scientists. Not surprisingly, this began almost immediately to produce a new narrative on the nature of science.

Scientific facts are in a literal sense manufactured. The resources used to make facts are locally available social, material, and symbolic capital. This capital is part of a system of shared norms, values, and beliefs and a more or less stable social structure. This structure can define a research team, a laboratory group, a group of scientists working within a large research facility, or any other community of scientists working on similar problems and guided by similar paradigms. Think of science as a labor process—a social practice—in which workers cooperatively process raw materials (e.g., glass) or refined materials (e.g., test tubes) and turn a disordered set of contingencies (ranging from scotch and duct tape, scraps of paper and metal, and assorted objects to symbols, from paper to money) into an ordered set of facts. A lab experiment (for example; not all science flows from experiments) is followed by a sequence of notes, papers, and publications in which sentences become increasingly mechanical and objective. The earliest writings tend to be seasoned with subjectivities— first-person, emotionally colored, rhetorical flourishes. By the time we reach the published description and interpretation of the experiment, the subjectivities, the flesh and blood, the sensual nature of experimental science have been progressively erased.

We hear and read a lot in science about "universal truths" as if these truths *are* universal. Scientific facts, however, are not immediately, necessarily, and naturally "universal"; they become universal (more or less true for scientists first and then wider and wider circles of lay people across regions, nations, cultures, and the world) through the activities of scientists nationally and internationally. Scientists travel about the world communicating with other scientists and, along with engineers, tourists, journalists, and other travelers, act as agents of professions and governments. Their mobility makes them ambassadors for the legitimacy of scientific facts.

The new sociologists of science write and speak about science and scientific facts as "socially constructed." This has fueled the "science wars," which are discussed in a separate entry. Such controversies are based on mistakes, misinterpretations, misunderstandings, and prejudices. For those who do not view social science as science, it is easy to dismiss ideas and findings, especially if they deal with a high-status profession and an important social institution such as science. Some of this is a more-or-less straightforward problem of scientists protecting their territory and their presumed jurisdiction over the analysis and theory of science.

When sociologists claim that science is socially constructed, many scientists hear them saying that science is arbitrary, subjective, and indeed not much more than a literary fiction. A fair, careful, and complete reading of what sociologists of science do and say, however, demonstrates that they consider themselves scientists, champions of science and its ways of knowing, doing, and finding. Society and culture are natural phenomena amenable to scientific study. Some critics claim that this leads to paradoxes when we study science as a natural phenomenon. If science is social, is not the sociology of science then also social? This is only a problem or paradox, the sociologists of science reply, if you assume that saying science is social is the same as saying it is arbitrary and even perhaps irrational. The paradox disappears once it is realized that the only way humans can reach true or false conclusions, the only way they can invent or discover, is through their collective efforts in social and cultural contexts, where biographies (individual and collective) and history intersect.

Sociologists do not claim jurisdiction over the subject matter of the sciences. They study the ways in which scientists produce, manufacture, and construct facts, and they can analyze those facts as social constructions. It is not, however, their job to decide based on sociological ideas whether the moon is made of green cheese or planetary materials. Contrary to the claims of some scientists and philosophers, sociologists of science do not deny reality, truth, or objectivity. They do argue, however, that we need to view these notions in a new light based on what we now know about society, culture, and the ways in which sight, perception, and the senses in general operate under the influence of social forces. In general, the sociological sciences have led us to the view that the self, the mind, and the brain are social phenomena.

When scientists say that there is a "reality out there," this should not be taken to mean there is a description of that reality that we can approach to closer and closer approximations. Few if any scientists (social, natural, or physical) or philosophers would dispute the idea that there is "a reality" (or that there are realities) "out there," outside of and independent of humans. There was something here or there before you were born, and there was something here or there before humans (or any other life forms) appeared on planet Earth. The question is not whether there is a "reality out there" but whether it is possible for us to know anything certain about that reality. Science does not (and cannot) give us closer descriptions of a "reality out there" but rather culturally and historically tailored descriptions of our experiences in that reality.

Finally, science—as the basic rationality of humans as adaptive animals—is at its best when it is not being directly or indirectly controlled by powerful interests supported by the policing power of state or religious institutions. This is in fact the case for the social institution of science—Science—which is tied to the institutional context that nourishes it. Modern Science, for example, is the science of modern industrial, technological Western society, even though it carries within itself the science that is integral to the human condition.

Let's look next at what is sometimes referred to as "the" hard case in the sociology of knowledge and science. Hard cases in this field are subjects that are considered on the surface invulnerable to sociological analysis. Scientific knowledge, logic, math, and God are classic hard cases. Traditionally, mathematics has defined the limits of the sociology of science.

Mathematics has been shrouded in mystery for most of its history. The reason for this is that it has seemed impossible to account for the nature and successes of mathematics without granting it some sort of transcendental status. Classically, this is most dramatically expressed in the Platonic notion of mathematics. Consider, for example, the way some scholars have viewed the development of non-Euclidean geometries (NEGs). Platonically inclined mathematicians and historians of mathematics have described this development as a remarkable and startling example of simultaneous invention (or discovery, if you are inclined in that direction) in two respects. First, they point out, the ideas emerged independently in Göttingen, Budapest, and Kazan; second, they emerged on the periphery of the world mathematical community.

There are a couple of curiosities here. In the case of non-Euclidean geometry, for example, even a cursory review of the facts reveals that NEGs have a history that begins with Euclid's commentators, includes a number of mathematicians over the centuries, and culminates in the works of the three men credited with developing NEGs: Lobachevsky, Riemann, and Bolyai. Moreover, far from being independent, all three mathematicians were connected to Gauss, who had been working on NEGs since at least the 1820s. One has to wonder why in the face of the facts of the case, mathematicians and historians chose to stress the "remarkable" and the "startling." Even more curious in the case of the sociology of knowledge is the fact that by 1912, several of the early social theorists had speculated on science and mathematics as social constructions, even linking the sociology of religion and the sociology of logic. This work, coincident with the emergence of the social sciences from about 1840 on, would fail to get picked up by the twentieth-century sociologists of knowledge, and science and would languish until the new sociologists of science went to work beginning the late 1960s.

It is interesting that a focus on practice as opposed to cognition was already adumbrated in Richard Courant and Herbert Robbins's classic text titled *What Is*

Mathematics? (1941). It is to active experience, not philosophy, they wrote, that we must turn to answer the question "what is mathematics?" They challenged the idea of mathematics as nothing more than a set of consistent conclusions and postulates produced by the "free will" of mathematicians. Forty years later, Davis and Hersh (1981) wrote an introduction to "the mathematical experience" for a general readership that already reflected the influence of the emergent sociology of mathematics. They eschewed Platonism in favor of grounding the meaning of mathematics in "the shared understanding of human beings." Their ideas reflect a kind of weak sociology of mathematics that still privileges the mind and the individual as the creative founts of a real objective mathematics.

Almost 20 years later, Hersh, now clearly well-read in the sociology of mathematics, wrote *What Is Mathematics, Really?* (1997). The allusion he makes to Courant and Robbins is not an accident; Hersh writes up front that he was not satisfied that they actually offered a satisfactory definition of mathematics. In spite of his emphasis on the social nature of mathematics, Hersh views this anti-Platonic anti-foundationalist perspective as a philosophical humanism. Although he makes some significant progress by comparison to his work with Davis, by conflating and confusing philosophical and sociological discourses, he ends up once again defending a weak sociology of mathematics.

There is a clear turn to practice, experience, and shared meaning in the philosophy of mathematics, in the philosophy of mathematics education, and among reflexive mathematicians. This turn reflects and supports developments in the sociology of mathematics, developments that I now turn to in order to offer a "strong programme" reply to the question "What is mathematics?"

We are no longer entranced by the idea that the power of mathematics lies in formal relations among meaningless symbols, nor are we as ready as in the past to take seriously Platonic and foundationalist perspectives on mathematics. We do, however, need to be more radical in our sociological imagination if we are going to release ourselves from the strong hold that philosophy has on our intellectual lives. Philosophy, indeed, can be viewed as a general Platonism and equally detrimental to our efforts to ground mathematics (as well as science and logic) in social life.

How, then, does the sociologist address the question, what is mathematics? Technical talk about mathematics—trying to understand mathematics in terms of mathematics or mathematical philosophy—has the effect of isolating mathematics from the turn to practice, experience, and shared meaning and "spiritualizing" the technical. It is important to understand technical talk as social talk, to recognize that mathematics and mathematical objects are not simply (to use terms introduced by the anthropologist Clifford Geertz) "concatenations of pure form," "parades of syntactic variations," or sets of "structural transformations." To address the question "what is mathematics?" is to reveal a sensibility, a collective formation, a worldview, a form of life. This implies that we can understand mathematics and mathematical objects in terms of a natural history, or an ethnography of a cultural system. We can answer this question only by immersing ourselves in the social worlds in which mathematicians work, in their networks of cooperating and conflicting human beings. It is these "math worlds" that produce mathematics, not individual mathematicians or mathematicians' minds or brains.

Mathematics, mathematical objects, and mathematicians themselves are manufactured out of the social ecology of everyday interactions, the locally available social, material, and symbolic interpersonally meaningful resources. All of what I have written in the last two paragraphs is captured by the shorthand phrase "the social construction of mathematics." This phrase and the concept it conveys are widely misunderstood. It is not a philosophical statement or claim but rather a statement of the fundamental theorem of sociology. Everything we do and think is a product of our social ecologies. Our thoughts and actions are not products of revelation, genetics, biology, or mind or brain. To put it the simplest terms, all of our cultural productions come out of our social interactions in the context of sets of locally available material and symbolic resources. The idea of the social seems to be transparent, but in fact it is one of the most profound discoveries about the natural world, a discovery that still eludes the majority of our intellectuals and scholars.

What is mathematics, then, at the end of the day? It is a human, and thus social, creation rooted in the materials and symbols of our everyday lives. It is earthbound and rooted in human labor. We can account for the Platonic angels and devils that accompany mathematics everywhere in two ways. First, there are certain human universals and environmental overlaps across biology, culture, space, and time that can account for certain "universalistic" features of mathematics. Everywhere, putting two apples together with two apples gives us phenomenologically four apples. Yet the generalization that 2 + 2 = 4 is culturally glossed and means something very different across the generations from Plato to our own era. Second, the professionalization of mathematics gives rise to the phenomenon of mathematics giving rise to mathematics, an outcome that reinforces the idea of a mathematics independent of work, space-time, and culture. Mathematics is always and everywhere culturally, historically, and locally embedded. There is, as the historian and mathematics teacher Oswald Spengler wrote early in the twentieth century, only mathematics and not Mathematik.

The concept-phrase "mathematics is a social construction" must be unpacked in order to give us what we see when we look at working mathematicians and the products of their work. We need to describe how mathematicians come to be mathematicians, the conditions under which mathematicians work, their work sites, the materials they work with, and the things they produce. This comes down to describing their culture—their material culture (tools, techniques, and products), their social culture (patterns of organization—social networks and structures, patterns of social interaction, rituals, norms, values, ideas, concepts, theories, and beliefs), and their symbolic culture (the reservoir of past and present symbolic resources that they manipulate in order to manufacture equations, theorems, proofs, and so on). This implies that in order to understand mathematics at all, we must carry out ethnographies—studies of mathematicians in action. To say, furthermore, that "mathematics is a social construction" is to say that the products of mathematics—mathematical objects—embody the social relations of mathematics. They are not freestanding or culturally or historically independent, Platonic objects. To view a mathematical object is to view a social history of mathematicians at work. It is in this sense that mathematical objects are real.

Arithmetic, geometry, and the higher mathematics were produced originally by arithmetical or mathematical workers and later on by professional mathematicians. Ethnographies and historical sociologies of mathematics must, to be complete, situate mathematics cultures in their wider social, cultural, and global settings. They must also attend to issues of power, class, gender, ethnicity, and status inside and outside of more-or-less well-defined mathematical communities.

If mathematics has been the traditional arbiter of the limits of the sociology of knowledge, logic and proof have posed even more formidable challenges to students of the hard case. In his study of religion, Emile Durkheim, one of the nineteenth-century founders of sociology, argued that God was a symbol of society and that when we worshipped God, we were in reality worshipping our own social group, our community. Religion then came into focus as a social institution dedicated to sustaining the social solidarity of communities. Religion and God, in other words (and from a Durkheimian perspective), are institutional and symbolic glues for holding societies together. It was Durkheim, indeed, who connected the sociology of God and the sociology of logic by demonstrating that God and logic are firmly grounded in our everyday earthly activities, products of our social lives. In this demonstration, he solved the problem of the transcendental. He interrogated the idea that there is a realm of experience that transcends time and space, history, society, culture, and human experience. By tackling this sense that there are things "outside" of ourselves, he put us on the path to understanding that this sense of outsideness is in fact our experience of society. So God, for example, is real but not in the commonsense way many or most people think God is real. Sociology corrected an error of reference. God was not real in the sense of being a real entity but rather real in the sense of being a real symbol.

Scientific facts, mathematics, logic, and proof pose the same sort of "God" problem. They have the appearance of being outside of us, but only until sociology comes along to ground them in our earthly and social realities. There is a philosopher I know who often writes the following phrase in large letters on the blackboard before he starts his lectures: LOGIC IS IRREFUTABLE. And so it seems; there is a force that ideas such as 1 + 1 = 2 exert on us, compelling us to come to the "right" conclusion. Consider, for example, the following set of statements (known in the technical vocabulary of logic as *modus ponens*): If A; and if A then B; then B. This says that every time you encounter B, A is always going be there too. Therefore, if you come across A, you can be certain that B is going to be there too. Problems arise for compelling "universals" such as God and *modus ponens* when we come across equally compelling alternatives, new Gods, Gods that die, Gods that get transformed in predictable ways as history unfolds and societies and cultures change, alternative logics. It is the case, for example, that for every logical rule you can identify, there is a system someone has proposed that rejects this rule, no matter how compelling it is.

Consider the universally valid logical form or argument known as the syllogism. This is often demonstrated as follows: All men are mortal; Socrates is a man; therefore Socrates is mortal. From the first two statements, known as premises, the third statement follows. You are—or at least you feel—compelled to reach that conclusion. The story could end right here, except that there is more than one form of syllogism. Essentially, syllogism means a three-step argument: A is B; C is A; therefore C is B. There are, however, two types of disjunctive syllogisms. P or Q; not P, therefore Q. For example: A–either the Yankees win or the Red Sox win; B–the Yankees win; C–therefore, the Red Sox do not win. This is known as an inclusive syllogism. The exclusive form looks like this: Either P or Q (exclusive); P, therefore, not Q. In an inclusive syllogism, P or Q must be true, or P and Q must be true. In an exclusive syllogism, one term must be true, and one term must be false; they cannot both be true, and they cannot both be false. We have already gotten too technical and complicated for the point I wish to make, and we could add to the complexity ideas about hypothetical syllogisms, categorical syllogisms, the syllogistic fallacy, the fallacy of propositional logic, and a variety of other fallacies and forms of logic. The point is that in the end, you have to choose an appropriate logic for a given situation and context from among the multitude of available logics. Without going into the technical details or definitions, consider a world in which the only logic we have is classical logic. "Logic is irrefutable" might then be a defensible claim. But we live in a world with multivalued logic, relevance and intuitionistic logic, second- and higher-order logics, linear and non-monotonic logics, quantum logic, and so on. Historically, some mathematicians and logicians are always going to feel uncomfortable with things that seem obvious and unchallengeable on the surface. This is the case, for example, with the Law of the Excluded Middle (LEM). LEM says "Either X or Y." Either X is true, or its negation is true; you cannot have it both ways. Some alternative logics have been created because mathematicians and logicians did not feel compelled by LEM.

Complexity and comparative analyses complicate our worlds, whether we are trying to figure out God, logic, or numbers. Numbers? Are there alternative numbers? Well, first remember that some of the early Greeks did not consider 1 a number, some said it was neither odd nor even but odd-even, and some did not consider 2 an even number. So even something as supposedly obvious as the answer to the question "What is a number?" can lead to complications. Again, without going into the details, consider the natural numbers (N: 0, 1, 2, 3, etc.); the integers (Z: . . . −2, −1, 0, +1, +2 . . .); the rational numbers (Q) or fractions and the real numbers (R), repeating decimals such as e and pi, which cannot be written as fractions; and finally (for the moment), C, the complex numbers. Many of you will have come across one or more of these numbers in your schoolwork. But there are other numbers that you are less likely to have encountered, such as quaternions, 4-dimensional numbers. Hey, you might say, why not go ahead to create 5-dimensional numbers, or 6- or 7-dimensional ones. Well, we have tried that. It turns out that 5- to 7-dimensional numbers are rather unruly. Eight-dimensional numbers, however, are rather well-behaved. These are known as Cayley numbers or octonions. We can create sedenions,

16-dimensional numbers, but they are again rather unruly. It is not outside the realm of the possible that some of the unruly systems might be made more well-ordered in some future application.

What is the point of multiplying all these complexities? The point is that universals are always going to turn out to be situated, local, and contingent. Here is one way to bring order to all this confusion, and we do this be turning to socio-logic—sociology. One can doubt any formally expressed number system, logic, or religion. You can doubt *modus ponens,* for example. But suppose we restate *modus ponens* (recall: If A; and If A then B; then B) as follows:

> If you (always as a member of some collectivity) accept A and "If A then B," and if you accept "if A" and "If A then B," you must or will accept "B."

This makes the compulsion a function of a shared culture.

We can adopt a similar approach to the compulsions of proofs. First, we notice that proofs change over time, and proofs that are acceptable in some periods and by some mathematicians are not acceptable at other times and by other mathematicians. Plato could prove $1 + 1 = 2$ in one line, simply by claiming that it was necessarily true by virtue of its independence of any preliminary act of construction. In other words, it is true outside of human time, space, history, and culture; or it is a priori. It took Leibniz about six lines to prove $2 + 2 = 4$. He had three givens, an axiom, and the proof line. This sort of simple addition became a product of a more complicated system in Peano's axioms. And then along came Bertrand Russell and Alfred North Whitehead and their multi-volume exercise in deriving all of arithmetic and mathematics from logic. Their goal was not to prove $1 + 1 = 2$, but the world of mathematics had become so advanced and complex by comparison with Plato's world that it took them all of volume 1 and about 100 pages into volume 2 to establish the foundation for proving $1 + 1 = 2$.

Once again, we can adopt a socio-logic (a sociology) to help us bring order to all of this. And we do it like this. First, we notice that mathematicians treat proofs as if they were real things in and of themselves. But the real world is a world of events and actions, so instead of talking about proofs (or numbers, or logic, or God), we could try talking about proof events or proving. "Proofs," then, are real experiences unfolding in time and place (situated), involving real people with specific and shared skills, norms, values, and belief. These people constitute a "proof community," and proving can occur only within proof communities. Proof A is going to make sense only in proof community A; proof B will compel only members of proof community B.

Proof outcomes are never simply "true" or "false." They are effective proving or proof events if the social context is appropriate, if there is an appropriate interpreter, and if there is an appropriate interpretation. In that case, and surprisingly—as the late Joe Goguen pointed out—almost anything can be a proof.

One consequence of the unfolding history of sociology and the sociology of knowledge and science has been the progressive rejection of transcendence. This has been a history of locating referents for experiences that seemed to come from outside of experience. One might ask how, if we are creatures of time and space, we could have knowledge of things and entities that are outside of time and space (such as God and numbers). Classical logic, it turns out, is actually situated in the material world and the rules that determine how things interact with each other and with their environment. In a world of cows and horses, it is easy to develop a generalization about gathering up two horses, gathering up another two horses, bringing them together, and recognizing that you have four horses. The next step is to represent the generalization, for example, as $1 + 1 = 2$. In a world of clouds, however, cloud A and cloud B might add up to cloud C (or AB) if the clouds were moving toward each other. This could, using the same notation we used with the horses give us $1 + 1 = 1$. Indeed there are algebras that are based on mathematical worlds in which $1 + 1 = 1$.

In other words, what sociology and the social sciences in general have done is given us a new logic alongside the logics that represent generalizations from the physical and natural world.

Science, math, proof, and logic—and God! There is an emerging battleground that may become as significant as the battleground that led to the success of the Copernican system over the Ptolemaic system in astronomy. The new battleground may be resolving itself into a conflict between science and religion. Science will prevail, or we will all die because it is at its roots the basic adaptive modality in human evolution; it is our species' core adaptive methodology. Religion will prevail too because it is a manifestation of our species' core requirement for moral order and community. As social science penetrates closer and closer to the core of general science, more knowledge and evidence will accrue that demonstrates that traditional religion is only one way of constructing a moral order and of grounding communities. The pathway to a new understanding of reality—in the terms adumbrated in Durkheim's sociology—will be cluttered with the waste of scholarly debates and warfare, and the final outcome could as easily be annihilation as a new world order. Proof and logic will be brought to bear on this battleground by all the combatants, and what they mean will continue to be transformed. Our future will unfold on this battlefield of symbols and guns and will unify us, destroy us, or leave us in the dark, urban, technified, and terrorized *Bladerunner* world portrayed in recent film and literature.

See also Education and Science; Math Wars; Science Wars.

Further Reading: Bauchspies, W., Jennifer Croissant, and Sal Restivo. *Science, Technology, and Society: A Sociological Approach*. Oxford: Blackwell, 2005; Bloor, David. *Knowledge and Social Imagery*. 2nd ed. Chicago: University of Chicago Press, 1991; Courant, Richard, Herbert Robbins, and Ian Stewart. *What Is Mathematics?: An Elementary Approach to Ideas and Methods*. 2nd ed. Oxford: Oxford University Press, 1996; Davis, P. J., and R. Hersh. *Descartes' Dream: The World according to Mathematics*. Mineola, NY: Dover, 2005; Davis, Phillip J., and Reuben Hersh. *The Mathematical Experience*. New York: Mariner Books, 1999; Geertz, Clifford. *The Interpretation of Cultures: Selected Essays*. New York: Basic Books, 1973; Hersh, Reuben. *What Is Mathematics, Really?* New York: Oxford University Press, 1999; Restivo, Sal. *Mathematics in Society and History*. Boston: Kluwer Academic, 1992.

Sal Restivo

MEDICAL ETHICS

Medical ethics, an offspring of the field of ethics, shares many basic tenets with its siblings: nursing ethics, pharmaceutical ethics, and dental ethics. The definition of medical ethics is itself an issue of some controversy. The term is used to describe the body of literature and instructions prescribing the broader character ideals and responsibilities of being a doctor. Recent sociopolitical and technological changes, however, have meant medical ethics is also involved with biomedical decision making and patients' rights.

In the first sense of the term, medical ethics consists of professional and character guidelines found in codes and charters of ethics (e.g., the American Medical Association Code of Medical Ethics); principles of ethics (e.g., autonomy, beneficence, nonmaleficence, and justice); and oaths (e.g., the Hippocratic Oath). These formal declarations have the combined effect of expressing an overlapping consensus, or majority view, on how all physicians should behave. It is common to find additional heightened requirements for certain specialties in medicine such as psychiatry, pain medicine, and obstetrics and gynecology. Moreover, these ethical norms tend periodically to shift as the responsibilities of good doctoring change over time. Such shifts can give rise to heated debates, especially when individuals maintain certain values that have been modified or rejected by the majority.

For example, in the 1980s the medical profession was forced to consider doctors' obligations in treating HIV/AIDS patients in a climate of discrimination. The American Medical Association (AMA) promulgated ethical rules requiring that physicians treat HIV/AIDS patients whose condition is within the physicians' realm of competence. When such rules are violated, boards of medicine, medical associations, hospital and medical school committees, and other credentialing agencies have the difficult task of reviewing alleged breaches and sanctioning misconduct. These professional guidelines begin to clarify the boundaries and goals of medicine as a social good. They attempt to ensure that medical practitioners act humanely as they fight and prevent diseases, promote and restore health, and reduce pain and suffering. The ethical customs found in codes, charters, principles, and oaths form the basis of an entire culture of medicine for the profession.

The practice of medicine is bound by ethical rules for an important reason. In order to fulfill their healing obligation, medical practitioners must often engage in risky procedures interfering with the bodies and minds of vulnerable individuals. Bodily interference, if unconstrained by certain legitimate guiding rules, can be nothing more than assault and battery. Patients must be assured that they will benefit from, or at least not be harmed by, doctors' care. The establishment of trust is crucial to this end, and once earned, trust marks the doctor–patient relationship.

Perhaps one of the most enduring doctrines in the history of medical ethics is the Hippocratic Oath. The oath dates back to the fourth century B.C.E. and forms the basis of Western medical ethics. It reflects the assumed values of a brotherhood of physicians who charge themselves to care for the sick under a pledge witnessed by the Greek deities. Of great interest to doctors at the time was distinguishing the genuine physician from the fraudulent practitioner. One way in which the oath furthers this goal is by prizing teachers and teaching, requiring that the physician hold his teacher in the "art" of medicine on par with his own parents. It also requires the physician to pledge to help the sick according to his skill and judgment and never do harm to anyone, never administer a deadly drug even when asked to do so, never induce abortion, and never engage in intentional misdeeds with patients (sexual or otherwise). It further requires the physician to keep secret all those things that ought not be revealed about his patients. The good physician, the oath concludes, may enjoy a good life and honored reputation, but those who break the oath shall face dishonor.

To this day, most graduating medical school students swear to some version of the Hippocratic Oath, usually one that is gender-neutral and that departs somewhat from the traditional prohibitions. The mandate of the oath is strong; it directs practitioners to desire what is best for the health of patients. A growing number of physicians and ethicists realize, however, that the Hippocratic Oath and similar ethical codes, though motivational, are inadequate when dealing with the novelties of current practice.

Medicine has recently undergone radical shifts in the scientific, technological, economic, social, and political realms, giving rise to artificial life-sustaining devices and treatments, legalized abortions, new artificial reproductive technologies, inventive cosmetic surgeries, stem cell and gene therapies, organ transplantation, palliative care, physician-assisted suicide, and conflicts of interest more powerful than anyone could have predicted just a few decades ago. Many matters previously thought of as "human nature" are continuously being recharacterized to reflect changing knowledge, scientific and otherwise. Medical ethics engages these debates and evaluates the correlative concerns over the definition of death, the moral status of the fetus, the boundaries of procreation and parenting, the flexibility of the concept of personhood, the rights and responsibilities of the dying, and the role of corporations in medicine.

Although physicians' codes play a crucial role in defining the broad parameters of ethical conduct in medicine, in the last few decades, sociopolitical demands and market forces have played a much larger role in both shaping and complicating ethics in medicine. Medical ethics then becomes a tool for critical reflection on modern biomedical dilemmas. Ethics scholars and clinical ethicists are regularly consulted when principles or codes appear inadequate because they prescribe unclear, conflicting, or unconscionable actions. Even for ethicists, it is not always obvious what "doing the right thing" means; however, many ethical dilemmas in medicine can be deconstructed using the theoretical tools of medical ethics and sometimes resolved by encouraging decision makers to consider the merits, risks, and psychosocial concerns surrounding particular actions or omissions.

To be sure, clinical ethicists usually do not unilaterally declare right and wrong. But they can ensure that all rightful parties have a fair and informed voice in the discussion of ethically sensitive matters. Medical ethics, as a clinical discipline, approaches decision making through formal processes (e.g., informed consent) and traditional theories (e.g., utilitarianism) that can enhance medical and ethical deliberation. The need for these processes and theories was not just a by-product of technological advances, but also a consequence of a movement that recharacterized the civil status of doctors and patients.

The American Civil Rights Movement of the 1950s and 1960s brought previously denied freedoms to people of color and reinvigorated the spirit of free choice. The unconscionable inferior treatment of marginalized groups was the subject of great sociopolitical concern. Significant legal and moral changes took place both in the ideology surrounding the concepts of justice and equality and in the rigidity of hierarchies found in established institutions of status such as churches, families, schools, and hospitals. Out of the movement came a refreshing idea of fundamental equality based on the dignity of each individual. In the decades that followed, strong criticism arose against paternalism—the practice of providing for others' assumed needs in a fatherly manner without recognizing individuals' rights and responsibilities. It was no longer acceptable for all-knowing physicians to ignore the preferences and humanity of patients while paternalistically doing what they thought was in their "best interests." Doctors were required to respect patients' autonomy, or ability to self-govern. With this recognition came a general consensus that patients have the legal and ethical right to make uncoerced medical decisions pertaining to their bodies based on their own values.

Autonomy, now viewed by many as a basic principle of biomedical ethics, often translates in practice into the process of "informed consent." Full informed consent has the potential to enrich the doctor–patient relationship by requiring a competent patient and a physician to engage in an explanatory dialogue concerning proposed invasive treatments. By law, physicians must presume that all patients are competent to make medical decisions unless they have a valid reason to conclude otherwise. If a patient is diagnosed as incapable of consenting, the patient's surrogate decision maker or "living will" should be consulted, assuming they are available and no other recognized exception applies. At its core, informed consent must involve the discussion of five elements:

1. the nature of the decision or procedure
2. the reasonable alternatives to the proposed intervention
3. the relevant risks, benefits, and uncertainties related to each alternative
4. an assessment of patient understanding
5. the acceptance of the intervention by the patient

A physician's failure to abide by this decision process can lead to ethical and legal sanctions.

Scholarly questions often arise regarding the diagnosis of incapacity, the determination of how much information must be shared, the definition of "understanding," and the established exceptions to informed consent (e.g., emergency, patient request not to be informed, and "therapeutic privilege"). It is important for informed consent to be an interactive process and not merely the signing of boilerplate forms. The latter does not take the interests of patients into account, it does not further the doctor–patient relationship, and it can result in future conflict or uncertainty if previously competent patients become incapacitated.

In addition to doctors, many other parties are involved in caring for the ill and facilitating medical decision making. Relatives, spiritual counselors, nurses, social workers, and other members of the health care team all help identify and satisfy the vital needs of patients. Informed consent is a process that can give rise to meaningful dialogue concerning treatment, but like some other tools of practical ethics, it alone may not provide the intellectual means for deeper reflection about values and moral obligations.

To this end, medical ethics makes use of many foundational theories that help situate values within wider frameworks and assist patients, families, and doctors with making ethical choices. These moral theories are typically reduced to three categories: the deontological (duty-based, emphasizing motives and types of action); the consequentialist (emphasizing the consequences of actions); and the virtue-based (emphasizing excellence of character and aspiration for the good life).

The most influential deontological theory is that of Immanuel Kant (1724–1804). Kant derived certain "categorical imperatives" (unconditional duties) that, in his view, apply to the action of any rational being. Generally speaking, the relevant imperatives are as follows: first, individuals have a duty to follow only those subjective principles that can be universalized without leading to some inconsistency; and, second, individuals must treat all rational beings as "ends in themselves," respectful of the dignity and integrity of the individual, and never merely as a means to some other end. Despite some philosophical criticism, Kant's revolutionary thoughts on the foundations of morality and autonomy are still very timely.

In contrast to Kantian deontology, an influential consequentialist theory is utilitarianism, which states that the moral worth of an action is determined solely by the extent to which its consequences maximize "utility." For Jeremy Bentham (1748–1832), utility translates into "pleasure and the avoidance of pain"; for John Stewart Mill (1806–73), utility means "happiness." Utilitarianism offers another popular way to conceptualize right and wrong, but it gives rise to the oft-asked question of how one might accurately calculate the tendency to maximize happiness.

Finally, virtue-based ethics, principally attributed to the Greek philosophy of Plato and Aristotle, generally holds that a person of good character strives to be excellent in virtue, constantly aiming for the *telos* or goal of greater happiness. In leading a virtuous life, the individual may gain both practical and moral wisdom.

These three basic ethical frameworks maintain their relevance today, inspiring many complementary models of ethical reasoning. For example, medical ethics has benefited significantly from scholarship in theological, feminist, communitarian, casuistic, and narrative ethics. These perspectives either critically analyze or combine the language of deontology, consequentialism, and virtue. Together, theories of ethics and their descendants provide some further means of deconstructing the ethically difficult cases in medicine, giving us the words to explore our moral intuitions.

Medical ethics is now often described within the somewhat broader context of bioethics, a burgeoning field concerned with the ethics of medical and biological procedures, technologies, and treatments. Though medical ethics is traditionally more confined to issues that arise in the practice of medicine, both bioethics and medical ethics engage with significant overlapping questions. What are the characteristics of a "good" medical practitioner? What is the best way to oversee the use and distribution of new medical technologies and therapies that are potentially harmful? Who should have the right and responsibility to make crucial moral medical decisions? What can individuals and governments do to help increase access, lower cost, and improve quality of care? And how can individuals best avoid unacceptable harm from medical experimentation? Patients, doctors, hospital administrators, citizens, and members of the government are constantly raising these questions. They are difficult questions, demanding the highest level of interdisciplinary collaboration.

In sum, since the days of Hippocrates, the medical profession has tried to live by the principle of *primum non nocere* (first, do no harm). This principle has been upheld by many attentive professionals but also betrayed by some more unscrupulous doctors. To stem potential abuses offensive to human dignity and social welfare, medical ethicists carefully consider the appropriateness of new controversial medical acts and omissions. They try to ensure that medical decision makers do not uncritically equate the availability of certain technoscientific therapies and enhancements with physical and psychosocial benefit. Doctors and patients can participate in a better-informed medical discourse if they combine the dictates of professional rules with procedural formalities of decision making, respecting the diversity of values brought to light. Through this deliberative process, individuals will be able to come closer to understanding their responsibilities while clarifying the boundaries of some of the most difficult questions of the medical humanities.

See also Health and Medicine; Health Care; Medical Marijuana; Research Ethics.

Further Reading: Applebaum, P. S., C. W. Lidz, and A. Meisel. *Informed Consent: Legal Theory and Clinical Practice.* New York: Oxford University Press, 1987; Beauchamp, Tom L., and James F. Childress. *Principles of Biomedical Ethics.* 5th ed. New York: Oxford University Press, 2001; Clarke, Adele E., et al. "Biomedicalization: Technoscientific Transformations of Health, Illness, and U.S. Biomedicine." *American Sociological Review* 68 (April 2003): 161–94; Daniels, N., A. Buchanan, D. Brock, and D. Wikler. *From Chance to Choice: Genes and Social Justice.* Cambridge: Cambridge University Press, 2000; Engelhardt, H. Tristram, Jr. *The Foundations of Bioethics.* 2nd ed. New York: Oxford University Press, 1996.

Joseph Ali

MEDICAL MARIJUANA

Whether marijuana should be made legally available for doctors to prescribe as a drug for treatment of certain medical conditions is hotly debated among politicians, lawyers, scientists, physicians, and members of the general public.

The cannabis plant (marijuana) has been cultivated for psychoactive, therapeutic, and nondrug uses for over 4,000 years. The primary psychoactive drug in the plant is tetrahydrocannabinol (THC)—a molecule that produces a "high" feeling when ingested and, as is most often the case with cannabis, when inhaled in smoke or vapor form. There are hundreds of other chemical components in marijuana, from Vitamin A to steroids, making it somewhat unclear how the human body will physiologically react to short-term and long-term use of the substance.

Supporters of medical marijuana argue that the drug is acceptable for medical treatment, citing reports and several scientific peer-reviewed studies. There has been considerable interest in the use of marijuana for the treatment of glaucoma, neuropathic pain, AIDS "wasting," symptoms of multiple sclerosis, and chemotherapy-induced nausea, to name a few conditions.

The Food, Drug, and Cosmetic Act—a key law used by the Food and Drug Administration (FDA) in carrying out its mandate—requires that new drugs be shown to be safe and effective before being marketed in the United States. These two conditions have not been met through the formal processes of the FDA for medical marijuana, and it is therefore not an FDA-approved drug.

Proponents of medical marijuana argue that the drug would easily pass the FDA's risk-benefit tests if the agency would give the drug a fair and prompt review. One significant hurdle to obtaining FDA approval is the fact that marijuana has been listed as a Schedule I drug in the Controlled Substances Act (CSA) since 1972. As such, it is considered by the U.S. government to have a "lack of accepted safety," "high potential for abuse," and "no currently accepted medical use." Schedule I drugs, however, have occasionally been approved by the FDA for medical use in the past, with significant restrictions on how they must be manufactured, labeled, and prescribed.

At present, the possession and cultivation of marijuana for recreational use is illegal in all states and in most countries around the world. Further, representatives of various agencies in the current U.S. federal government have consistently stated that there is no consensus on the safety or efficacy of marijuana for *medical* use, and without sufficient evidence and full approval by the FDA, the government cannot allow the medical use of a drug that may be hazardous to health. Some say that the availability of various other FDA-approved drugs, including synthetic versions of the active ingredients in marijuana, make the use of marijuana unnecessary. They claim furthermore that marijuana is an addictive "gateway" drug that leads to abuse of more dangerous drugs and that it injures the lungs, damages the brain, harms the immune system, and may lead to infertility. The use of marijuana for some medical purposes is allowed in Canada, however, though under strict Health Canada regulations.

Proponents maintain that the approved synthetic versions of marijuana are not chemically identical to the actual plant and therefore not as medically beneficial. They further argue that many of the claims of harm either have not been shown to be true or are not at all unique to marijuana, but are comparable to the potential side effects of a number of alternative drugs currently on the market. They insist that the U.S. federal government is setting unfair standards for medical marijuana because of sociopolitical rather than scientific reasons. They point to a respected scientific report published in 1999 by the U.S. Institute of Medicine (IOM) and commissioned by the U.S. government through a $1 million grant, which recommends that under certain conditions marijuana be made medically available to some patients, even though "numerous studies suggest that marijuana smoke is an important risk factor in the development of respiratory disease."

Despite a broad federal stance in opposition to the distribution, possession, and cultivation of marijuana for *any* drug-related use, many U.S. states have enacted their own "medical use" laws. Currently 12 states have approved the medical use of marijuana for qualified patients. The level of permissibility for marijuana use in state laws varies. Some states, such as California, allow doctors to prescribe marijuana very liberally, whereas others, such as New Mexico, allow access to medical marijuana only for patients suffering pain as a result of a few specific conditions. The enactment of medical marijuana state statutes that conflict with the federal Controlled Substances Act has given rise to lawsuits brought by both sides in the controversy.

The issue has gone so far as to reach the U.S. Supreme Court in the case of *Gonzales v. Raich*. In that 2005 case, the Supreme Court ruled that Congress has the authority to *prohibit* the cultivation and use of marijuana in California and across the United States, despite laws in California allowing the use of medical marijuana. The court did not require California to change its laws, however. As a result, both the California medical-use statutes and the conflicting federal laws remain in force today. Some doctors in California continue to prescribe medical marijuana through the state's program, and the federal government's Drug Enforcement Administration (DEA) continues to enforce the federal statute in California against those who choose to prescribe, possess, or cultivate marijuana for medical use. The issue remains largely undecided in law.

In *Gonzales v. Raich,* the Supreme Court did state that Congress could change the federal law to allow medical use of marijuana, if it chose to do so. Congress has voted on several bills to legalize such use, but none of these bills has been passed. Most recently, a coalition has petitioned the U.S. government to change the legal category of marijuana from "Schedule I" to a category that would permit physicians to prescribe marijuana for patients they believe would benefit from it. Given recent trends, it is unlikely that the current federal government will respond favorably to this petition; it is equally unlikely, however, that supporters of medical marijuana will be quick to abandon this controversial battle.

See also Drugs; Medical Ethics; Off-Label Drug Use.

Further Reading: *Controlled Substances Act,* U.S. Code Title 21, Chapter 13; *Federal Food, Drug, and Cosmetic Act,* U.S. Code Title 21, Chapter 9; *Gonzales v. Raich* (previously *Ashcroft v. Raich*), 545 U.S. 1 (2005); Joy, Janet Elizabeth, Stanley J. Watson, and John A. Benson, *Marijuana and Medicine: Assessing the Science Base.* Institute of Medicine Report. Washington, DC: National Academies Press, 1999.

Joseph Ali

MEMORY

Conflicts about memory are part of the larger battleground involving the brain and mind. The major controversies in the general and particular arenas of brain and mind studies arise within the brain sciences and at the intersection of the brain sciences and the social sciences. Philosophers were the original arbiters of systematic approaches to mind and brain; later, psychologists became as or more important overseers of this territory. Today, although philosophers and psychologists still play key roles in this field, cognitive scientists and neuroscientists are increasingly the key players on this field of inquiry. There is, it should be noted, much fluidity across these disciplines, and it is sometimes difficult to put a disciplinary label on a particular researcher, methodology, model, or theory.

At the most general level, the battleground here is one that was set up by social theorists and philosophers in the nineteenth century. It amounts to asking how much of mentality is inside us, and indeed inside the brain, and how much is outside of us, in our experience and behavior. In brief, the questions fueling this battleground are these: what is memory, where is memory, and how do we "get" at what and where using our current research methods and technologies?

Within memory studies as a province of the brain sciences, one of the main controversies pits advocates of laboratory research against naturalistic students of memory in everyday life. The traditional storehouse model, which treats memory as a matter of storage and retrieval, has not fared well in light of developments in memory research over the last few decades. Memory seems less and less like something we can think of using filing-cabinet and filing-system analogies but more and more like something that requires a new and elusive approach. Staying with the conventional framework, the main alternative to the storehouse model is based on a correspondence metaphor. In this approach the person remembering, the "subject," plays a more active role than in the storehouse model. If we allow subjects to freely report memories, we get better results in terms of accuracy than if we constrain them according to the requirements of laboratory experiments or even of behaviors in natural settings. Some researchers believe that focusing on a variety of metaphors will turn out in the long run to be important in the comparative assessments of the laboratory–versus–everyday, natural settings debate and may even resolve this debate and lead to a more firmly grounded general theory of memory.

Models of memory that focus on storage are then distinguished as either multi-storage or uni-storage. In a multi-storage model, for example, the researcher would distinguish sensory memory, short-term memory, and long-term memory. Some mode of attentiveness is assumed to move sensory memory into short-term memory, and some form of "rehearsal" transfers the content into long-term memory. Remembering is the process of retrieving memories from long-term memory. Clearly, the terms *short-term memory* and *long-term memory* refer to two different types of "storage containers" in the brain.

Critics argue that the multi-storage model is too simplistic. One of the alternatives is the so-called working memory model. Just as in the case of the correspondence metaphor, the alternative to the more traditional idea is more active, more "constructive," one might say. This is important because it feeds directly into the more recent social science models of memory, especially those that stress a social construction approach. In the working memory model, attention is conceived as a complex process metaphorically embodied in a "central executive." The central executive funnels information to and through a phonological loop and a visio-spatial sketchpad.

Following the original development of the working model, the concept of an episodic buffer was introduced. This buffer was needed to integrate information across temporal and spatial dimensions in order to allow for the holistic rendering of visual and verbal stories. The buffer may also be the mechanism needed to link the early stages of remembering to long-term memory and meaning.

The emerging emphasis on the correspondence metaphor draws our attention to accuracy and fidelity in remembering, and this has become a part of the increasingly volatile debates about false memories. Do children remember traumas and abuses (sexual traumas are of particular concern) because the events happened or because they are encouraged to recall events that never happened given how psychotherapeutic, legal, and other interrogations operate in conjunction with the child's memory apparatuses? The controversies over repressed memories and forced recall have provoked numerous research efforts to tease out the way memory works as a brain function (hypothetically, internally and in a way that can be isolated and demarcated) and as a contextual phenomenon that has situational determinants. At this point, we can say that the debate is about whether memories are fixed or flexible. Some of today's sociologists of science might put the question this way: Like scientific facts, truths, and logics, are memories situated? That is, does what we remember or can remember depend on the social and cultural context in which we are prompted to remember? These are the basic factors that have led to the memory wars. Psychologists, lawyers, politicians, social scientists, ethicists, parents, and children are among the key combatants on this battleground. It is no coincidence that the memory wars are products of the late twentieth century, which saw the rise of the science wars and the broader culture wars. These "wars" are signs of a general global paradigm shift that is impacting basic institutions across the globe and fueling conflicts between fundamentalists, traditionalists, and nationalists on the one side and agents of a science-based global restructuring of our ways of life and our everyday worldviews on the other side.

In now seems clear that we need to distinguish "real" memories from "false" memories, memories of events that really occurred from memories that are products of suggestion and other modes of manufacturing. This is not as simple as it might sound. Imagine that we experience something and that that experience enters our long-term memory (disregarding for the moment whether this is a storage process, a constructive one, or something entirely different). The first time we recall that experience, we are as close to remembering what actually happened as possible. The next recall, however, is not bringing up the original experience but our first recall of that experience. Each time we retrieve the event, we retrieve the last recollection. As we build up these multilayered levels of recollection, it is more than likely that our memories are going to be, to some degree and in some respects, corrupted. The point is that remembering is not as straightforward as our experience of remembering "clearly" may suggest, though this is a matter of degrees for any given memory, person, and situation.

There is experimental evidence that the same techniques that can help people recover repressed memories can also intentionally or unintentionally implant memories. Clinical scientists are at odds with academic scientists on the issue of repression, especially "massive" repression. Judges and lawyers tend to side with the academic scientists against the clinical scientists. The political and legal contexts of most of these debates surely cannot facilitate a reasoned scientific way of getting to the heart of the matter. Such an approach, however, may be constrained by the volatility of the knowledge and information markets as great strides are made in our understanding of the brain as a neuroscience object on the one hand and the social brain on the other. One of the features of our current world is that information and knowledge are growing and changing at very rapid rates by any measure you wish to use, and whatever the underlying causes for these world-historical dynamics, they are making it next to impossible to settle controversies quickly. This is one of the reasons, indeed, that we have so many battlegrounds in science, technology, society, and culture.

Within neuroscience, one of the important approaches to understanding memory involves the development of a neuronal model of memory. Neuronal models are very popular in the brain sciences today, reflecting in part a theoretical orientation but perhaps even more strongly the development of technologies for studying the brain at the micro-level of neurons as well at the level of brain modules and units of localization for specific tasks. Neuroscientists assume that working memory resides in the prefrontal cortex. This is the part of the brain assumed to be involved in planning, controlling, organizing, and integrating thoughts and actions. By studying neuronal activity in this part of the brain in healthy individuals as well as in individuals who exhibit mental health problems, neuroscientists hope to unravel the mechanisms of reason and what goes wrong when the control functions of the prefrontal cortex fail.

Another of the emerging battlegrounds in the brain, mind, and memory wars involves conflicts between philosophers, psychologists, cognitive scientists, and neuroscientists, who tend to view the brain and the individual person as freestanding, context-independent entities, and social scientists and some neuroscientists who view the brain and the person as social constructions, social things, context-dependent entities. This battleground, by contrast to others discussed in this reference set, is still more or less being carried on under the radar of media and public awareness. Nonetheless, there is an approach to memory that depends more on social theories of mind and brain than on traditional brain sciences approaches.

Ideas about mind and brain as topics of social and cultural investigation are not new. They were part of the tool kits of intellectuals and scholars who forged the social sciences beginning in the 1840s and continuing on into the early decades of the twentieth century. It has taken some recent developments in the social sciences, developments that began to take shape in the late 1960s, to recover some of these traditions and to begin to shape them into a credible theory about thinking and consciousness. Consider then how a social theory of mind and brain might affect how we think about memory. No one is prepared to deny that in our time and in our society thinking is experienced as occurring inside us and (perhaps for most people) inside people's heads. Nor is it controversial that some thinking at least goes on outside the presence of others. It may not be so obvious, unless you attend closely to what is going on, that thoughts—especially those we experience without others about—tend to be ephemeral. They will rapidly evaporate if they are not recorded or reported. This is true of experiences outside the presence of others. It is only less true in the presence of others because of the immediacy of the opportunity to rehearse the experience or the thought. Think about why people will get up in the middle of the night to jot down an idea or interrupt a lunch meeting to jot down an idea or diagram on a napkin or call a friend to tell him or her about a new idea. The reason is that they recognize the phantom nature of these ideas and that the ideas will evaporate out of consciousness unless we actively work to create a memory. Think about how many Post-its you or your friends have covering refrigerators. Is memory inside our heads, or is it outside on our refrigerators?

The social theory of memory is not a well-developed field except in the area of cultural remembering and repressing. Many studies have been carried out in recent years on culture and memory at the macro-historical level, especially, for example, involving remembrance among survivors of the Holocaust. I want to illustrate some ways in which memory could be approached using the tools and methods of the social sciences, however. This work is still quite exploratory, and there has not been much if any clinical or empirical research to ground the perspective. Nonetheless, it is one of many areas of social science theory and research that are transforming the way we think about the world around us.

I will introduce you to this way of thinking by posing some questions:

Consider whether we remember all at once. Without reflection, it may seem obvious that we do, and this is certainly consistent with the idea of remembering as information processing, storage, and retrieval. Do we really remember all at once, or do we remember in steps? We may experience remembering as virtually instantaneous, but that might be an illusion. This idea follows from the conjecture that thinking in general and remembering in particular are much more interactive, much more matters of manufacturing, much more processes. It may be that as we remember, we begin with a provocation that leads to a part of the memory we are seeking to reconstruct; that remembering triggers the next partial memory, and so on. Consider what happens when you write down directions for someone or draw the person a map. Do the directions come to you all at once and present you with a picture that you then see in your mind and copy down on paper? Or does each movement from memory to paper trigger the next move and so on until you have all the directions down?

Some neuroscientists have built the idea of "rehearsal" into their theories and models of memory. Suppose that rehearsal is not quite what they think, but something more pervasive, more active, and more constant. What if we remember by constantly rehearsing narratives of experiences in some sub-sub-vocal way? We know that we subvocalize when we think. It might be that there is some mechanism that allows us to rehearse our experiences on numerous channels simultaneously. These channels vary in their capacities to sustain our rehearsals. On some channels, our narrative rehearsals dampen quickly and fade away. On other channels our narratives are constantly boosted, enhanced, sustained. I imagine these channels have relevancy valences (RV). Stories fade, fade in and out, and sometimes disappear on low-RV channels. High-RV channels keep the story going. Remembering is in this model a matter of tuning into one of these channels, focusing our attention on it, and elevating it into our awareness so that we can think about it or verbalize it. If this is a reasonable pathway toward a social theory of memory, it needs to take account of the "fact" that the channels are not independent of one another or of current consciousness experience. Of course, this theory requires a mechanism we have not yet been able to identify.

The point of this idea is to help change the terms of the current rules for thinking about mind, brain, thinking, and memory. Clearly, neuroscientists as well as social scientists recognize that our prevailing individualistic, psychologistic models, theories, and basic assumptions are producing and sustaining more problems than they are solving. Social science models, though necessarily more speculative at this stage, may be important provocations to bringing an important battleground into the range of the media's and the public's radar.

See also Artificial Intelligence; Brain Sciences; Mind.

Further Reading: Bal, Mieke, Jonathan Crewe, and Leo Spitzer, eds. *Acts of Memory: Cultural Recall in the Present.* Hanover, NH: University Press of New England, 1999; Brothers, Leslie. *Friday's Footprint: How Society Shapes the Human Mind.* Oxford: Oxford University Press, 2001; Connerton, Paul. *How Societies Remember.* Cambridge: Cambridge University Press, 1989; Kandel, Eric R. *In Search of Memory: The Emergence of a New Science of Mind.* New York: Norton, 2007; Star, Susan Leigh. *Regions of the Mind: Brain Research and the Quest for Scientific Certainty.* Stanford: Stanford University Press, 1989; Stewart, Pamela J., and Andrew Strathern, eds. *Landscape, Memory and History: Anthropological Perspectives.* London: Pluto Press, 2003.

Sal Restivo

MIND

There have been many debates on the nature of "mind," and although our understanding of it has evolved over time, disagreements remain. Advances in science and technology have spawned an interest in studying the mind using knowledge gained from computer science. A variety of computational models have been proposed to explain how the mind works. Other models have been developed in the biological and neurosciences. Social theories of mind that are very different from physical and natural science approaches also line the landscape of mind studies.

The *Oxford English Dictionary* defines *mind* primarily as "the seat of consciousness, awareness, thought, volition, and feeling." In the seventeenth century, René Descartes (1596–1650) argued that the mind is something distinct from the body. His philosophical dualism was inscribed in the slogan "Cogito, ergo sum." The neuroscientist Antonio Damasio calls this "Descartes' error." This reflects an evolving interdisciplinary interrogation of mind–body dualism. Perhaps mind and body are not two separate things, substances, or natural kinds. Perhaps mind is no more substantial than the soul.

For a long time after Descartes, dualism took center stage, and scholars distinguished between the mind and the body (and brain). Other schools of thought, some predating Descartes, held different points of view. Idealism, for example, is closely linked to theology and the idea of the soul. It states that the world is only mind; we only have access to sensations and our interpretations of them (e.g., thoughts, feelings, perceptions, ideas, or will). As well, around the time that Descartes posited mind–body dualism, Thomas Hobbes and others advocated a materialist cosmology that sharply contrasted with idealism, claiming that all reality is matter. The roots of materialism go as far back as ancient Greek, Chinese, and Indian philosophy.

From materialism emerged behaviorism. Behavioral scientists accused Descartes of contributing to the dogma of the ghost in the machine, where the mind is the ghost and the body is the machine. To them, the mind is part of the body, or an aspect of behavior.

Functionalism also emerged from materialism. Its advocates believe that there is nothing intrinsically biological to the mind and that any system that manipulates symbols according to a given set of programmed rules will have a mind. For functionalists, the mind is a computer, dependent on the activities of the brain. Just as the computations of the computers are not reducible to any one part of the computer, so are mental states not reducible to any specific brain location.

Some functionalists believe that mental states are functional states of the computational machine (machine functionalism). This view allows for the possibility of a greater level of abstraction than biological models in studying the functioning of the mind. It also permits multiple realizability.

Multiple realizability suggests that state X can be achieved by many means, rather than only one. For example, one can feel pain in one's finger for different, unrelated reasons, yet the pain may still feel the same. Thus the state of pain can be achieved by many means; therefore, pain has multiple realizability. In the same way, according to machine functionalism, the mind can be created by many means.

Opponents of functionalism say that it cannot account for qualitative aspects of conscious experience, such as the experience of seeing colors or that of feeling pain. A system could have the same functional states as those of a conscious system and yet experience none of the feelings of a conscious system. In other words, a system can pass the Turing test (see sidebar) based on syntax and yet be unconscious of semantics.

TURING TEST

Can computers think? In 1950 Turing proposed an imitation game whose results should be more pertinent than the answer to the afore-stated question. The imitation game proposes that if a human being can be misled to believe a computer to be another human being, then, for all that we are concerned, that computer can be said to think. The imitation game takes place in an isolated room in which a participant sits. In another room, there is another person (A) and a computer (B). The participant is asked to determine which of A and B is the person and which is the computer. The participant is allowed to interact via written language with both A and B. Turing suggested that in the future, science and technology should be so advanced as to permit the creation of a computer able to simulate the coherent and pertinent language of a human being. Turing suggested that the participant should be incapable of telling the person and the computer apart and could very well end up thinking that B, the computer, is the person. According to Turing, this would be evidence enough to consider the computer a thinking machine.

This argument is another version of the Chinese room (see sidebar), a thought experiment put forth by John Searle, professor of philosophy at the University of California at Berkeley, whose main interests include philosophy of mind and artificial intelligence. If the Chinese room argument holds true, and functionalism cannot account for qualitative aspects of conscious experiences, then this in itself demonstrates that qualitative states are not identical to functional states; the mind is not the same as a computer. Thus, functionalism is false.

Yet another more recent theory rooted in materialism is identity theory, the earliest theory driven by the advances in neuroscience. The founders of the identity theories are Ullin T. Place, John Jamieson Carswell Smart, and Herbert Feigl. Identity theorists believe that mental states are brain states, literally, and they too endorse multiple realizability.

In a parallel series of developments, computer science has encouraged cognitive scientists to use new computer models to explain the mind. Yet the historic influence of debates on the nature of mind continues into the present, as researchers debate how these models should be used. Should they be taken to reflect the actual organization of the mind, or perhaps of the brain? In these modern theories, one can still feel the influence of the mind-body dualism problem, Descartes' error.

The classical theory of cognition is Computational Theory (CT). Its premise is that cognition is computation. Both processes are considered to be semantic. That is, they are both considered to be dependent on meaning. Computational theory seeks to build models of the mind based on the semantic and syntactic structures of our symbols for the inputs from the world (e.g., words) that are considered to foment mind processes.

Computational theory is based on the concept of a cognitive architecture whose main feature is its ability to allow cognitive representations. Cognitive architecture thus resembles the central processing unit of a computer. It fixes the nature of the symbols to be used by the system as well as the operations possible

CHINESE ROOM ARGUMENT

Among the most important criticisms of the Turing Test was the Chinese room argument, by John Searle. Opposing functionalism, Searle argues that syntax and semantics are quite distinct and that one cannot replace the other. He presented four formal premises to explicate his argument, of which the second premise is supported by the Chinese room thought experiment. The premises are the following:

1. Brains cause minds.
2. Syntax is not sufficient for semantics.
3. Computer programs are entirely defined by their formal, or syntactical, structure.
4. Minds have mental contents; specifically, they have semantic contents.

The Chinese room argument was presented for the first time in 1980. One is to imagine a room in which a person A, who speaks and understands no Chinese, sits. Person A has detailed instructions on how to create meaningful Chinese responses to Chinese inputs. Person A has no understanding of the language manipulated; person A has only been given mechanical instructions on how to process Chinese messages. Person A could receive any Chinese message and reply with an appropriate Chinese response. Searle points out that under these conditions, a Chinese person B standing outside the Chinese room could never know that the person inside the Chinese room does not understand one word of Chinese. Person A would reply in a normal manner to person B and could pass the Turing Test. Yet person A could hardly be said to understand Chinese (semantics); person A only has a grasp of the syntax. Person A merely converted input X into Y, following a series of mechanical instructions.

for these symbols. Just as computers have hardware and software, so the mind has a cognitive architecture and cognition. The cognitive architecture is impenetrable and unchangeable, just as is hardware. Cognition can undergo modifications, but this will never have any effect on the cognitive architecture. Because of these similarities between cognition and computers, symbolic computation is thought to be an ideal tool to study the organization of the mind.

An important point of CT is that it allows for a clear distinction between cognitive and noncognitive processes. CT makes explicit symbolic computation a requisite for cognition. Symbolic computation is any process or manipulation of symbols with semantic interpretations that is described by a non-semantic (syntactical) cognitive architecture. An explicit symbol is one of whose existence or presence we are aware. For computationalists, learning, for example, is a cognitive process. It is a case of explicit symbol manipulation in syntactically legal ways. However, changing one's belief about X by taking a pill would not involve manipulation of any explicit symbol; it could not therefore be an example of "learning."

A critical and philosophically relevant implication of CT, however, is that it also gives the mind multiple realizability; the human mind is but a type of computational device whose symbols, arranged and manipulated in the way that they are, lead to what we consider "mind." This suggests that there is nothing intrinsically biological and unique about the human mind because there are infinite possible ways to create mind. According to CT, as long as the physical system is programmed in the appropriate manner, the nature of the system is irrelevant. Any machine, any computer, if programmed rightly, will have a mind.

In the 1980s, connectionism became very popular among cognitive scientists. This theory agrees with CT, in that content characterizes both computation and cognition, but it seeks to build a computational model that will accurately reflect the actual organization of the brain and, as a result, explain the workings of the mind. Its models focus on neural networks and not so much on the semantics of inputs. Connectionist models of cognition are based on neural networks of units (nodes) all connected together in various patterns, determined by the weights (strengths) of the connections. They thus account for graceful degradation of function, spontaneous generalization (generalizing from vague cues), and so forth.

Connectionism considers both explicit and tacit symbols to be involved in mental processes. Moreover, processing can occur at sub-symbolic levels. Connectionism understands subconscious processes (e.g., Pavlov conditioning) to be cognitive events, whereas CT does not.

Unlike CT, however, connectionism treats cognitive processes to be distributed. This permits explanation of the holistic representation of data (rather than unit-by-unit identification of a whole.) Parallel Distributed Processing (PDP), endorsed by many connectionists, claims that all incoming information (input) is processed in parallel. Thus, the parts and the whole of an object are processed simultaneously.

These types of models also do an excellent job of explaining the graded notions of category that hold our human minds. Indeed, it is quite impossible to pinpoint definite categories of our view of the world, a CT requirement. How is one to define a dog? At which point is it not a dog, but a wolf? It may be difficult, even impossible, to come up with a final and finite definition for a dog, yet most of us will be certain we could distinguish between the two animals. Connectionism proposes that we hold no finite, whole symbolic notions of what a dog is and what a wolf is. Rather, we maintain statistical connections among varying units of relative importance to the inputs in question, which make us say in any particular instance "this is a dog" instead of "this is a wolf."

Social theories of mind have been around since the emergence of classical social theory in the nineteenth century. Social theorists from Durkheim, Nietzsche, and Marx to Mead and C. Wright Mills and most recently R. Collins and S. Restivo have sought the basis of mind in social relationships and networks and communication systems rather than in the brain. Restivo and others have argued that the mind is not an entity at all but a secular version of the soul, a concept without a natural world referent. For classical and contemporary developments in this area, see the readings by Restivo and Bauchspies, Collins, and Valsiner and van der Veer.

See also Brain Sciences; Memory.

Further Reading: Chalmers, David J. *Philosophy of Mind: Classical and Contemporary Readings*. New York: Oxford University Press, 2002; Clapin, Hugh. "Content and Cognitive

Science." *Language & Communication* 22, no. 3 (2002): 232–42; Collins, Randall. *The Sociology of Philosophies.* Cambridge, MA: Harvard University Press, 1998; Damasio, Anthonio. *Descartes' Error.* New York: G. P. Putnam's Sons, 1994; Gregory, Richard L. *Mind in Science: A History of Explanations in Psychology and Physics.* New York: Cambridge University Press, 1981; Restivo, Sal, with Wenda Bauchspies. "The Will to Mathematics: Minds, Morals, and Numbers" (revised). In "Mathematics: What Does It All Mean?" ed. Jean Paul Van Bendegem, Bart Kerkhove, and Sal Restivo, special issue. *Foundations of Science* 11, no. 1–2 (2006): 197-215. "O arbítrio da matemática: mentes, moral e números." [Portuguese translation.] *BOLEMA* 16 (2001): 102–24; Valsiner, Jaan, and Rene van der Veer. *The Social Mind: Construction of the Idea.* Cambridge: Cambridge University Press, 2000.

Sioui Maldonado Bouchard

MISSILE DEFENSE

Ever since the advent of long-range weapons, militaries have been concerned with defending themselves against objects falling from the sky. Developing technologies in the 1950s brought a new threat in the form of ballistic missiles. Governments and their armed forces sought defensive measures, culminating recently in the United States in a National Missile Defense (NMD) program. There are three main concerns with NMD: destabilization, functionality, and who should be in charge of decisions about its development and deployment.

The first attempt at missile defense in the United States came in the late 1950s with the Nike-Zeus interceptor. Because the United States lacked advanced guidance technology, the only reasonable path to interception lay in arming the defensive missile with a nuclear warhead. This system was unsuccessful and was replaced in 1961 by the Ballistic Missile Boost Interceptor (BAMBI). Housed in satellite platforms, BAMBI would intercept enemy missiles shortly after launch (the "boost" phase) by deploying a large net designed to disable intercontinental ballistic missiles (ICBMs). Again, because of technical difficulties, it was never deployed.

In 1963, U.S. Defense Secretary Robert McNamara unveiled the Sentinel program. This program differed from its predecessors by layering defensive missiles. Made up of both short- and long-range interception missiles and guided by radar and computers, the system would protect the entire United States from a large-scale nuclear attack. Political concerns about the destabilizing influence of this system, along with the technical difficulties involved in tracking and intercepting incoming ICBMs, ensured that the Sentinel fared no better than its predecessors.

In 1967 the Sentinel was scaled back and renamed Safeguard. With this reduction in scale, the entire United States could not be protected, and Safeguard was installed only around nuclear missile sites. This enabled launch sites to survive a first strike and then retaliate. For the first time in U.S. missile defense theory, survival of retaliatory capability outweighed the defense of American citizens.

While the United States worked at developing NMD systems, the USSR did the same. It became obvious to the two superpowers that this could escalate into a defensive arms race. In an effort to curb military spending, the two countries agreed in 1972 to limit their defensive systems, creating the Anti-Ballistic Missile (ABM) treaty. Under this agreement, each country could deploy one defensive system. The United States chose to defend the Grand Forks Air Force base in North Dakota, and the USSR chose Moscow.

In 1983 President Reagan revived the NMD debate by announcing the Strategic Defense Initiative (SDI), known derisively as "Star Wars." Although previous missile defense systems had used ground-based control systems, Star Wars called for an elaborate series of nuclear-pumped X-ray laser satellites to destroy enemy missiles. This program would provide complete protection to the United States in the event of an all-out attack by a nuclear-armed adversary. Unfortunately, the technical problems were too great, and with the collapse of the USSR and the end of the Cold War, the program was canceled.

Today, SDI has morphed into NMD. This project is less ambitious than SDI, and its goal is the defense of the United States against nuclear blackmail or terrorism from a "rogue" state. The system consists of ground-based interceptor missiles in Fort Greely, Alaska, and at Vandenberg Air Force Base in California. As of 2005, there have been a series of successful test launches from sea- and shore-based launchers against a simulated missile attack.

As with its predecessors, there are three current concerns with the NMD program: destabilization, functionality, and who should be in charge of decisions about its development and deployment.

Under the doctrine of Mutually Assured Destruction (MAD), both sides avoided launching missiles because the enemy would respond in kind. Neither side could win; therefore, neither would go to war. Developing an effective NMD would eliminate the retaliatory threat, destabilizing the balance of power by making a nuclear war winnable and thus increasing the chance one might occur. Even the fear that one side might field such a system could cause a preemptive strike.

The desire for a successful NMD assumes a system that works. To date, missile defense systems have had numerous technical problems and have never achieved true operational status. Critics of NMD argue that this current system will fair no better than others, whereas supporters claim that the successful tests of the past few years show that the technology is viable. It remains to be seen how the system performs under actual battle conditions and thus whether it is, in the end, functional.

Finally, there is the question of who is in charge. Given post-9/11 security issues, the main concern is defending against launches from countries that have possible links to terrorists. As the developer of NMD, the United States wants the final say in its deployment and use. Unfortunately, to maximize interception probabilities, NMD requires sites in other countries, mostly members of the North American Treaty Organization (NATO). Poland and the Czech Republic, because of their position along possible missile flight paths, figure prominently in U.S. strategies. The current plan calls for up to 54 missiles to be based in Poland, and the controlling X-band radar would be sited in the Czech Republic. Negotiations are ongoing.

These and other NATO countries, however, believe participating in NMD makes them into potential targets of both terrorists and countries unfriendly to NATO. They feel they should have the authority to launch missiles in their own defense, should the need arise. Understandably, after all its investment, the United States feels otherwise. This remains an ongoing debate, though the United States likely will retain control over its launch sites.

NMD is still very much an unproven system. Despite over 50 years of work, the probability of successful ballistic missile defense remains low. Add to this the concerns over destabilization, and the future of the system is far from certain. It remains to be seen if the NMD is the final answer to the United States' missile defense problems or if it will become just another in a long list of failed or cancelled projects.

See also Asymmetric Warfare; Nuclear Warfare; Warfare.

Further Reading: Carus, Seth W. *Ballistic Missiles in Modern Conflict.* New York: Praeger, 1991; Daalder, Ivo H. *The SDI Challenge to Europe.* Cambridge, MA: Ballinger, 1987; Mockli, Daniel. "US Missile Defense: A Strategic Challenge for Europe." *CSS Analyses for Security Policy* 2, no. 12 (2007): 1–3; Snyder, Craig, ed. *The Strategic Defense Debate: Can "Star Wars" Make Us Safe?* Philadelphia: University of Pennsylvania Press, 1986.

Steven T. Nagy